CSR, Sustainability, Ethics & Governance

Series Editors

Samuel O. Idowu, London, United Kingdom
René Schmidpeter, Cologne Business School, Köln, Germany

More information about this series at
http://www.springer.com/series/11565

Mia Mahmudur Rahim • Samuel O. Idowu
Editors

Social Audit Regulation

Development, Challenges and Opportunities

Editors
Mia Mahmudur Rahim
QUT Business School
Queensland University of Technology
Brisbane
Queensland
Australia

Samuel O. Idowu
London Guildhall Faculty of Business and Law
London Metropolitan University
London
United Kingdom

ISSN 2196-7075 ISSN 2196-7083 (electronic)
CSR, Sustainability, Ethics & Governance
ISBN 978-3-319-15837-2 ISBN 978-3-319-15838-9 (eBook)
DOI 10.1007/978-3-319-15838-9

Library of Congress Control Number: 2015939894

Springer Cham Heidelberg New York Dordrecht London
© Springer International Publishing Switzerland 2015
This work is subject to copyright. All rights are reserved by the Publisher, whether the whole or part of the material is concerned, specifically the rights of translation, reprinting, reuse of illustrations, recitation, broadcasting, reproduction on microfilms or in any other physical way, and transmission or information storage and retrieval, electronic adaptation, computer software, or by similar or dissimilar methodology now known or hereafter developed.
The use of general descriptive names, registered names, trademarks, service marks, etc. in this publication does not imply, even in the absence of a specific statement, that such names are exempt from the relevant protective laws and regulations and therefore free for general use.
The publisher, the authors and the editors are safe to assume that the advice and information in this book are believed to be true and accurate at the date of publication. Neither the publisher nor the authors or the editors give a warranty, express or implied, with respect to the material contained herein or for any errors or omissions that may have been made.

Printed on acid-free paper

Springer International Publishing AG Switzerland is part of Springer Science+Business Media (www.springer.com)

Foreword

Corporate social responsibility is increasingly prevalent in companies globally, and consumers are demanding that businesses become more transparent in their actions. Although often difficult, a systematic approach to progress measurement is extremely valuable for both companies and stakeholders. It is therefore of great importance that current literature focuses on effective tools and strategies for the widespread implementation of social auditing. Social auditing must be used as a way to promote transparency and accountability and not as a marketing mechanism internally controlled by the institution. The purpose of social auditing is not to enhance superficial corporate image. It should include all significant environmental and social data and serve as an evaluation system from which further progress stems.

Consumers are no longer blindly purchasing but are demanding more and more information about social and environmental impacts (both positive and negative) at all stages of a product or service life cycle. Social auditing empowers consumers by offering a comparison of different companies' performance with regard to responsible economic development. The enlightened customer can use this information to influence responsible behavior among industry practices by either supporting those doing good or opposing those underperforming. Thus, enabling the public to *vote with their Euro and Dollar* in an informed way, the stakeholders in general could also use this information to decide whether or not to withhold the license to operate which all companies desperately crave for.

Social auditing is also an extremely valuable tool for businesses. Companies can use it to evaluate the extent to which they have met their CSR goals and the effect specific actions have had on their own performance, therefore offering a critical evaluation of which actions are most efficient and effective at producing social added value and which have failed to make a big difference. This gives companies a better idea of which projects to continue investing resources in and which to revise.

Recent CSR debates have established that corporate social and environmental programs should aim to consistently support societal objectives that promote the long-term health of the planet and its inhabitants. This can be ensured through a comprehensive social auditing approach.

Social auditing within CSR is a critical component of sustainable development. Companies cannot predict the best route to take if they do not track where previous paths have led them. As society continues to battle challenges such as the mass extinction of species globally, resource scarcity, and climate change, society cannot afford to let the capacity and innovative expertise of businesses go to waste.

I congratulate Mia Rahim and Samuel O Idowu for furthering the CSR discussion and continuing to bridge the gap between academia and practice with this book. It not only addresses the role of business in society but also explores this valuable tool that can be used by companies, stakeholders, and governments to solve critical global issues in effective and efficient ways. Thank you for this further milestone of CSR literature and all the fruitful discussions on the future of CSR we have.

Cologne, Germany   René Schmidpeter

Preface

Our world has continued to develop its understanding and practice of corporate social responsibility (CSR) even during the heat of the recent global financial crisis which shook our world to a breaking point and made things almost impossible for all. There are many compelling tools which have been devised and continued to be used to improve corporate activities and performance in the field of CSR. Irresponsible practices and scandalous activities have been exposed through fatal accidents, by research studies and the media in different factories that supply merchandise to retailers worldwide. Problems in the factories of many companies in emerging economies that operate in the supply chain sector have meant that responsible actions were necessary to avert the occurrence and reoccurrences of unimaginable disasters that could ensue in the supply chain sector and those sectors that rely on it for their own operational activities. Responsible social auditing of what goes on in the sector is indeed a welcome corporate action which stakeholders including many nongovernmental organizations (NGOs) and those sectors whose survival depends on the supply chain sector would applaud.

Modern corporate entities which aspire to be perceived by all as socially responsible are embedding different socially responsible activities into their strategies. CSR reports have been issued by corporate entities worldwide for more than two decades. The quality of these reports has continued to improve year in, year out. Many multi-stakeholder organizations have emerged over the course of time to provide needed guidelines and directions and ensure that corporate entities that aspire to make a positive difference in their social, economic, and environmental impacts on our world are aware of what they should do and how they should do them. See, for example, organizations such as the Social Accountability International (SAI), the AccountAbility (AA), the Global Reporting Initiative (GRI), and the Ethical Trading Initiative (ETI), just to mention a few, which have continued to guide corporate entities with usable standards and guidelines. That we are still having serious issues in this sector is beyond belief, but the fact remains that we still have a lot to contend with in regard to this aspect of CSR which is why this book on *Social Audit Regulation* is now being added to the literature.

It is hoped that contributors' attempts in the different chapters of this book by 26 scholars who work in this area and are based in 11 countries around the globe would improve our readers' understanding of how corporate entities in different economies globally are faring in the field of social audit and those issues that surround it.

London, Uk Samuel O. Idowu
Köln, Germany Mia M. Rahim
Winter 2014

Acknowledgements

We are extremely grateful to all our contributors who have made the publication of this book a reality. There are also others who are not featured directly in the book as contributors but have played some equally active parts in ensuring the publication of the book. We would also like to show our gratitude to them.

We are equally grateful to the publishing team at Springer headed by the Senior Editor Christian Rauscher, Barbara Bethke, and other members of the publishing team who have supported this publication and all our other projects. We are deeply grateful to our respective families for their support. Thanks to our colleagues and friends who have supported us in seeing through this publication.

Finally, we apologize for any errors or omissions that may appear anywhere in the book; no harm was intended to anybody.

Contents

Social Audit: A Mess or Means in CSR Assessment?............... 1
Mia M. Rahim and Victor Vicario

New Challenges for Internal Audit: Corporate Social Responsibility
Aspects... 15
Adriana Tiron-Tudor and Cristina Bota-Avram

The Development of Integrated Reporting and the Role
of the Accounting and Auditing Profession...................... 33
Dominic S.B. Soh, Philomena Leung, and Shane Leong

United States Accounting Firms Respond to COSO Advice
on Social Audit, Sustainability Risk and Financial Reporting........ 59
Katherine Kinkela

Social Audit Regulation Within the NGO Sector: Practices
of NGOs Operating in Bangladesh and Indonesia................. 79
Vien Chu and Belinda Luke

Social Audit for Raising CSR Performance of Banking Corporations
in Bangladesh... 107
Md. Tarikul Islam

Corporate Social Responsibility Assurance: Theory, Regulations
and Practice in China... 131
Yuyu Zhang and Lin Liao

Social Audit: Case Study of Sustainable Enterprise Index-ISE
Companies.. 155
Dalia Maimon and Cristiana Ramos

Corporate Climate Change-Related Auditing and Disclosure
Practices: Are Companies Doing Enough?....................... 169
Shamima Haque

Social Audit in the Supply Chains Sector 187
Samuel O. Idowu

AA1000: An Analysis of Accountability and Corporate Social Responsibility in the Contemporary Context 201
Priscila Erminia Riscado

History and Significance of CSR and Social Audit in Business: Setting a Regulatory Framework 217
Anjana Hazarika

Defining a Methodology for Social Audit Based on the Social Responsibility Level of Corporations 257
Adriana Tiron-Tudor, Ioana-Maria Dragu, George Silviu Cordos, and Tudor Oprisor

Social Audit Failure: Legal Liability of External Auditors 281
Larelle Ellie Chapple and Grace Y. Mui

Fostering the Adoption of Environmental Management with the Help of Accounting: An Integrated Framework 301
A.D. Nuwan Gunarathne

Index 325

About the Editors

Samuel O. Idowu Is a senior lecturer in Accounting and Corporate Social Responsibility at the London Guildhall Faculty of Business and Law, London Metropolitan University, where he was course organizer for Accounting Joint degrees, course leader/personal academic adviser (PAA) for students taking Accounting Major/Minor and Accounting Joint degrees, and currently course leader for Accounting and Banking degree. Samuel is a professor of CSR and Sustainability at Nanjing University of Finance and Economics, China. He is a fellow member of the Institute of Chartered Secretaries and Administrators, a fellow of the Royal Society of Arts, a liveryman of the Worshipful Company of Chartered Secretaries and Administrators, and a named freeman of the City of London. Samuel has published about fifty articles in both professional and academic journals and contributed chapters in edited books and is the editor in chief of two major global reference books by Springer—the *Encyclopedia of Corporate Social Responsibility* (ECSR) and the *Dictionary of Corporate Social Responsibility* (DCSR)—and he is a series editor for Springer's CSR, Sustainability, Ethics, and Governance books. Samuel has been in academia for 27 years winning one of the Highly Commended Awards of Emerald Literati Network Awards for Excellence in 2008 and 2014. In 2010, one of his edited books was placed in 18th position out of 40 top Sustainability books by Cambridge University Programme for Sustainability Leadership. He has examined for the following professional bodies—the Chartered Institute of Bankers (CIB) and the Chartered Institute of Marketing (CIM)—and has marked examination papers for the Association of Chartered Certified Accountants (ACCA). His teaching career started in November 1987 at Merton College, Morden, Surrey; he was a lecturer/senior lecturer at North East Surrey College of Technology (Nescot) for 13 years where he was the course leader for BA (Hons) Business Studies and ACCA and CIMA courses. He has also held visiting lectureship posts at Croydon College and Kingston University. He was a senior lecturer at London Guildhall University prior to its merger with the University of North London, when London Metropolitan University was created in August 2002. He has served as an external examiner to a number of UK universities including the University of Sunderland; the University of Ulster, Belfast, and Coleraine, Northern Ireland; and Anglia Ruskin University, Chelmsford. He is currently an external examiner at the University of Plymouth; Robert Gordon University, Aberdeen, Scotland; and Teesside University, Middlesbrough, UK. He was also the treasurer and a trustee of Age *Concern*, Hackney, East London, from January 2008 to September 2011. He is a member of the Committee of the Corporate Governance Special Interest Group of the British Academy of Management (BAM). Samuel is on the Editorial Advisory Boards of the Management of Environmental Quality Journal, International Journal of Business Administration, and Amfiteatru Economic Journal. He has been researching in the field of CSR since 1983 and has attended and presented papers at several national and international conferences and workshops on CSR. Samuel has made a number of keynote speeches at international conferences and workshops.

Mia M. Rahim Is a lecturer in law in the QUT Business School at the Queensland University of Technology, Australia. Previously, he taught law units at Macquarie University where he was pursuing his Ph.D. with a Research Excellence Scholarship. He did his LLB with honors and LLM from Dhaka University; LLM in International Economic Law from Warwick University as a Chevening Scholar; and MPA from LKY School of Public Policy with NUS Graduate Scholarship. Before he joined academia, he was a lawyer and Joint District and Sessions Judge in Bangladesh. He also worked for the Law Commission and the High Court of Bangladesh.

Mia has contributed to internationally reputed conferences and journals; some of the journals he has been published in are the Journal of Business Ethics, Law and Financial Market Review, Common Law World Review, Competition and Change, and Australian Journal of Asian Law. A special issue of the Transnational Corporation Review was published under his editorship. Palgrave Macmillan, Gower, and Springer have also published his recent book chapters, encyclopedia, and dictionary entries. Two of his research monographs were published by Springer in 2013–2014. He is a member of the Center for Legal Governance of Macquarie University and Technical Committee for Asian Consumer Protection Research Network.

About the Contributors

Cristina Bota-Avram Is a lecturer at the Department of Accounting and Audit, Faculty of Economics and Business Administration, Babeş-Bolyai University. Her research activity and teaching interests lie in the areas of audit, corporate governance, public governance, accounting ethics, managerial accounting, public accounting, and other governance-related topics. Her area of interest for the postdoctoral research is related to the various issues connected to the relationship governance audit. She holds a Ph.D. in economic sciences, in the field of accounting and audit, her Ph.D. thesis being entitled *"Internal audit in enterprises."* Her doctoral thesis aimed to address a very exciting theme through the growing interest and topicality from day to day, especially in the context of an increased volatility of economic and financial environment. This paper mainly deals with the internal audit of private enterprises, through a comparative approach, from both perspectives, theoretical and practical, with the internal public audit of entities from public sector. Also, during the period 2010–2013, she successfully achieved a postdoctoral research project entitled *"Developing an integrated framework of good audit practices in the context of corporate governance for the Romanian economic entities,"* as part of the research project POSDRU/89/1.5/S/59184 "Performance and excellence in postdoctoral research within the field of economic sciences in Romania," Babeş-Bolyai University, Cluj-Napoca, and Reading University UK being partners within the project.

Larelle Chapple Is a professor at the QUT Business School, Queensland University of Technology, Brisbane, Australia. Her teaching and research interests are primarily corporate law, corporate governance, and audit. She has over 25 refereed journal articles in top-ranked regional journals in accounting, finance, and commercial law. Her research record relates to studies in corporate accountability, reporting, and corporate transactions, and she favors a mixed-methods approach, that is, qualitative analysis of policy and legal documents and statistical analysis of accounting and market measures and data. Professor Chapple's contribution to this monograph on social audit is an example of the former research style, combining skills and literature from the accounting and commercial law disciplines to inform policy debates about regulation of social audit. She is currently joint editor of the Accounting Research Journal.

Vien Chu Is a Ph.D. candidate and tutor in the School of Accountancy at Queensland University of Technology, Brisbane, Australia. Her research interests include accountability in the third sector, social enterprise, and micro-enterprise development. Prior to joining academia, Vien worked as a financial and general manager for international private sector companies in Vietnam.

George Cordoş Is a Ph.D. candidate at Babeş-Bolyai University, Faculty of Economics and Business Administration, Accounting and Audit Department, Cluj-Napoca, Romania. He graduated in Accounting within the Faculty of Economics and Business Administration, and, afterward, he obtained a master's degree in managerial accounting, auditing, and controlling and also in audit and financial management of European funding. During the 2-year master's program, he started research in the audit field. He is currently writing his Ph.D. thesis on the topic of audit reporting changes, pushed by regulating bodies and standard setters such as IAASB or the European Commission. His research highlights how the current form of the audit report affects users' decisions of investment; how the audit report can be improved and revised in this direction, in order to include more relevant information for users; and how the expectation gap can be reduced. He is thoroughly following the topics of his interest, has published a series of papers on audit reporting, and has attended a series of conferences and seminars. In recent studies, he focused on a content analysis of comment letters submitted by interested parties, regarding changes proposed by the IAASB 2013 Exposure Draft. Being a 1st year Ph.D. candidate, he has attended several workshops on the topic of academic writing and research techniques. His research relates to social audit insights and the integration process into standard reporting practices, as well as information disclosure.

Ioana Dragu Is a Ph.D. candidate at Babeş-Bolyai University, Faculty of Economics and Business Administration, Accounting and Audit Department, Cluj-Napoca, Romania. She graduated in Finance and Banking, English Line of Study within Faculty of Economics and Business Administration, and then followed an MBA of 2 years in managerial accounting, audit, and control. She started her research in corporate reporting and disclosure when she was still a student. Currently she is writing her thesis on Integrated Reporting as the mixture between financial and nonfinancial information brought together, including in her topic sustainability and corporate social responsibility disclosure practices. She published a series of papers on reporting practices and disclosure regarding financial and nonfinancial information. Her research interests incorporate social audit insights and their integration into standard reporting practices and information disclosure.

Nuwan Gunarathne Is a lecturer at the Department of Accounting, University of Sri Jayewardenepura, Colombo Sri Lanka. He is an associate member of the Chartered Institute of Management Accountants (CIMA-UK) and the Institute of Certified Management Accountants (CMA-Sri Lanka) and a Chartered Global Management Accountant (CGMA). He also holds a postgraduate diploma in marketing from the Chartered Institute of Marketing (CIM-UK). After completing his bachelor's degree in business administration with a first-class standing from the University of Sri Jayewardenepura, Nuwan completed his master's degree in business administration from the Postgraduate Institute of Management of University of Sri Jayewardenepura with a merit pass. He currently teaches cost and management accounting, strategic management accounting, sustainability management accounting, and contemporary issues in management accounting in undergraduate and postgraduate degrees. He has presented papers in conferences in Finland, Australia, Sri Lanka, and Indonesia. He has served as a special guest editor of the Journal of Accounting Panorama and a guest reviewer of Issues in Social and Environmental Accounting journal. With wide-ranging research interests in environmental and sustainability management accounting, waste accounting, accounting education, and integrated reporting, Nuwan has published papers in Journal of Accounting and Organizational Change, Journal of Certified Management Accountants, Journal of Accounting Panorama, and Professional Manager Magazine. He is a committee member of the Environmental and Sustainability Management Accounting Network (EMAN)-Asia Pacific (AP) and the country representative of the Sri Lanka Chapter of EMAN-AP.

Shamima Haque Is a lecturer in the School of Accountancy, Queensland University of Technology, Brisbane, Australia. Since completing her Ph.D. in accounting in 2012, Shamima has maintained her research focus on corporate social and environmental accounting and accountability, building on her work on climate change issues but extending it to corporate attitudes and practices related to bribery and human rights issues.

Anjana Hazarika Is currently working as an assistant professor and assistant director at Centre for Law and Humanities, Jindal Global Law School, O. P. Jindal Global University, Sonepat, Haryana, India. She is also serving as the editor of the Journal of Global Studies (a monthly international peer-reviewed research journal, published from Delhi). She also publishes an online newsletter named "Pragati" on behalf of the Centre for Law and Humanities of the University. Her research and writing focus on sociology, corporate social responsibility, social audit, globalization, and social conflict. She has recently completed a book, "Corporate Social Responsibility: A Liberal Democratic Perspective," to be published by next year. Among her published book chapters, "Ethical CSR-Competency, Community and Consumer Driven-An Indian Experience" in "Corporate Social Responsibility and Sustainable Development" is noteworthy.

Tarikul Islam Is a Ph.D. student at QUT School of Accountancy since July 2014. Before joining QUT as a Ph.D. student, he was an employee of two universities—Khulna University and Jahangirnagar University in Bangladesh in various positions, and he is currently on study leave as an associate professor of Finance and Banking in Jahangirnagar University. His research interests include CSR and corporate governance. He can be reached at imtarikul@gmail.com.

Katherine Kinkela Is an assistant professor at the Hagan School of Business of Iona College in New York, USA. She has JD and LLM degrees from Fordham University School of Law. She teaches courses in financial and managerial accounting principles, intermediate accounting, and taxation. She has published in refereed journals and presented papers at national and international conferences on accounting. She has worked at KPMG LLP and law firms with experience in consulting on accounting, taxation, and related legal matters. This paper won the best in conference in the Academic and Business Research Institute AABRI 2014 summer conference held in Honolulu, Hawaii, USA, and was published in the Journal of Finance and Accountancy in October 2014.

Shane Leong Is an early career research fellow with Macquarie University's Department of Accounting and Corporate Governance. He holds a Ph.D. in accounting, awarded by Macquarie University in 2013, and has published articles in the *Journal of Business Ethics* and *Journal of Cleaner Production*. His research area is social and environmental accounting. Shane is especially interested in exploring how social and environmental accounting disclosures can advance sustainability and enhance corporate accountability.

Philomena Leung Is professor and head of the Department of Accounting and Corporate Governance in Macquarie University. Philomena's research interests include auditing, ethics, corporate governance, and accounting education. Philomena is Chair of Education Committee of the Institute of Internal Auditors in Australia. She co-convenes a think tank group comprised of the heads of accounting in Australia and New Zealand [Chairs of Accounting and Finance Forum (CAFF)]. Philomena has led major accounting research projects on immigrant accountants, gender issues, ethical problems, and internal auditing and was the lead researcher commissioned to develop a model for ethics education in accounting for 160 countries under the auspices of the International Federation of Accountants. She is joint editor for *Managerial Auditing Journal* and a lead author for the text *Modern Auditing and Assurance Services in Australia*. Philomena has a Ph.D. (RMIT) and a master of accounting and finance (Glasgow). She was an auditor with one of the international audit firms in Hong Kong before joining academia and has held senior academic roles since 1976, in Hong Kong, Melbourne, and Sydney.

Lin Liao Is an associate professor in accounting at Southwestern University of Finance and Economics, China. He received his Ph.D. from the University of New South Wales, Australia. His research areas include financial accounting, auditing, carbon accounting, and corporate governance. He has been a member of the CPA Australia since 2008.

Belinda Luke Is an associate professor in the School of Accountancy at Queensland University of Technology. Her research interests include accountability in the public sector and third sector. Belinda's research has a strong case study-based focus, and she has received two finalist awards from the Academy of Management Dark Side Case study competition. Prior to joining academia, Belinda worked as a senior tax and human resources manager for PricewaterhouseCoopers in Australia, Ukraine, Papua New Guinea, and China.

Dalia Maimon Is the head of the Laboratory of Social Responsibility (LARES) and full professor at Institute of Economics since 1972 and academic coordinator of MBA in sustainability and third sector and MBA in economics and sustainability management. She teaches evolution of the sustainability concept and sustainable business. She has published 9 books in environmental and social responsibility. She holds a post Ph.D. in economics from Sorbonne Paris 7 (1995), Ph.D. in environmental economics at Ecole des Hautes Etudes In Sciences Sociale Paris (1989), master's degree in production engineering from the Graduate School and Research in Engineering (COPPE) (1979), and MBA in economics from UFRJ (1972). She was the coordinator of a Project with the UNDP-UN Development Report and participated actively in the NGO group for ISO 26000. She was for 9 years a member of the follow-up of UNCED (UNESCO). She was a consultant for the World Bank and Inter-American Development Bank in the area of Social Technologies. She was an active participant at the UN Conference Rio+20, Johannesburg 2002, and Rio 92. She was the president of NGOs ProNatura and SIGA (member of INEM—International Network on Environmental Management). Her research interests include sustainable business and responsible social investment. She won the Prize Beija Flor's Riovoluntário in 2009.

Grace Y. Mui Is the senior consultant at Thye and Associates, Malaysia. She received her Ph.D. in accounting from The University of Queensland, Australia, in 2010. She has taught courses in financial accounting, management accounting, electronic commerce, and research methods at undergraduate and postgraduate level. She has designed and facilitated corporate training programs for government, private, and not-for-profit organizations. Her research interests include auditor expertise, fraud detection, and risk management. In recent years, she has received funding from the Institute of Internal Auditors Malaysia and the Malaysian Ministry of Higher Education.

Tudor Oprisor Is a Ph.D. candidate at Babeş-Bolyai University, Faculty of Economics and Business Administration, Accounting and Audit Department, Cluj-Napoca, Romania. He graduated in Accounting within the Faculty of Economics and Business Administration, and, afterward, he obtained a master's degree in managerial accounting, auditing, and controlling. During the 2-year master's program, he was employed by a business investment company, and he conducted a series of financial analyses on public-listed companies. He is currently writing his thesis on Integrated Reporting within the public sector, with emphasis on how integration can be defined and implemented in public sector reporting, which are the users of these reports, and to which extent can the disclosure level be improved as a result of integrated reporting. He is also interested on the regulating activities conducted by the governing bodies and professional organizations regarding the Integrated Reporting initiative in both the private and public sector. In this respect, he has written a series of articles on the way comment letters submitted by significant worldwide stakeholders have contributed to the improvement of the Integrated Reporting Framework. He is closely following the topics and has attended a series of conferences and seminars on public sector accounting and integrated reporting. His research interests also relate to human capital, environmental and sustainability reporting, public sector economics, as well as social audit insights and the integration process into standard reporting practices, as well as information disclosure.

Cristiana Ramos Sadly, Cristiana passed away in Brazil in December 2013 according to her coauthor professor Dalia Maimon. We take this opportunity to send our condolences to her family, friends, and colleagues.

Priscila Erminia Riscado Is an adjunct professor of Political Theory in the Institute of Education in Angra dos Reis, in Federal Fluminense University (IEAR/UFF) in Brazil. She holds a Ph.D. in political science from the graduate program in political science from the Federal Fluminense University (PGCP/UFF). She is a researcher at the Center for Political Culture (NCPAM) in Federal University of Amazonas (UFAM) in Brazil. She has been dedicated to research on the analysis and implementation of Public Policy and Social Responsibility in Business Programs. Her research areas are in implementation and evaluation of programs in the social responsibility area and implementation and evaluation of public policies. She has professional experience in the area of political science, acting on the following topics: democracy, society public policy, social responsibility, company, entrepreneur, and the environment.

Dominic Soh Is a lecturer in the Department of Accounting and Corporate Governance at Macquarie University, Australia. He has previously worked in auditing, management accounting, and in a technical role at the Institute of Internal Auditors—Australia (IIA). Dominic actively researches in internal and external auditing and assurance issues, in particular, those relating to environmental, social, and governance. He maintains close ties with professional bodies as a member of the IIA's Education Committee and in his involvement in the Audit and Assurance Module of Chartered Accountants Australia and New Zealand's Chartered Accountants Program. He is also a coauthor of *Auditing and Assurance: A Case Studies Approach*, published by LexisNexis, a text widely used in a number of leading Australian universities.

Adriana Tiron-Tudor Is a Ph.D., Full Professor, at Babeş-Bolyai University, Faculty of Economics and Business Administration, Accounting and Audit Department, Cluj-Napoca, Romania, where she coordinates doctoral students in accounting and is actively involved in the development of the Romanian accounting profession. A graduate of Babeş-Bolyai University with a Ph.D. in accounting, Mrs. Tiron-Tudor received her expert accountant designation from CECCAR in 1995. She has lectured extensively at a national level on financial management, on internal audit and control topics for public entities, and on public sector accounting standards. In 2006, she contributed to the implementation of the accrual accounting system in Romania by being involved in the training of practitioners, especially for local governmental administrations. Since 2008, she has also been a member of the Fédération des Experts Comptables Européens (Federation of European Accountants, or FEE) Public Sector Committee. She became a member of the International Public Sector Accounting Standards Board (IPSASB) in January 2012. She was nominated by the Body of Expert and Licensed Accountants of Romania (CECCAR). She has also written several publications relating to government accounting and standard setting in the public sector.

Victor Vicario Joined the School of Accountancy in 2014 as a research assistant focusing on corporate social responsibility (CSR) and as sessional academic in business and corporations law. Previously, he completed his LLB at Bond University and LLM in Intellectual Property Law at Queensland University of Technology. Before commencing his master's degree, Victor completed his postgraduate degree in legal training at Bond University and was admitted as a solicitor in the Supreme Court of Queensland in 2012.

Yu Yu Zhang Is a lecturer in accounting at Queensland University of Technology, Australia. She received her Ph.D. from the University of New South Wales. Her research interests center on the economics of auditing, sustainability audit, financial institutions, the audit market, and financial crisis.

Social Audit: A Mess or Means in CSR Assessment?

Mia M. Rahim and Victor Vicario

1 Introduction: What Is Social Audit?

Since the initial recognition of Corporate Social Responsibility (CSR) in the 1960s, companies worldwide have been placed under increasing scrutiny with regards to their level of social performance; namely their positive or negative impact on stakeholders, the community and the environment. Slowly, a notion developed that, by comparing different companies' sustainability performance, stakeholders would be able to influence the social responsibility of businesses either by pledging their support to, and investing in good performing companies, or alienating and withdrawing from poorly performing companies.

Throughout the years companies have organically developed their own codes of conduct and they have gradually collated and released CSR reports voluntarily in order to respond to the increasing pressure from stakeholders to improve their social, environmental and ethical standards. However, while corporate governance, and the use of CSR reports, has grown exponentially throughout the years, the same cannot be said about the manner in which the data is gathered and processed. The vital importance of ensuring "quality" CSR reports has attracted substantial attention from scholars, who, for the past four decades, have been trying to determine whether or not companies are effectively managing to meet stakeholder demands. To establish this, focus has been directed towards the level of social accounting and more specifically, about the efficacy of social "audits" being conducted by companies themselves (Gray et al., 1996). The data, presentation and practice of social audits varies so significantly between corporations, industries and jurisdictions that it has proven very difficult for any accurate and meaningful analysis to take place. While various attempts have been made to establish a universal standard for social

---

M.M. Rahim (✉) • V. Vicario
School of Accountancy, QUT Business School, Queensland University of Technology,
2 George St, Qld 4001 Brisbane, Queensland, Australia
e-mail: mia.rahim@qut.edu.au

© Springer International Publishing Switzerland 2015
M.M. Rahim, S.O. Idowu (eds.), *Social Audit Regulation*, CSR, Sustainability,
Ethics & Governance, DOI 10.1007/978-3-319-15838-9_1

auditing, variations among corporations of the notion, interpretation, preparation and implementation of social audits have resulted in some barriers to such system (Owusu & Frimpong, 2012; Perrini, 2006). Nonetheless, social audit may be broadly defined as a way of analysing, measuring and reporting an organisation's social and ethical performance by scrutinising its nonfinancial activities which, directly or indirectly, impact stakeholders (Ghonkrokta & Lather, 2007). In other words, it is a means of formally measuring and recording the level of a company's social and environmental performance with regular monitoring through the collection of data from interviews, documents and inspections gathered within an organisation (Björkman & Wong, 2013).

Similarly to its definition, the real purpose of social audit is also interpreted differently depending on the corporation or jurisdiction. For example, one justification for a corporation to gather sufficient information about its own social performance, would be to directly determine the extent to which it is able to meet the values and objectives it has committed itself to (Björkman & Wong, 2013; Locke, Qin, & Brause, 2007). Alternatively, a company with high CSR performance may release the collated social performance data to its stakeholders so that they may compare such data with equivalent data from other companies, possibly raising the profile and popularity of the well performing company. Lastly, social audit is seen a means of assurance, relied upon by governments, to ensure that companies are collecting social and environmental data and disclosing it in a satisfactory manner.

Social auditing facilitates a transparent control and monitoring mechanism of individual companies. This allows stakeholders to evaluate a corporation's social performance against particular standards or expectations, ultimately taking the role of instruments of social accountability for an organisation similarly to a financial audit (Hess, 2001; Kurian, 2005; Timane & Chavan, 2012). A financial audit provides verification of the financial statements provided by a company and provides an assurance that the financial statements are true and accurate, thus increasing the value and credibility of the statements. Similarly, a social audit can verify the CSR of a company by demonstrating how social and environmental programs are being carried out. If these actions reflect social, environmental and community objectives, stakeholders will have an increased confidence in the company and its values. Also, like financial audits, the purpose of social audits is not to place judgement on the performance of a company (Kurian, 2005), but instead, to focus on the data verification, and evidence gathering, for all significant assertions in the report.

The manner in which social and environmental information is gathered and processed in a social audit varies significantly. Data may be collected and verified by the company itself, by a hired external and independent consultant, by a NGO representative or even by a dedicated government entity (Courville, 2003). Each of these different bodies has started to recognise the importance and value of social audits and is promoting their uptake.

2 Importance of Social Audit

Though still in its infancy (Dando & Swift, 2003), social auditing plays a fundamental role in upholding Corporate Social Responsibility and is proving direct and indirect benefits to both the company and its stakeholders (Miles, Hammond, & Friedman, 2002; Owusu & Frimpong, 2012). Reviewing the socially responsible practices and impact on stakeholders, and comparing the level of social performance in relation to pre-set social, environmental and community goals (Ghonkrokta & Lather, 2007), provides valuable information and insight to a company. This, in turn, leads to a better self-assessment and establishment of the strengths and deficiencies present within a corporate strategy. Such knowledge, allows companies to implement improvements leading to a more efficient social performance, which in turn leads to the enhancement of a company's image (Humble, 1975; Kok et al., 2001). In fact, where a company has a good social performance, a social audit has the potential to safeguard its image in the case of a particular event leading to negative publicity, or simply, enhance a company's reputation, image and relationship with stakeholders by demonstrating its social performance and its commitment to social objectives (Owusu & Frimpong, 2012). Furthermore, regular yearly social audits allow companies to compare their own social performance over time as well as against external norms and standards and competing companies (The Seep Network, 2008).

In addition to allowing companies to increase their image and performance by providing valuable data, social audits play also an important role in increasing an organisation's transparency and accountability to its stakeholders (US Aid, 2008), much like a financial audit does. They inform the community, the public, as well as other organisations, about the allocation of companies' resources invested in the organisation itself: such as the sustainability of the company, the treatment of its employees and the impact on the environment. This, in turn, stimulates healthy competition between companies to increase their social performance as stakeholders and investors rely on the social responsibility reports to establish whether a corporation is achieving the goals it has set itself and how it is performing against other companies. This leads companies to a race of maintaining the best reputation and thus maintaining or gaining an increased market share. Transparency in fact, has become a key element required by stakeholders when reviewing CSR reports. It falls in line with two renowned theories, namely stakeholder theory and legitimacy theory. According to stakeholder theory, (Roberts, 1992; Roberts & Mahoney, 2004) since corporations are actively taking advantage of, and relying upon, social and environmental resources, stakeholders have the right to be informed about their actions (Owusu & Frimpong, 2012). Legitimacy theory (Deegan, Rankin, & Tobin, 2002; Magness, 2006) emphasises the existence of a relationship between a company and the society it directly or indirectly affects, as well as the responsibility on the organisation's part to disclose its overall impact on such society. This theory emphasises the importance of community expectations in ensuring the survival of an organisation (Owusu & Frimpong, 2012). In Chapter "Corporate climate change-related auditing and disclosure practices: are companies doing enough?", Shamima Haque extends this issue taking climate change and its impact assessment for the society at large; it explores how the climate change related 'disclosures' of corporations can serve as a tool for auditing corporate accountability practice to climate change.

Both the stakeholder theory and the legitimacy theory are commonly referred to when promoting the uptake of social reporting (Laan, 2009). Although corporate social and environmental reporting is mostly a voluntary practice, increasingly, organisations are being legally required to disclose information about their interactions with, and impacts on, society. This transition has been supported by a range non-government organisations (NGOs) and regulatory agencies as well as ethical or socially responsible investment fund managers (Laan, 2009).

In addition to allowing companies to analyse and determine the practicality and efficiency of their corporate social and environmental strategy and providing stakeholders with accurate non-financial data, social audits also play an important role in helping governments monitor companies and hold them accountable when breaching certain social, ethical and environmental standards. In fact, social audits are a very important means of assessing the success or failure of a particular CSR regulation (Ghonkrokta & Lather, 2007) and thus serve the purpose of meeting regulatory requirements (Darnall, Seol, & Sarkis, 2009; Owusu & Frimpong, 2012). This is a growing trend, which relies heavily on the assurance that social accounts are audited by qualified and objective social auditors which are independent from management and with no vested interests in the outcome of the audit (The Seep Network, 2008). However this is not the norm and very often, social audits are either conducted by the companies themselves or by an external accountant paid by the company being audited, which could potentially threaten the unbiased nature of the report.

3 Development of Social Audit

While the term "social audit" first appeared in the 1940s and 1950s its notion started to receive significant attention by academic scholars and the business industry in general in the 1970s (Fetyko, 1975; Hess, 2001), mainly as a response to another notion which emerged in the 1960s and started to gain momentum (Hess, 2001), namely corporate social responsibility (Courville, 2003; Fetyko, 1975). As companies started to make reference to their social performance, issues were raised by interest groups in relation to their failure to disclose balanced and accurate information as companies were deemed to have too much control on what information was disclosed and how the information was presented. This called for an independent formal standardised analysis and presentation of information coupled with a lack of management discretion, which ultimately led to the development of social audits (Fetyko, 1975). However, this initial rapid level of interest slowed down during the 1980s (Hess, 2001; Gray et al., 1996), perhaps attributed to seemingly too good stock market results, removing from companies mind the interest of undertaking their own social performance in exchange with the sole pursuit of profits (Henriques, 2000). Others pointed the finger towards the recession of the early eighties in the US (Hess, 2001). Regardless of both contradicting alternatives, the adoption of social audits significantly increased in the mid 1990s (Gray et al., 1996) led by a number of ethically-oriented companies (Henriques, 2000), NGOs, national governments and socially responsible investment funds.

In the past two decades, social auditing has been consistently adopted across different industries and jurisdictions (Owusu & Frimpong, 2012), driven by companies' own regulatory requirements and pursuit of benefits derived from the portrayal of a positive image. However, while traditionally the role of social audits was perceived to be mainly of a monitoring nature, this notion found itself always losing against the more established corporate goals such as the pursuit of profits (Spira & Page, 2002). This has mostly been attributed to the fact that social auditing has developed as a voluntary notion and corporate auditors have no obligation to report to stakeholders on a company's social impact on society. Although the concept of social auditing is valuable, it appears not to be sufficient to fulfil its intended purposes.

While throughout the years certain corporations have significantly improved their social reporting practices, concerns have been raised with regards to the actual credibility of corporate self-regulation (Courville, 2003). Pressure from outside groups and stakeholders has led many corporations to release social responsibility reports voluntarily, but the format, content and detail included in the data is often unregulated. This has raised arguments that social audits undertaken with the sole purpose of meeting outside pressures or advertising the company's good deeds are not meaningful (Fetyko, 1975). The lack of regulation allows the reporting companies to decide what information to disclose and in what manner to present such information (Gray, 2001; Laan, 2009). Instead, it would be better to provide assurance to stakeholders in relation to the accuracy and adequacy of a social responsibility report through the development of a regulatory system which establishes a standard for auditors, the type of relationship they may have with the audited company and the manner in which social audits should be conducted (Dando & Swift, 2003). This would make social audits far more valuable as they could be used to verify the validity and accuracy of the information in a CSR report.

Over the years there has been a number of "green wash" scandals uncovered which have understandably reduced stakeholders' trust and confidence in the honesty and reliability of CSR reports. This has led to an increased need for social reporting and auditing regulation.

4 Current Trend of Social Audit

Recent years have seen a rise in the general uptake of CSR reporting. This rise has been attributed partly to the voluntary commitment of companies, but, more importantly, it has been linked to the constant increase of legislation and regulation and a direct stakeholder action, which has taken place across the world (Jaramill &

Altschuller, 2013). Countries have begun to realise, that solely relying on market forces to increase corporate social reporting, and, more importantly, social performance, is not bringing the anticipated results. In fact, a lack of mandatory obligations to conduct business activities in a socially responsible manner has the tendency to lead companies to avoid focusing too much attention to, or neglecting altogether any CSR related matters as they do not directly provide an enhancement of profit. However, the imposition of regulations relating to CSR and the monitoring and enforcement of such regulations is clearly challenging and not necessarily the only method to compel companies to behave in a socially responsible manner.

A balanced solution has emerged where governments regulate the disclosure of CSR reports which ensures companies disclose all of their CSR activities, whatever their perceived impact. This could have the desired effect of placing pressure on the company to try to increase their CSR performance to attract investors and raise its profile and reputation above that of its competitors. This notion has gained traction in light of numerous international corporate scandals throughout the years which have produced a growing view that, not only must companies be held accountable for their actions and repair all damages caused by such actions, but that they should also actively provide some direct benefits back to the community. This would not necessitate that a company abandons its main pursuit for profits, but, instead, that a company elaborate "a comprehensive corporate strategy with a balanced business perspective" (Sy, 2013). Furthermore, companies are now starting to realise the potential gain of adopting corporate best practices that go beyond what is necessary for legal and regulatory compliance as a good public image is a crucial marketing asset and its importance just cannot be underestimated (Jaramill & Altschuller, 2013). The most successful attempts to regulate non-financial reporting of companies across all sectors of the economy have taken place in the European Union (EU), in particular, the United Kingdom (UK), France and Denmark, where companies have traditionally been pioneers in the regulation of CSR reporting. In Chapter "Social audit failure: Legal liability of external auditors", Ellie Chapple and Grace Mui focus on the development of financial report auditing standards and legal liability of auditors to raise issues about potential legal liability for social auditors under the current English and Australian case laws. However, many countries, such as the United States (US), have expressed their reluctance to move away from the voluntary model, the main concern being that over-regulation could have a negative impact on financial markets (Tschopp, 2005). In Chapter "United States Accounting Firms respond to COSO Advice on Social Audit, Sustainability Risk and Financial Reporting", Katherina Kinkela and Iona College address the current situation in the United States with particular focus on a recent paper from the Committee of Sponsoring Organisations of the Treadway Commission (COSO) providing guidance on internal controls of corporations.

While CSR has mainly developed within the largest developed countries, it is also becoming a significant topic in emerging countries. In Chapter "Corporate Social Responsibility Assurance: Theory, Regulations and Practice in China", Yuyu Zhang and Lin Liao focus on providing an insight to the development, theories, regulation and challenges of CSR assurance in China, identifying the stakeholders and analysing the strengths and weaknesses of the major social assurance providers. Further insight in the current CSR trend in emerging countries is expanded upon by Dalia Maimon in Chapter "Social Audit: Case Study of Sustainable Enterprise Index-ISE companies", with her in depth study of the incorporation of social responsibility into the audits of Brazilian companies, exploring the trends and addressing the effectiveness of state regulation as well as by Vien Chu and Belinda Luke which instead in Chapter "Social audit regulation within NGO sector: Practices of NGOs operating in Bangladesh and Indonesia" will focus on investigating the benefits and constraints of social audit regulation of NGOs operating in Bangladesh and Indonesia. An additional overview of social audit in Bangladesh will be undertaken by Tarikul Islam, in Chapter "Social Audit for Raising CSR Performance of Banking Corporations in Bangladesh", which will focus on social audit practices in the national banking sector, in particular in relation to the level of stakeholder uncertainty in regard to bank claims.

However, while both alternative models might lead to a larger uptake of CSR reporting, only the mandatory model has the potential to ensure the quality and consistency of the report. One widely shared concern is that in the absence of regulation, private companies would be reluctant to go beyond their duties and, instead, would address their attention mainly on their own financial issues and pursuit of profits without working towards a socially optimal level of disclosure. It is clear that a voluntary system of reporting has not produced the desired results of influencing companies to significantly focus on their CSR behaviours and actions and that some form of regulation is required to push companies to collate and release accurate and audited non-financial information. While various arguments have been raised in favour (Lennox & Pittman, 2011), Admati & Pfleiderer, 2000; Lambert, Leuz, & Verrecchia, 2007) and against (Lennox & Pittman, 2011; Leuz & Wysocki, 2008; Sunder, 2003) mandating disclosure of accounting information, obliging companies to have their financial statements audited would ultimately be the best way to ensure that outsiders have access to reliable accounting information (Lennox & Pittman, 2011). Otherwise, companies would lack sufficient incentives to provide social and environmental information on a voluntary basis (Lennox & Pittman, 2011).

Currently, most CSR reports vary significantly in format, length and content. While several key international guidelines are available such as the Organisation for Economic Co-operation and Development (OECD), Guidelines for Multinational Enterprises (OECD, 2014), the Global Reporting Initiative Reporting Guidelines, the International Finance Corporation's Performance Standards and the Accountability's AA1000 Assurance Standards, their voluntary adoption does not ensure that they are used at all or closely follow the guidelines. Of these guidelines, the Accountability's AA1000 Assurance Standards have provided a significant development by creating a standardised system for social and environmental auditing (Graham & Woods, 2006). The AA1000S Assurance Standard provides guidance that can be used by stakeholders and regulating entities to judge the quality of a social audit, setting a standard to which social auditors must uphold. The AA1000S Assurance Standard sets out a variety of requirements such as the need for the auditor to establish the extent to which the reporting company has disclosed adequate and timely information about its activities, performance and impacts on the community and the environment, as well as to determine whether stakeholders concerns have been dealt with (Dando & Swift, 2003). The AA100 Assurance Standards will be further addressed in more detail by Pricila Erminia Riscado, in Chapter "AA1000: an analysis of accountability and corporate social responsibility in the contemporary context". It will discuss the debate over private accountability, focusing on the standard AA 1000 as it aims to disseminate the central position focused on accountability. An additional international perspective is provided by Dominic Soh, Philomena Leung and Shane Leong in Chapter "The Development of Integrated Reporting and the Role of the Accounting and Auditing Profession", which focuses on the recent release of the International Integrated Reporting (IR) Framework in December 2013. It will address the growing trend to improve social reporting as a result of increasing stakeholder demands and regulatory initiatives.

5 Effectiveness of Social Audit

The level of effectiveness of a social audit relies on a number of factors. Scholars have expressed their concern with regards to the lack of legislation regulating social auditors and its effect on the level of uniformity in the findings of social audits (Barrientos & Smith, 2007; Björkman & Wong, 2013). Another criticised aspect has been the level of secrecy generally present with audits, as well as the lack of consistent data collection and analysis, limiting the range of discussion which would potentially be able to solve certain inefficiencies (Björkman & Wong, 2013). Furthermore, social audits tend to be too short and lack sufficient detail to be able to be used to identify particular code violations (Björkman & Wong, 2013).

Such concerns have fuelled ongoing discussions regarding the best structure for audits to follow, for example in relation to the manner and form, as well as the best individuals to undertake such audits (Björkman & Wong, 2013; Locke et al., 2007). The latter being a particular delicate and much focused aspect because of the general lack of guarantee of impartiality on behalf of auditors when assessing the data from a company, especially if this company is the employer of the auditor (Björkman & Wong, 2013; Locke et al., 2007; O'Rourke, 2002; Pruett, 2005). This was witnessed in a report on the examination of labour standards in China and Korea by Prycewater House Coopers (PwC). The report found that "significant and seemingly systematic biases" were present in the auditors' methodologies, which then raises questions as to the possibility of a company being truly unbiased and independent in its auditing process (Graham & Woods, 2006; O'Rouke, 2000). In a further study, Dara O'Rourke found significant problems with PwC's social auditing methods, one of the most important being the management bias in the audit process. The study found that the vast majority of information was gathered from managers, and only a minority from actual employees (Courville, 2003). These issues would be avoided or reduced if the audit was undertaken by an external independent entity otherwise referred to as third party auditors. This system would be an improvement if compared to the in-house audit but it still does not guarantee to solve all the issues of bias raised in the previous method. In fact, the objectivity of the audit may still be a matter of concern where external auditors are being paid directly by the audited company or may develop long-standing financial relationships with companies and then favour such companies to safeguard their relationship (O'Rourke, 2002). In these instances, auditors may compromise their impartiality and unbiased independence in order to please their clients and ensure an ongoing business relationship (Graham & Woods, 2006). This in turn, would severely impact on the total credibility of those social audits. With reference to the concerns regarding internal and external auditors, the only solution would be the use of a truly independent auditor (Björkman & Wong, 2013; Locke et al., 2007) In order to improve the current social and environmental auditing system there would be a need for the existing principles of the practice to be revised and re-thought (Boiral & Gendron, 2010; Kemp, Owen, & van de Graaff, 2012). The social audit would need to be carried out by an independent entity to the company being audited. The 'external social audit' process would need to be thorough and overreaching, and include investigations, unannounced visits, interviews and monitoring of aspects of organisational activity (Gray, 2000; Pruett, 2005; Mamic, 2005; Locke et al., 2007; Barrientos & Smith, 2007).

Independent third party audits provide an important tool for the regulation of the disclosure of accounting information (Barton & Waymire, 2004; Lennox & Pittman, 2011). Lack of regulation has the tendency to allow companies to focus on their own costs and profit maximisation, while neglecting the accuracy and quality of their social responsibility reports. This has been attributed to the fact that compared to the firm-level benefits society-level benefits are simply smaller (Lennox & Pittman, 2011). However, an increase in government monitoring and regulation does have some drawbacks. It is costly, time consuming and difficult to enforce effectively. In all countries there are gaps in regulatory enforcement due to lack of government capacity (Graham & Woods, 2006). A solution would be to establish a set of standards for social and environmental monitoring mimicking those in financial auditing, and by doing so, removing discretion from auditors and allowing stakeholders to have easy access to the final data (Dando & Swift, 2003). To this end, in Chapter "History and Significance of CSR and Social Audit in Business: Setting a Regulatory Framework", Anjana Hazarika presents a regulatory framework for CSR and social audit based on international guidelines likme the UN, Global Compact, Organisation for Economic Cooperation and Development (OECD) International Labour organisation (ILO) and Global Reporting Initiative (GRI).

While industries have been very wary of governments using too much impositions and controls, it has become evident that, in relation to social audits, companies are reluctant and unlikely to release adequate and comprehensive data "unless their reporting is mandated by the government" (Graham & Woods, 2006). In a voluntary system, if one company publishes detailed information regarding its social and environmental programs, while others either don't, or use CSR reports to disclose information to their personal advantage, then the most honest company would likely suffer the most (Graham & Woods, 2006) and gain no commercial advantage from their truthful disclosure. A company should not disregard its pursuit for profit maximisation as well as the protection of its interests. However it does owe a duty to stakeholders by accurately demonstrating how well it is performing socially and environmentally, whilst it is pursuing its financial goals. As such, it is important that a company must continuously take into account and respond to public expectations, while still maintaining a level of freedom to determine what is socially relevant and how to achieve its social goals (Fetyko, 1975).

Throughout the book, a number of chapters will provide a series of suggestions in relation to the best regulatory framework to be implemented. In Chapter "Social Audit in the Supply Chains Sector", Samuel Idowu describes the impact of the social audit in the supply chain, assesses the most popular guidelines for this audit and suggests the core requirements of an effective social audit. Adopting a slightly different approach, in Chapter "Fostering the Adoption of Environmental Management with the Help of Accounting: An Integrated Framework", Nuuan Gumarathne provides an integrated gradual framework initially driven by internal or external compliance that would facilitate the adoption of environmental management with the help of accounting. In Chapter "Social audit regulation within NGO sector: Practices of NGOs operating in Bangladesh and Indonesia", Vien Chu and Belinda Luke argue that there is a limited real change in corporate action if there is no government regulation proposing a radical approach such as mandatory monitoring and disclosure requirements necessary to ensure corporate accountability in relation to climate change. They proceed to propose a series of mandatory monitoring procedures and disclosure requirements.

Further advice is provided by Adriana Tiron–Tudor et al. in Chapter "Defining a Methodology for Social Audit Based on the Social Responsibility Level of Corporations", which proposes a framework based on a disclosure checklist simulated on a range of corporate social responsibility set of information, relying upon a sample of annual reports from the most socially responsible corporations worldwide. Finally, in Chapter "New Challenges for Internal Audit: Corporate Social Responsibility Aspects", Adriana Tiron-Tudor and Cristina Bota-Avram attempts to provide a framework defining the role of auditors in corporate social responsibility, focusing on the importance of internal social audit. They provide an assessment of the manner in which ethical environmental and social performance is reported and establishes a recommended procedure to increase the contribution of internal social audit in corporate social responsibility.

6 Conclusion

Even after decades of development, current social and environmental auditing practices are still in their infancy. Doubts still exist regarding the capability of the accounting profession to handle the situation especially when companies and their approaches to social auditing vary so significantly between them (Tipgos, 2000). Whilst concrete attempts have been made to implement social accounting procedures amongst an increasing number of international firms, the absence of guidance present in the field has undermined many of these attempts. The lack of a reliable monitoring system, in turn, is threatening the potential benefits gained from the current increasing trend of adopting CSR reporting practices, by not being able to guarantee the level of quality of CSR reports. The most effective way of achieving a soial auditing system which can be trusted and relied upon by stakeholders is to implement specific regulation and guidelines which help to ensure a satisfactory collation of data as well as a high level of impartiality by the auditor. Furthermore, in addition to making sure that social audits are conducted effectively, it is equally important to ensure that regulation would guarantee stakeholders full access to this information. Without such a guarantee, there would be a significant risk that the benefits of implementing a social auditing system would not be achieved. In order for social audits to really be effective they must not be hidden and obscured but should instead be released and disclosed in an adequate and clear manner (Graham & Woods, 2006).

As there is currently an increasing worldwide trend to implement CSR legislation it appears that now would be the best time to seriously consider the implementation of social auditing legislation with the desired effect of ensuring a substantial improvement in global CSR.

References

Admati, A., & Pfleiderer, P. (2000). Forcing firms to talk: Financial disclosure regulation and externalities. *Review of Financial Studies, 13*(3), 479.
Barrientos, S., & Smith, S. (2007). Do workers benefit from ethical trade? Assessing codes of labour practice in global production systems. *Third World Quarterly, 28*(4), 713.
Barton, J., & Waymire, G. (2004). Investor protection under unregulated financial reporting. *Journal of Accounting and Economics, 117*, 38.
Björkman, H., & Wong, E., (2013). *The role of social auditors—A categorization of the unknown* (master thesis, UPPSALA University).
Boiral, O., & Gendron, Y. (2010). Sustainable development and certification practices: Lessons learned and prospects. *Business Strategy and the Environment, 20*, 331.
Courville, S. (2003). Social accountability audits: Challenging or defending democratic governance. *Law & Policy, 25*(3), 269.
Dando, N., & Swift, T. (2003). Transparency and assurance: Minding the credibility gap. *Journal of Business Ethics, 44*, 195.
Darnall, N., Seol, I., & Sarkis, J. (2009). Perceived stakeholder influences and organizations' use of environmental audits. *Accounting, Organizations and Society, 34*, 170.
Deegan, C., Rankin, M., & Tobin, J. (2002). An examination of the corporate social and environmental disclosures of BHP from 1983–1997: A test of legitimacy theory. *Accounting, Auditing & Accountability Journal, 15*(3), 312.
Fetyko, D. F. (1975). The company social audit. *Management Accounting, 56*(10), 645.
Ghonkrokta, S. S., & Lather, A. S. (2007). Identification of role of social audit by stakeholders as accountability tool in good governance. *Journal of Management Research, 7*(1), 18.
Graham, D., & Woods, N. (2006). Making corporate self-regulation effective in developing countries. *World Development, 34*(5), 868.

Gray, R., Owen, D., & Carol, A. (1996). *Accounting and accountability: Changes and challenges in corporate social and environmental reporting.* London: Prentice Hall.
Gray, R. (2000). Current developments and trends in social and environmental auditing, reporting and attestation: A review and comment. *International Journal of Auditing, 4*, 247–268.
Gray, R. (2001). Thirty years of social accounting, reporting and auditing: what (If anything) have we learnt? *Business Ethics: A European Review, 10*(1), 9.
Henriques, A. (2000). Social auditing. *Quality Focus, 4*(2), 60.
Hess, D. (2001). Regulating corporate social performance: A new look at social accounting, auditing, and reporting. *Business Ethics Quarterly, 11*(2), 307.
Humble, J. (1975). *The responsible multinational enterprise.* London: Foundation for Business Responsibilities.
Jaramill, G., & Altschuller, S. (2013). *Expert Q&A on trends in corporate social responsibility* (Reuters, November 2013) Practical Law The Journal. online at http://www.csrandthelaw.com/wp-content/uploads/2013/12/Nov2013_OfNote1.pdf.
Kemp, D., Owen, J. R., & van de Graaff, S. (2012). Corporate social responsibility, mining and "audit culture". *Journal of Cleaner Production, 24*.
Kok, P., et al. (2001). A corporate social responsibility audit within a quality management framework. *Journal of Business Ethics, 31*, 285.
Kurian, T. (2005). *Social Audit* (Centre for good governence, 2005) online at http://www.sasanet.org/documents/Tools/Social%20Audit.pdf.
Laan, S. (2009). The role of theory in explaining motivation for corporate social disclosures: Voluntary disclosures vs 'solicited' disclosures. *Australasian Business and Finance Journal, 3*(4), 15.
Lambert, R., Leuz, C., & Verrecchia, R. (2007). Accounting information, disclosure, and the cost of capital. *Journal of Accounting Research, 45*(2), 385.
Lennox, C. S., & Pittman, J. A. (2011). Voluntary audits versus mandatory audits. *The Accounting Review, 86*(5), 1655.
Leuz, C., & Wysocki, P. (2008). Economic consequences of financial reporting and disclosure regulation: A review and suggestions for future research. (Working paper, University of Chicago, 2008)
Locke, R. M., Qin, F., & Brause, A. (2007). Does monitoring improve labor standards? Lessons from Nike. *Industrial and Labor Relations Review, 61*(1), 3.
Magness, V. (2006). Strategic posture, financial performance, and environmental disclosure: an empiric al test of legitimacy theory. *Accounting Auditing and Accountability Journal, 19*(4), 540.
Mamic, I. (2005). Managing global supply chain: The sports footwear, apparel and retail sectors. *Journal of Business Ethics, 59*, 81–100.
Miles, S., Hammond, K., & Friedman, A. (2002). *Social and environmental reporting and ethical investment, ACCA Research Report No.77* (London: Certified Accountants Educational Trust).
O'Rouke, D. (2000). *Monitoring the monitors: A Critique of PricewaterhouseCoopers PWC Labour Monitoring.* (Massachusetts Institute of Technology, September 28th).
O'Rourke, D. (2002). Motivating a conflicted environmental state: Community driven regulation in Vietnam. In A. P. J. Mol, & F. H. Buttel (Eds.), *The environmental state under pressure. Social Problems and Public Policies Series.* Amsterdam: Elsevier.
Organisation for Economic Co-operation and Development. (2014). OECD Guidelines for Multinational Enterprises (2011 Edition) Online at: .http://www.oecd.org/daf/inv/mne/48004323.pdf;
Owusu, C. A., & Frimpong, S. (2012). Corporate social and environmental auditing: Perceived responsibility or regulatory requirement? *Research Journal of Finance and Accounting, 3*(4), 47.
Perrini, F. (2006). The practitioner's perspective on non—financial reporting. *California Management Review, 48*(2), 73.

Pruett, D. (2005). *Looking for a quick fix: How weak social auditing is keeping workers in sweatshops*. Amsterdam: Clean Clothes Campaign.

Roberts, R. W. (1992). Determinants of corporate social responsibility disclosure: An application of stakeholder theory. *Accounting, Organizations and Society, 17*(6), 595.

Roberts, R., & Mahoney, D. L. (2004). Stakeholder conceptions of the corporation: Their meaning and influence in accounting research. *Business Ethics Quarterly, 14*(3), 399.

Spira, L. F., & Page, M. (2002). Risk management—The reinvention of internal control and the changing role of internal audit. *Accounting, Auditing & Accountability Journal, 16*(4), 640.

Sunder, S. (2003). Rethinking the structure of accounting and auditing. *Indian Accounting Review, 7*(1), 1.

Sy, M. V. U. (2013). *Asia Pacific Industrial Engineering and Management System Drivers of corporate social responsibility leading to sustainable development*. Philippines: University of San Jose-Recoletos.

The Seep Network. (2008). *Social Performance Map* (The Seep Network, 2008) online at http://www.setoolbelt.org/system/files/resources/130_835.pdfonline;

Timane, R., & Chavan, M. (2012). A study of stakeholder engagement in social audit. *International Journal of Research in It & Management, 2*(12).

Tipgos, M. A. (2000). A case against social audit. *Management Accounting, 58*(5), 23.

Tschopp, D. (2005). Corporate social responsibility: A comparison between the United States and the European Union. *Corporate Social-Responsibility and Environmental Management, 12*(1), 55.

US Aid. (2008). *Social audit tool handbook: Using the social audit to assess the social performance of microfinance institutions*. (U.S. Agency for International Development April 2008) online at http://www.microlinks.org/sites/microlinks/files/resource/files/ML5896_social_audit_handbook.pdf.

New Challenges for Internal Audit: Corporate Social Responsibility Aspects

Adriana Tiron-Tudor and Cristina Bota-Avram

1 Introduction

Corporate social responsibility and accountability at corporate level are some of the topics with an increasing interest from both researchers and practitioners. According to Gray, Owen, and Maunders (1987), corporate social responsibility can be defined as the process of providing information designed to accomplish social accountability. No doubt, there is general consensus about the fact that business should no longer operated in a vacuum, but within a social environment, with an increasing focus for achieving principles of good social, ethical and environmental practice (King, 2002). Even more, the economic value added provided by organisations should no longer be the main criteria for evaluation of organisational performance, but also its impact on environment and value or contributions to social issues are more and more relevant.

In this context, a major challenge would be for internal audit to provide its value added by improving the value of the company and firm performance in terms of good social, ethical and environmental practices. Starting from the following principle issued by Ridley (2008) for all internal auditors that *"Internal auditing has a responsibility to contribute to the processes of assessing reputation risks and advising at all levels in their organizations on how reputation can be managed and enhanced through good corporate responsibility practices"*, this chapter proposes an examination of the arguments that justify the necessity for internal audit to play a significant role in corporate social responsibility. Emphasis is placed on the need for internal auditors to be sensitive to the complex of their social responsibilities and challenges; trying in the same time to highlight the main actions through

---

A. Tiron-Tudor (✉) • C. Bota-Avram
Faculty of Economics and Business Administration, Babes-Bolyai University, Al. Densusianu 24, Cluj-Napoca, Romania
e-mail: adriana.tiron.tudor@gmail.com

internal audit can contribute to the ensuring of corporate accountability in terms of societal issues.

2 Corporate Social Responsibility: Conceptual Approaches and International Development

The interest in corporate social responsibility increased over the last two decades, while topics such as environmental and social responsibilities at corporate level are now even of greater significance at the global level (Maignan & Raltson, 2002). The concept of corporate social responsibility (CSR) has been evolving quite dynamically during the last few decades. Thus, if Nobel-Prize winning economist Milton Friedman (1970) clearly stated his vision about CSR, in his article *The Social Responsibility of Business is to Increase Its Profits* (Friedman, 1970 cited by Sprinkle & Maines, 2010), today, the academic literature on CSR contains a large spectrum of conceptual approaches of CSR, the interaction between CSR and other areas such corporate governance, accounting and audit being largely analysed. Even more, according to the study achieved by McKinsey (2006, cited by KaKabadse & Kakabadse, 2007), among 4,238 executives in 116 countries, more than 84 % no longer accepted the opinion shared by Nobel Laureate Milton Friedman (1970), that major goal of business is only to increase its profits. The participants to this study also admitted that now companies have to face with a larger and increasing spectrum of organisational risks, especially in terms of reputational and minimisation of shareholder value, therefore, now companies should be more careful about the economic, social, political and environmental impact of their actions.

One definition of CSR is given by the European European Commission (2010) where this concept was defined as follows:

> Corporate social responsibility is a concept whereby companies integrate social and environmental concerns in their business operations and in their interaction with their stakeholders on a voluntary basis (European Commission, 2010).

Davis (1973) cited by Sprinkle and Maines (2010) attached to corporate social responsibility the following definition:

> the firm's considerations of, and response to, issues beyond the narrow economic, technical, and legal requirements of the firm to accomplish social [and environmental] benefits along with the traditional economic gains which the firm seeks (Davis, 1973 cited by Sprinkle & Maines, 2010).

According to the World Business Council for Sustainable Development (2010), corporate social responsibility or sustainability could be defined as the continuing commitment of organizations to behave in an ethical manner and trying to provide a valuable contribution to the economic development and social prosperity, looking for the enhancement of quality of life for the employees and their families as well as the local community and society. To achieve this objective in developing their businesses, companies should *"innovate, adapt, collaborate and execute"*. To perform these activities as best as possible, companies should form close partnerships with other businesses, governmental agencies, academic and non-governmental organisations *"in order to get it right for all"* (The World Business Council for Sustainable Development, 2010). This definition is in line with the previous conceptual approach of CSR concept developed by the World Business Council for Sustainable Development in its previous paper—Corporate Social Responsibility: Making good Business Sense, where the corporate social responsibility was defined as follows:

> the commitment of business to contribute to sustainable economic development working with employees, their families, the local community and society at large to improve their quality of life (World Business Council for Sustainable Development, 2000).

In the vision of the Government of Canada, this concept of CSR should be understood as:

> the way firms integrate social, environmental and economic concerns into their values, culture, decision-making, strategy and operations in a transparent and accountable manner and thereby establish better practices within the firm, create wealth, and improve society (IIA, 2010a).

Corporate Social Responsibility was defined by the Global Reporting Initiative (GRI) 2002 "Sustainability reporting guidelines" as a significant part of sustainability and sustainability is *"one of the three ideas that are playing a pivotal role in shaping how business and other organizations operate in the twenty-first century"* (GRI, 2002), while the other two ideas are accountability and governance. Looking for a more comprehensive understanding of "sustainability" concept, United Nations Global Compact provides in its paper some clarifications about this concept:

> Throughout this report, we use the term "sustainability" to encompass environmental, social and corporate governance issues, as embodied in the United Nations Global Compact's Ten Principles. These ten principles cover areas of "human rights, labour, the environment and anti-corruption" (United Nations Global Compact, 2010 cited by Ridley, D'Silva, & Szombathelyi, 2011).

Another interesting and complex perspective on this CSR concept was given by Rayner (2003), who develops the definition of CSR around some major elements such as:

- Companies should operate in a way that goes beyond fundamental legal compliance to larger areas, which include social and environmental aspects.
- In designing corporate strategy, organisations should take into account the impact, but also the value-added contributions to society and environment, by paying attention for minimising negative influences and maximising positives ones.
- The risks addressed and evaluated by company should also consider social, ethical and environmental risks.

- Transparency in disclosing the responsibilities and respect for human rights related to different internal or external stakeholders of the organisation.
- Companies should consider an appropriate answer to the needs and expectations of various stakeholders.
- All the elements mentioned above should be correlated in an integrated framework where elements like corporate strategy, corporate governance, management decisions and reporting systems are interrelated.

There are numerous terms used for CSR, but the most common term used in addition to CSR is "corporate sustainability". The principles incorporated in this is similar to those connected with CSR and are derived from the following areas: ethics, governance, transparency, business relationships, financial return, community involvement, product value, employment practices and environmental protection (Epstein, 2008 cited by Sprinkle & Maines, 2010). The overlapping of corporate social responsibility issues with all aspects of sustainability is well argued by Ridley et al. (2011), who stated that the commitment of organisations for ensuring sustainability generates a strong impact and influence on all strategic and operational practices of the organisations.

One thing is clear—organizations are no longer only accountable to shareholders in terms of ensuring increasing of financial return on their capital investment. The spectrum of corporate responsibilities is now much more complex, including also social and environmental responsibilities about the potential impact of their operations and actions on stakeholders such as employees, existing and prospective customers, suppliers, governmental and non-governmental agencies and society at large.

In spite of the many conceptual approaches of CSR, one could identify some common elements around which corporate social responsibilities should be designed. These could include social, environmental and financial issues, which for major international companies require them to develop a "Triple Bottom Line" reporting strategy which encompasses relevant areas to be covered by the strategy. Thus, the interest of companies in their reporting is moved forward from the needs of shareholders to the interests of all the stakeholders who might be influenced by the organisation's business policies and actions. While looking for a global positive impact for the entire society in achieving business purposes, the responsibility of companies has to be focused more significantly on how to impregnate the business operations with the commitment for social values, without significant impacts on its profitability, looking at the same time to minimize its potential negative impacts on the stakeholders.

This is no doubt, a complex corporate strategy which should incorporate corporate responsibility issues that should organised around major objectives of CSR, as depicted in Fig. 2.1. The responsibility for ensuring that CSR objectives are established, the management of risks, the measurement of performance and

Fig. 2.1 CSR objectives

appropriate monitoring and reporting of companies' actions belong to management, while, on the other hand—

> CSR activities are pervasive throughout the organisation; thus every employee has a responsibility for ensuring the success of CSR objectives (IIA, 2010a).

Considering the increasing interest in CSR or corporate sustainability, at international level many set of principles, guidelines and standards have been developed by various governmental, non-governmental and professional organisations from around the world. Thus, while some address particularly social and ethical aspects of CSR/corporate sustainability, others adopt a broader perspective where aspects like employees, human rights, the environment and corporate governance issues are included and interrelated (Ackers, 2008b). A synthesis of these major international guidelines in terms of ensuring sustainability is provided in Table 2.1.

3 Role of Internal Audit in Corporate Social Responsibility

In this context of growing significance paid for corporate social responsibility and sustainability, one issue that needs more analysis is about the role of the audit profession in providing assurance on CSR topics. As Ackers (2009) admitted:

> the auditors' role as a CSR assurance provider is expected to become increasingly more important .... This increased demand for assurance services will require the global audit profession's paradigm to be re-examined to include competence in contextual accounting and auditing.

There is a large amount of research papers and reports of various professional organisations and researchers dedicated to issues related to corporate social responsibility and how to provide its assurance, but in spite of the fact that latest research

Table 2.1 CSR/Corporate sustainability—a synthesis of international guidelines

Title	Description
ISO 26 000	ISO (The International Organisation for Standardisation) issued the ISO 26000 standards in order to provide voluntary guidance on corporate social responsibility. It clearly states the importance of CSR to the sustainability of the organisation (ISO, 2010).
AA1000AS Assurance Standard	It is a free open-source set of internationally recognised standards specifically designed to provide assurance in terms of sustainability. Within this set of standards there are presented the principles that should define the robustness of the assurance process (AccountAbility, 2008).
Caux Round Table Principles for Business	It represents a set of principles developed by business leaders, with a greater focus on business conduct, community involvement and corporate governance, while aspects such environmental and human rights are less promoted.
Global Reporting Initiative (GRI) – Sustainability Reporting Guidelines	It is a framework of principles and guidance together with disclosures and indicators for voluntary use by organisations in reporting the performance achieved in terms of sustainability.
The International Standards on Assurance Engagements 3000 (ISAE 3000)	It is also known as the "Assurance engagements other than audit or reviews of historical financial information", published in 2005 by the International Federation of Accountants (IFAC). This standard provides guidance to the audit profession about the principles and procedures that have to be followed when performing non-financial assurance.
Organisation for Economic Cooperation and Development's Guidelines for Multinational Enterprises	These OECD principles and standards are voluntary and they are endorsed by over 33 countries focusing on responsible business conduct by multinationals covering human rights and environmental issues (OECD, 2008).
Principles for Global Corporate Responsibility: Benchmarks for measuring business performance	This set of performance standards and expectations for corporate behaviour was issued by Interfaith Center on Corporate Responsibility (ICCR), including a wide area of 60 issues that ICCR considers fundamental in order to ensure responsible corporate responsibility, including the environment, employees and corporate governance aspects.
Social Accountability (SA) 8000	It represents a set of standard containing nine principles focusing on labour and human rights for international companies.

(continued)

Table 2.1 (continued)

Title	Description
United Nations Global Compact (UNGC)	It is a voluntary corporate citizenship standard covering ten principles in the field of human rights, labour and environment. It, also, includes specific practices for determining the organisations to act with respect to both internal corporate practices and complementary external public policy actions (UNGC, 2004).

Source: adapted after Ackers (2008b)

trends show that CSR assurance has become an increasing field of interest in auditing research, still it would appear that little scientific research has been done on the role of the internal audit function (Ackers, 2008a).

The Institute of Internal Auditors (IIA) notes that as *"the internal audit profession is the global voice, recognized authority, and acknowledged leader"* the IIA has argued that the major objectives of internal audit through the definition assigned to this function:

> Internal auditing is an independent, objective assurance and consulting activity designed to add value and improve an organization's operations. It helps an organization accomplish its objectives by bringing a systematic, disciplined approach to evaluate and improve the effectiveness of risk management, control, and governance processes (The Institute of Internal Auditors, www.theiia.org).

Even more, the IIA's Standard 2130—Figure 2.2 on Governance states that internal audit has to provide an effective contribution to improving organizations' controls in managing the risks within the processes of governance, operations and information systems.

The study released by PriceWaterhouseCoopers (PWC, 2011) about the state of internal audit profession emphasized that corporate social responsibility and sustainability have become growing areas of internal audit interest, while 35 % of the survey respondents clearly stated that their internal audit plan explicitly takes corporate social issues into consideration. No doubt, the internal audit techniques and approaches are continuing to evolve thus highlighting the internal audit function to be as best as possible strongly implicated in the process of better understanding the risks and exposures that might negatively influence the corporations. According to the respondents of the study of PWC (2011), some examples of internal audit approaches about corporate social responsibility issues could be used to:

- Elaborate an inventory of regulatory reports and public statements in the field of corporate social responsibility
- Provide assistance in the process of development of policies and procedures for review and approval of public statements about corporate social responsibility.
- Provide assistance for the development of regulatory compliance programs in terms of social responsibility issues.

The major goal of internal audit is to provide an assessment of the internal control environment, to undertake complex reviews of organisations' compliance with the applicable legislation and regulatory framework. Therefore, the planning

> *The internal audit activity must assist the organization in maintaining effective controls by evaluating their effectiveness and efficiency and by promoting continuous improvement.*
>
> *2130.A1 – The internal audit activity must evaluate the adequacy and effectiveness of controls in responding to risks within the organization's governance, operations, and information systems regarding the:*
> - ✓ *Achievement of the organization's strategic objectives;*
> - ✓ *Reliability and integrity of financial and operational information;*
> - ✓ *Effectiveness and efficiency of operations and programs;*
> - ✓ *Safeguarding of assets; and*
> - ✓ *Compliance with laws, regulations, policies, procedures, and contracts.*

Fig. 2.2 IIA's Standard 2130. *Source*: The IIA International Standards for the Professional Practice of Internal Auditing (Standards)

of internal audit activities should necessarily take into account social and environmental aspects, including the fact that the spectrum of CSR risks is properly mitigated. In the vision of Sawyer, Dittenhofer, and Scheiner (2003) who note that the involvement of internal audit in corporate social responsibilities should provide at least the following benefits:

- Internal audit should provide assistance with legislative and regulatory compliance, including social and environmental aspects.
- Internal audit should be in a position of being able to identify the possible problematic areas, which could generate significant remediation costs and penalties, and also potential litigation against the company, which could negatively affect the further financial return of investor, but also its reputation.
- Internal audit should contribute to building a strong organisational image of the company, because of the assurance provided about the compliance with legal requirements and ethical business practices.
- Because of the internal audit's contribution to the effective self regulation, the relationship with regulatory authorities could be strongly improved.

Discussing the way many companies are achieving their CSR responsibilities, Jenkins (2001) states that only effective monitoring and independent verification could provide real assurance that a company fully meets certain ethical and social standards. Such an independent monitoring and effective assessment can be provided by the internal audit function.

Analysing the role that internal audit could play in this regards, Ackers (2008a) develops a study among a sample of 40 representative South African enterprises (with a response rate of 30 %). Their findings suggest that 91.7 % of respondents have a CSR policy which includes aspects such as: corporate social investment, sustainable development, human resource development, employee well-being, occupational health and safety, education, environmental management, community involvement, corporate citizenship, community Health & Welfare and energy efficiency. Considering the role of internal audit in CSR, 33.3 % of respondents stated the role played by internal audit in CSR, while under the clear requirement of compliance with internal audit IIA's standards, only 50 % reported compliance with IIA Standards 2130 on Governance, while 41.7 % of the respondents shared the opinion that internal audit could play a more active role in corporate social responsibility and governance issues. Because of the relatively small sample used in this study, the conclusions derived from the analysis of data results cannot be generated.

One relevant model for internal auditing to provide value-added to sustainability was developed by IIARF (2006), where it presented some of the key roles that internal auditors could play in providing assurance for an effective sustainable corporate framework (IIARF, 2006 cited by Ridley et al., 2011):

- Assistance for management in designing/implementing their company's sustainability management systems
- Training for employees in achieving their sustainability actions,
- Performing limited scope audits requested by top management,
- Achieving supply chain and compliance audits,
- Coordinating audit actions supplied by external providers.

Despite the existence in the scholarship literature of a limited amount of research dedicated to the role of internal audit in CSR, there is a general and common view that internal audit should have a distinct role to fulfill in CSR. Otherwise, if internal auditors are inactive from this perspective, they could be positioned in a dangerous area of not achieving their primary objective of delivering value added services about the assurance that should provide effective risk mitigation.

4 Ways to Increase the Contribution of Internal Audit in Corporate Social Responsibility

Ackers (2008b) notes that the audit process of sustainability/CSR must take into consideration all the dimensions of sustainability (social, environmental and economic) and should include at least following major steps:

- Define the audit program objective and scope.
- Select the representative business units to be included in the audit, specifying the frequency of the audits, their planning and the methodology used for establishing the plan of audit work.
- Select the audit protocols, checklist, procedures and guidelines to be used during the audit.
- Define the pre-audit activities such as meeting activities, file reviews, interviews and travel logistics.
- Establish procedures for audit reporting and document management. These procedures should indicate the elements that should be included within the

audit report, timing and distribution, but also instructions on how documents generated during the audit will be managed.
- Establish procedures for post-audit corrective actions and follow-up of findings. These procedures should indicate the responsibilities and the mechanisms to track the recommendations made by auditor in order to correct any deficiency identified.
- Establish quality assurance processes to be incorporate into the audit process.

The role of internal audit in providing assurance about CSR issues is also one of the major topic of interests of the Institute of Internal Auditors (IIA) which is the international professional body, its supremacy being recognised all over the world in coordinating the development and progress of the internal audit profession. The IIA organised a series of courses dedicated to the role of internal audit in CSR areas—"Corporate social responsibility: opportunities for internal audit" (IIA, 2010b), which has provided a significant framework about the important questions that every internal auditor should frequently be asking the organisations they perform their audit function as depicted in Fig. 2.3.

Internal auditors are in the position of being involved in corporate social responsibility at different levels. One of such position is as a result of them—internal auditors being asked to provide assistance and advice to management on how to manage CSR activities. Another situation might be when internal auditors are called upon to audit CSR programs and objectives providing assurance on how CSR objectives and responsibilities were achieved. But regardless, the position supplied by the internal auditor, there are a variety of risks related to CSR issues that need to be well-known and understood by the internal audit team in order to be included in the audit plan and procedures.

The subject of internal audit's role in the auditing CSR activities and providing assurance in this field was, also, one of the major interest topics for The Institute of Internal Auditors (IIA) which has developed a practice guide, where significant and valuable guidelines are offered to internal auditors when they are called on to evaluate corporate social responsibility or sustainable development (IIA, 2010a). There are a lot of various risks associated with CSR activities, but the responsibilities for performing an appropriate assessment of these risks and implementing the control activities to manage those risks belong to management. The primary role of the internal auditor will be to provide an assurance about the effectiveness of these control activities designed to manage the risks related to CSR. A valuable synthesis of main categories of risks that could influence corporate social responsibility of sustainable development activities is provided by this IIA's guide (IIA, 2010a), which is presented below in the Table 2.2.

According to the IIA's practice guide (IIA, 2010a); an internal auditor could adopt one of the next approaches in auditing their company's CSR activities and their related controls. The internal auditor could:

a. Separate audits of each element included in CSR activities, that could further be developed into audits of matters at the corporate head office, subsidiaries and with external business relationship. Generally, these CSR activities should

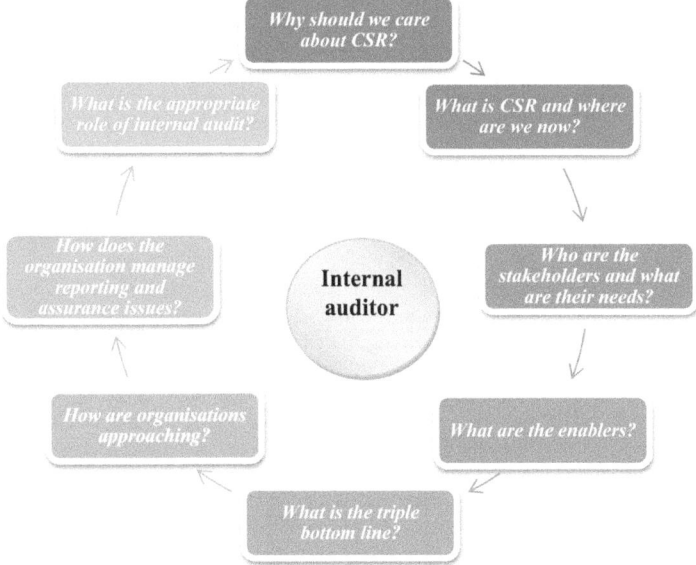

Fig. 2.3 Questions for internal auditor. *Source*: Adapted after Ridley, D'Silva, Szombathelyi, (2011) citing IIA (2010a, b)

include at least following elements: governance; community investment; environment; ethics; health, safety and security; transparency; working conditions and human rights.
b. Develop the audit of CSR activities by referring to each significant stakeholder group involved or affected by CSR activities that could also further be developed into audits of matters at the corporate head office, subsidiaries and with external business relationship. The main groups of stakeholders influenced by CSR activities are customers; employees and their families; the environment; local communities; shareholders/investors; suppliers, non-governmental organisations and activist groups.
c. Group the related subjects as follows :

- Workplace—which could include elements such as: health and safety; environmental management practices; ensuring diversity and equality; training and development; ethics; governance and human rights.
- Marketplace—ensuring the appropriate safety and quality of product and services; responsible advertising and sales; responsible supply chain management; adequate practices and procedures in product development and testing; disclosure practices and privacy.
- Environment—responsibility for natural resources such as air, water, land, waste, animals and energy use and compliance with regulatory framework related to them.

Table 2.2 The categories of risks affecting CSR activities and objectives

Category of risks	Description
Reputation	If there are violations of law or principles, errors or omissions in disclosed CSR information, under-performance compared with objectives/targets, or the appearance of indifference to social issues, then the organization's brand or reputation could be seriously damaged. By adopting behaviour in a socially responsible manner and showing availability for involving stakeholders in business decisions that might affect them, the companies could have the opportunity to improve their reputation or their image.
Compliance	There are a lot of regulations and legislative framework related to the environment, health and safety, employment, governance, conflict of interest, preventing fraud, political contributions that organizations have to comply with. But due to the extent, complexity and quantity of these regulations, companies may fail in ensuring the compliance with those legislative frameworks. These compliance risks are also given by the contractual obligations with third parties such as customers, non-profit organisations, and employees and from voluntary choice to adopt some standards. Also, for companies operation in multiple countries, the compliance risk increases due to the diversity and complexity of various regulations that should be complied with.
Liability	Liability risks are given by the situations when activists or specific interest organizations may proceed to legal actions for fighting against alleged harm done by the organization.
Operational	Operational risks derive from CSR pressures on the organisation's manufacturing processes, products, services and impact on the environment. In the vision of IIA's other examples of potential risks scenarios include: under-performance of other targets due to inappropriate CSR strategies, or over-emphasis on CSR strategies; failure to integrate CSR objectives into processes, or to educate staff appropriately; failure to develop well-controlled systems for CSR initiatives; risk associated with reporting CSR activities and results (e.g. incomplete disclosure of information and reporting related to CSR activities and responsibilities). Also, applying same rules or standards related to CSR activities for organisations operating in various countries may be difficult or at least challenging.
Stock Market	If the organisation does not qualify for various social responsible investment funds, they may be faced with the risks of losing investors or their number of investors to be limited.
Employment market	No doubt, employees are more willing to work for companies that respect their rights, show a culture of integrity and express their availability to contribute at solving of social and community concerns.
Sales Market	If the organisations are known in the group of companies socially responsible, than they have the chance to contribute to the increasing of their sales and to mitigate the risks of boycott of its products and services by its customers, due to environmental or social reasons.
External Business Relationships	The compliance with CSR terms, conditions, principles, laws or objectives should also be reflected by the organisation' business relationships with customers, suppliers or business partners or associate.

Source: IIA (2010a)

- Community—philanthropy, local economic support, capacity building, volunteerism and stakeholder engagement.

d. Audit the system of internal controls over risk management, recording, measuring and reporting of CSR activities applied within each department, function or activity included in the audit plan (or in other words, including in the audit of each department/ function/activity an audit objective related to CSR activities).
e. Conduct Assurance audits of the way that company provides public information about their approaches and results in achieving CSR activities and objectives (or in other words, the audit of reporting of financial and nonfinancial information related to CSR elements).
f. Audit third parties for compliance with contractual clauses, including compliance with CSR terms and conditions.

IIA (2010a) provides, also, some general considerations that should be taken in account by the internal auditors in developing their internal audit program, which could include the potential answers to a list of questions.

- Taking in consideration the disclosure of CSR information produced by the organizations. Internal auditor should look for answers to questions like: is there any consistency in the messages inserted in public reports, speeches, and presentations on the organisation's website? How is process of disclosure of CSR information and activities controlled?
- Has the organization made a decision to align its reporting process of its CSR activities and information with one of the recognised at international level reporting standards and guidelines in the field of corporate social responsibility?
- How is the organisation communicating its CSR strategies and priorities? Internal auditor should find answers to questions such as: How are CSR strategies and priorities be incorporated into the process of decision making and approval? When there are conflicting objectives, which elements take precedence?
- Which elements of CSR responsibilities and strategies are included in the organisational structure? Is the responsible position in the organisational structure occupied by qualified staff with experienced and necessary competences?
- Is the organization a signatory to voluntary standards of performance? If not, then why not? Were the standards adopted by management, or by the board? How are they integrated into management practices? How is compliance monitored within the organization?
- How does the organization manage the process of compliance with local and international laws? Does the organization meet standards required for inclusion in environmental or social investment funds? If not, then why not?
- It is there any risk that the company's reputation could be negatively influenced by the external business relationships that involve environmental or human rights issues?
- Can the CSR activities of customers adversely impact the organization's reputation? Would the organization refrain from selling products to organizations

with irresponsible or unsustainable practices? Does it provide programs to encourage or facilitate customers to be responsible with its products?
- How well controlled are the mechanisms put in place for capturing information about CSR activities and developing and reporting performance metrics? What instruments are used, and are there adequate control activities in place to ensure complete, accurate and timely information?
- It is there a procedure to follow in publishing a CSR report?

The Internal auditor should also look for answers to the following questions:

– Is the disclosure process for CSR activities and results as rigorous as for Financial reporting?
– Does this CSR report contain clear messages that are aligned with the company's vision and commitments?
– Does it contain balanced reporting (mixed presentation of good elements with less positive ones), performance measures, and trends?
– Is the CSR report written in a manner that allows the reader to understand the issues and the organization's accountabilities?
– Is there any possibility to compare the organisation's CSR disclosed program with others programs reported by other companies?
– Has the organisation received recognition or awards for compliance with international standards or guidelines in the field of CSR?
– How do independent organisations perceive the company's CSR report in this field of activity?
– Has the organisation asked for feedback about any of CSR issues or activities? Were these feedback given in a timely and effective manner? Are these feedback adequately disclosed to prove the company's preoccupation in the field of CSR?

Using these considerations and answers to the previous questions, in order to proceed to the audit of CSR activities/sustainability development, internal auditor should establish the company's audit objectives which could include the following elements:

- Based on audit evidence obtained by applying appropriate audit techniques, the internal auditor should deeply understand the company's risk identification, response plans and internal control activities to mitigate the identified risks in the field of CSR's programs, activities and responsibilities.
- Proceeding to test and analyse a sample of CSR programs in order to identify the degree of compliance with policies and procedures established by management in this field, but also the degree of compliance with international standards and guidelines that company has shown adherence.
- Identifying weaknesses in the control mechanisms and activities implemented in the field of CSR, and by making recommendations for enhancing the ability of the organization to fulfill its duties in terms of corporate social responsibility, represents now, one of the ways by which the internal auditor could provide real added-value to the company.

Considering these audit objectives, the internal audit program must be carefully planned, taking into consideration the following steps:

1. **Preparatory phase,** where the following procedures should be followed:
   - Determine the time period to be examined.
   - Obtain and review the applicable policies and procedures followed by company in the field of corporate social responsibility programs.
   - Identify and develop a risk profile related to CSR objectives and activities.
   - Identify strengths and weaknesses in this risk profile related to CSR programs.
   - Plan meeting, where the scope, approach, period, areas included and adequate auditee contacts are established.

2. **Testing phase**, where the following procedures should be followed:
   - Apply the audit procedures and techniques on the activities included in the CSR program, considering particularly the weaknesses identified previously, when CSR risk profile was identified.
   - Make a synthesis of the main deficiencies found
   - Make recommendations to improve the quality of internal controls in the field of CSR to mitigate CSR specific risks. These recommendations could start from the review of any known best practice for corporate social responsibility and their adaptation to the specific of audited company.

3. **Closing phase**, where the following procedures should be followed:
   - Prepare preliminary draft of the CSR internal audit using the standard format. Ensure that all findings are well argued.
   - Discuss the main findings with the representative of the audited entity in field of CSR. Also, ensure that any recommended action has been discussed with these representatives.
   - Issue preliminary report to management. At this time, management and representative should express their agreement about their availability to implement proposed recommendations and any suggested action has been identified and agreed in the internal audit report.
   - After all internal audit report content has been validated and agreed, the final version of the report should be issued.

Conclusions
No doubt, companies worldwide are increasingly preoccupied with how their activities and actions affect the environment and social welfare, while employees, consumers, governmental agencies, investors, non-profit organisations and other stakeholders groups are requesting that organisations act in

(continued)

a socially responsible manner (a view also shared by other researchers such as Sprinkle & Maines, 2010).

From their position and the requirement of absolute independence of internal auditors, the internal audit should be seen to be effective. From this statement, one conclusion is clear the internal audit is one of the features which demonstrate a company's independence that could provide an independent assessment of the ethical, environmental and social performance of a company, without affecting the financial expectations of the investment community. Internal auditors have a competitive advantage, as one of the main pillars of audit function, because they have the necessary resources and capacities to obtain a complete understanding of corporate responsibility and sustainability development. This could assist the function to develop into a value-added centre by proposing consistent recommendations to improve the organisation's ability to give an appropriate response to its obligations and responsibilities in the field of corporate social responsibility.

References

AccountAbility. (2008). *AA1000 series of standards*. Retrieved from www.accountability.org/standards/
Ackers, B. (2008a). *Corporate Social Responsibility—an internal audit perspective*, (Part 1), Internal Audit Adviser, October-November, http://www.academia.edu/1060355/Corporate_Social_Responsibility_An_Internal_Audit_Perspective
Ackers, B. (2008b). *Corporate Social Responsibility—an internal audit perspective*, (Part 2), Internal Audit Adviser, December, http://www.academia.edu/1529876/Corporate_Social_Responsibility_An_Internal_Audit_Perspective_IA_Adviser_-_Part_2_# .
Ackers, B. (2009). Corporate social responsibility assurance: how do South African publicly listed companies compare? *Meditari Accountancy Research, 17*(2), 1–17.
Davis, K. (1973). The case for and against business assumption of social responsibilities. *Academy of Management Journal, 16*(2), 312–323.
Epstein, M. J. (2008). *Making sustainability work: Best practices in managing and measuring corporate social, environmental, and economic impacts*. San Francisco: Berrett-Koehler Publishers, Inc.
European Commission. (2010). Corporate social responsibility (CSR). Retrieved from http://ec.europa.eu/enterprise/policies/sustainable-business/corporate-social-responsibility/index_en.htm
Friedman, M. (1970). The Social responsibility of business is to increase its profits. *New York Times Magazine*. Retrieved from http://www.colorado.edu/studentgroups/libertarians/issues/friedman-soc-resp-business.html.
Global Reporting Initiative. (GRI). (2002). "Sustainability reporting guidelines." *G2 guidelines*.
Gray, R. H., Owen, D., & Maunders, K. (1987). *Corporate social reporting; accounting and accountability*. Hertfordshire, UK: Prentice Hall.
IIA. (2010a). Evaluating corporate social responsibility/Sustainable Development, IPPF – Practice Guide, February. Retrieved from http://www.iia.nl/SiteFiles/10038_PRO-Corporate_Social_Responsiblity_PG-FNL%5B1%5D.pdf

IIA. (2010b). Course—*Corporate social responsibility: Opportunities for internal audit.* Retrieved from https://na.theiia.org/training/courses/Pages/Corporate-Social-Responsibility-Opportunities-for-Internal-Audit.aspx

IIARF. (2006). *Sustainability and internal auditing.* Altamonte Springs, FL: The Institute of Internal Auditors Research Foundation.

ISO. (2010). *ISO 26000: Guidance on social responsibility.* Retrieved online at www.iso.org/iso/catalogue_details?csnumber=42546.

Jenkins, R. (2001). *Corporate codes of conduct: Self-regulation in a global economy.* Technology, Business and Society, Programme Paper No. 2, April 2001. United Nations Research Institute for Social Development (UNIRISD), Geneva, Switzerland

KaKabadse, A., & Kakabadse, N. (2007). *CSR in practice—delving deep.* New York, NY, USA: Palgrave Macmillan.

King, M. E. (2002). *The King Report on Corporate Governance for South Africa 2002.* Parklands, South Africa.

Maignan, I., & Raltson, D. A. (2002). Corporate Social Responsibility in Europe and the US: Insights from businesses' self-presentations. *Journal of International Business Studies, 33*(3), 497. 3rd Quarter.

Organisation for Economic Co-operation and Development –OECD. (2008). *OECD guidelines for multinational enterprises.*

PWC. (2011). *State of internal audit profession study.* accessible on-line at http://www.pwc.com/en_US/us/internal-audit/publications/assets/state-of-internal-audit-profession-study-2011.pdf.

Rayner, J. (2003). *Managing reputational risk—Curbing threats, leveraging opportunities.* Chichester, England: John Wiley & Sons.

Ridley, J. (2008). *Cutting edge internal auditing.* Chichester, UK: Wiley.

Ridley, J., D'Silva, K., & Szombathelyi, M. (2011). Sustainability assurance and internal auditing in emerging markets. *Corporate Governance, 11*(4), 475–488.

Sawyer, L. B., Dittenhofer, M. A., & Scheiner, J. H. (2003). *Sawyer's internal auditing—The practice of modern internal auditing* (5th ed.). Altamonte Springs, FL: Institute of Internal Auditors.

Sprinkle, G. B., & Maines, L. A. (2010). The benefits and costs of corporate social responsibility. *Business Horizons, 53*, 445–453.

United Nations Global Compact. (2010). *A new era of sustainability.* New York, NY: United Nations.

United Nations Global Compact—UNGC (2004). The ten principles. 2004. Retrieved from www.unglobalcompact.org/aboutthegc/thetenprinciples

World Business Council for Sustainable Development. (2000). *Corporate social responsibility: Making good business sense.*

World Business Council for Sustainable Development (2010). *Vision 2050—The new agenda for business.* Retrieved from http://www.wbcsd.org/pages/edocument/edocumentdetails.aspx?id=219&nosearchcontextkey=true

The Development of Integrated Reporting and the Role of the Accounting and Auditing Profession

Dominic S.B. Soh, Philomena Leung, and Shane Leong

1 Introduction

Corporate reporting has undergone significant expansion in the last decade as organisations increasingly report on matters outside of their financials, both voluntarily and as a result of growing mandatory requirements internationally. Recent studies have pointed to the increasing uptake of wider corporate reporting around environmental, social and governance (ESG) issues, finding that over 80 % of Fortune Global 500 companies in 2010 (Mori Junior, Best, & Cotter, 2014) and 93 % of Fortune Global 250 companies in 2012–2103 (KPMG, 2013) issued ESG reports. In recognition of the benefits of reporting ESG information and the limitations of a voluntary approach to reporting, regulators internationally have increasingly turned to mandating the disclosure of ESG information, with 72 % of ESG reporting policies in 2013 mandatory compared to 58 % of policies in 2006 (KPMG et al., 2013). In the EU, for example, the European Parliament's Directive on disclosure of non-financial and diversity information will require companies to "disclose information on policies, risks and results as regards environmental matters, social and employee-related aspects, respect for human rights, anti-corruption and bribery issues, and diversity on boards of directors" (European Commission, 2014, p. 1).

While the consistent approach in the EU has been hailed as a "smart policy approach", the existing legislative framework for ESG reporting internationally has been labeled as "fragmented and heterogeneous" (GRI, 2013b, p3).[1] This has in

---

[1] Refer to KPMG, UNEP, GRI & UCGA (2013) for a comprehensive overview of the global inventory of ESG reporting policies and guidelines.

D.S.B. Soh (✉) • P. Leung • S. Leong
Department of Accounting and Corporate Governance, Faculty of Business and Economics, Macquarie University, Sydney NSW 2109, Australia
e-mail: dominic.soh@mq.edu.au

turn resulted in increasingly complex, disjointed corporate reporting that does not necessarily serve the needs of and accountability to various stakeholders. It has been argued that due to a lack of standards, regulations and uniform accounting schemes, contemporary sustainability accounting is better described as "a weak approximation of the triple bottom line" (Ngwakwe, 2012, p. 29). Sustainability reporting still suffers from the inescapable fact that reporting a company's social, environmental, economic and ethical performance actually provides little information on planetary sustainability. As Gray (2010, p. 48) observes, "any simple assessment of the relationship between a single organisation and planetary sustainability is virtually impossible. The relationships and interrelationships are simply too complex."

Led by the formation of the International Integrated Reporting Council (IIRC), recent years saw a trend towards the practice of Integrated Reporting (IR) as a means of disclosing financial and ESG information and their relationships in an integrated way to providing a holistic view of an organisation and its value creation process (EY & BCCCC, 2014; KPMG et al., 2013). In explaining the relationship between integrated reporting and sustainability reporting, the GRI (2013a, p. 85) in its *G4 Sustainability Reporting Guidelines* suggests that:

> Integrated reporters build on sustainability reporting foundations and disclosures in preparing their integrated report. Through the integrated report, an organization provides a concise communication about how its strategy, governance, performance and prospects lead to the creation of value over time. Therefore, the integrated report is not intended to be an extract of the traditional annual report nor a combination of the annual financial statements and the sustainability report. However, the integrated report interacts with other reports and communications by making reference to additional detailed information that is provided separately.

This chapter examines the accounting and auditing profession's involvement in IR development. This was facilitated through analysing public submissions regarding key aspects of IR sent to the IIRC by numerous international and national accounting and auditing professional bodies at various developmental stages of the International <IR> Framework. The chapter draws on these responses to highlight significant issues of IR and challenges to the profession.

The remainder of this chapter is structured as follows. Section 2 provides a brief overview of the development of the International <IR> Framework. Section 3 discusses the involvement of the accounting and auditing profession using a regulatory theory perspective and develops our research question. Section 4 presents the results of the study. Section 5 discusses the novelty of IR, the challenges for the accounting and auditing profession in maintaining its relevance in the evolving corporate reporting landscape, and concludes the chapter.

2 An Overview of the Development of Integrated Reporting

The IR initiative can be traced to The Prince's Accounting for Sustainability Project (A4S) set up by The Prince of Wales in 2004 with the aim of "developing practical tools and approaches in order to embed sustainability into mainstream decision-making, accounting and reporting" (Fries, McCulloch, & Webster, 2010, p. 30).[2] A4S developed a connected reporting framework in 2007 and reporting guide in 2009, which explained and illustrated ways in which diverse areas of organisational performance could be presented in a connected, integrated way with traditional accounting information, thereby providing greater insight into an organisation's strategic objectives and its sustainability.

In 2009 it was proposed and confirmed in multi-stakeholder meetings, convened by A4S and the GRI in 2009, that an international body should be created to develop a generally accepted connected and integrated reporting framework (Fries et al., 2010). The formation of International Integrated Reporting Committee[3] was announced by A4S and the GRI in August 2010, bringing together an international cross section of leaders from the corporate, investment, accounting, securities, regulatory, academic and standard-setting sectors, as well as civil society.

Following its inception, the IIRC moved quickly in its aim of establishing a "globally accepted framework for accounting for sustainability" (A4S & GRI, 2010, p. 1). In September 2011, its initial discussion paper, *Towards Integrated Reporting: Communicating Value in the 21st Century*, was released for exposure until December 2011, receiving 214 submissions. This formed the basis for the outline of the International <IR> Framework and the Prototype Framework, respectively released in July 2012 and November 2012.

Between March and July 2013, a series of five background papers for IR (on business model, capitals, connectivity, materiality and value creation) were released by Technical Collaboration Groups, predominantly consisting of professional accounting bodies, accounting firms and the World Intellectual Capital Initiative. The Consultation Draft of the International <IR> Framework was made public by April 2013, receiving 359 submissions. The development process culminated in the release of the International <IR> Framework in December 2013. Table 1 provides a timeline of the development of IR.

The International <IR> Framework (hereafter the Framework) sets forth fundamental concepts aimed at explaining to providers of financial capital how an

---

[2] It is worth noting that the first usage of 'integrated report' is often attributed to Novo Nordisk's 2004 Annual Report (refer to Dey and Burns (2010) for a discussion on Novo Nordisk's approach to integrated reporting). The King Code of Governance for South Africa (King III) published in 2009 has also required companies listed on the Johannesburg Stock Exchange to issue an integrated report annually (or explain where not issued).

[3] The International Integrated Reporting Committee was renamed the International Integrated Reporting Council in 2011.

Table 1 Timeline of the development of integrated reporting

Date	Event
2004	The Prince's Accounting for Sustainability Project (A4S) established
2007	A4S's Connected Thinking Framework launched
2008	A4S Forum consisting or international organisations and 19 international accounting bodies started
2009	A4S's *Connected Reporting—A practical guide with worked examples* released
Sep–Dec 2009	Proposal to convene International Integrated Reporting Committee to create generally accepted connected and integrated reporting framework
Aug 2010	International Integrated Reporting Committee established
Sep–Dec 2011	Discussion paper, *Towards Integrated Reporting: Communicating Value in the 21st Century*
Jul 2012	Draft Outline of the Integrated Reporting Framework issued
Nov 2012	Prototype of the International <IR> Framework
Mar–Jul 2013	Background papers for IR released
Apr–Jul 2013	Consultation draft of the International <IR> Framework exposure period
Dec 2013	International <IR> Framework released

organisation creates value over time, by measuring and providing information on changes in, and interconnectivity of, various stocks of capitals utilised by the organisation in its operations (IIRC, 2013b). It adopts a principles-based approach to reporting intended to strike a balance between "flexibility and prescription that recognizes the wide variation in individual circumstances of different organizations while enabling a sufficient degree of comparability across organizations" (IIRC, 2013b, p. 4).

IR is defined as "a process founded on integrated thinking that results in a periodic integrated report by an organization about value creation over time and related communications regarding aspects of value creation", while an integrated report is described as "a concise communication about how an organization's strategy, governance, performance and prospects, in the context of its external environment, lead to the creation of value in the short, medium and long term" (IIRC, 2013b, p. 33).

The following section outlines the accounting profession's involvement in the development of the Framework and develops the research question for this study.

3 Examining the Accounting Profession's Involvement in Developing Integrated Reporting

The accounting profession (accounting firms, professional bodies and standard setters) has played a significant role in the development of IR. The profession has undertaken a lead role in developing a majority of the background papers and been strongly represented on the IIRC's Board, Council and Working Group.

The influence of the profession can further be seen in how the IIRC has signed Memorandums of Understanding (MOUs) with IFAC and the IFRS (International Financial Reporting Standards) Foundation (for the International Accounting Standards Board (IASB)), with the former (IIRC & IFAC, 2012, p. 2) stating that "the support of the accounting profession is materially relevant and beneficial to the development and implementation of <IR>." The President of the World Business Council for Sustainable Development declared that "accountants would save the world" at the Rio+20 United Nations Conference on Sustainable Development, a position which has since been subsequently reiterated (Bakker, 2013). The involvement of the profession in the IR development prompts us to analyse the extent of acceptance within professional bodies and the possible impact of IR on the profession.

This chapter examines the profession's perceptions on key aspects of IR and the Framework by drawing from the submissions of ten accounting and auditing professional bodies (hereafter respondents), to the IIRC's discussion paper and consultation draft. We also examine the profession's influence on IR development using a regulatory theory perspective. Assessing the likelihood of capture provides important contextual insight into the potential future impact of IR on the profession. We adopt the method utilised by Chalmers, Godfrey, and Lynch (2012) in relation to water accounting standards by using the lens of three regulatory theories, namely public interest theory, private interest theory and regulatory capture.

As Chalmers et al. (2012) observe, public interest theory views regulation as a means of protecting society from market failure; private interest theory views regulation as rules that have potentially been influenced by interest groups trying to benefit themselves; while regulatory capture theory is a specific form of private interest theory which suggests that the very groups to which regulation is applied often subvert regulation for their own purposes. Ultimately, it is likely that regulation is affected by many influences. Each of the three regulatory theories provides a different perspective on how regulation develops. Using all three therefore permits a more complete analysis of the forces influencing the development of the <IR> Framework.

We analyse responses from peak representative bodies representing a broad range of accounting and auditing professionals internationally. Submissions were analysed based on the key themes and issues in the Framework. The Appendix shows the list of respondents and the submissions analysed. The following research question is posed:

RQ How did the accounting and auditing profession perceive key aspects of the International <IR> Framework during its development?

4 Results

Sections 4.1–4.6 present the findings from a content analysis of the accounting and auditing profession's (respondents') comments on key aspects of the <IR> Framework and their support for IR. The key aspects discussed are: the Framework's guiding principles, business model reporting, use of multiple capitals, content elements and focus on investors. A discussion on the profession's influence on the development of the Framework, followed by a critical discussion of the evolution of IR, the implications of the findings and a conclusion are provided in Section 5.

4.1 Principles-Based Approach and the Guiding Principles

While there was broad support for a principles-based approach, respondents noted that the language used in the consultation draft was somewhat prescriptive (ACCA, 2013; AICPA, 2013) and that terms such as 'requirements' and 'comply with' are inconsistent with a principles-based approach. The case was also made by a number of respondents for a market-led approach in determining specific content (and their measurement), rather than reliance on an overly-prescriptive approach in some aspects of the Framework, despite recognition that it was high-level and conceptual in nature (AICPA, 2013). This would ensure that innovation would not be stifled (ACCA, 2013; AICPA, 2013).

The Framework also provides seven guiding principles that underpin the preparation and presentation of an integrated report as described in Table 2.

Respondents generally agreed that the guiding principles provided in the discussion paper were appropriate in forming the foundation for improving corporate reporting (AICPA, 2011; CPA Australia, 2011; FEE, 2011; IIA, 2011). However, some concerns were raised particularly in relation to future orientation and responsiveness and stakeholder inclusiveness. Responses to each guiding principle are briefly considered below.

4.1.1 Strategic Focus and Future Orientation

The inclusion of future orientation as a guiding principle was questioned (ICAEW, 2011), with suggestions that it should only go as far as requiring a commitment from organisations to disclose information relevant to users in understanding and analysing the organisation's future value creation potential, rather than requiring

Table 2 Guiding principles in the International <IR> Framework

Guiding principle	Description
Strategic focus and future orientation	An integrated report should provide insight into the organization's strategy, and how it relates to the organization's ability to create value in the short, medium and long term, and to its use of and effects on the capitals
Connectivity of information	An integrated report should show a holistic picture of the combination, interrelatedness and dependencies between the factors that affect the organization's ability to create value over time
Stakeholder relationships	An integrated report should provide insight into the nature and quality of the organization's relationships with its key stakeholders, including how and to what extent the organization understands, takes into account and responds to their legitimate needs and interests
Materiality	An integrated report should disclose information about matters that substantively affect the organization's ability to create value over the short, medium and long term
Conciseness	An integrated report should be concise
Reliability and completeness	An integrated report should include all material matters, both positive and negative, in a balanced way and without material error
Consistency and comparability	The information in an integrated report should be presented: (a) on a basis that is consistent over time; and (b) in a way that enables comparison with other organizations to the extent it is material to the organization's own ability to create value over time

Source: International Integrated Reporting Council (IIRC) (2013), *The International <IR> Framework*

disclosures of projections on expected future performance (AICPA, 2011). This continued to be a concern at the consultation stage as respondents indicated that this guiding principle appeared to require organisations to disclose forecasted or projected performance (AICPA, 2013), despite the liability implications and assurance challenges this would raise (IFAC, 2013).

4.1.2 Stakeholder Relationships

The responsiveness and stakeholder inclusiveness guiding principle in the discussion paper was raised as an issue of concern, as it was perceived to place undue emphasis on disclosures related to relationships, and being contradictory to the focus on investors (ACCA, 2011). It was suggested that this principle would be better characterised as 'relevance' to key concerns of all stakeholders (ACCA, 2011; AICPA, 2011).

The principle was subsequently renamed 'stakeholder responsiveness' in the consultation draft, but continued to receive limited support from respondents. It was suggested that information on stakeholder relationships be disclosed only to the extent that such relationships were relevant to providers of financial capital. As the

ICAEW (2013, p. 10) observed, "Understanding of the capitals on which a firm depends may also be enhanced by engagement with stakeholders, but this does not require responsiveness to stakeholders as an objective in its own right". The relevance of stakeholder engagement in ensuring the reliability of integrated reports was also questioned (ACCA, 2013).

4.1.3 Connectivity of Information

Connectivity of information was noted as the most important guiding principle (IIA, 2011). Although there was general support for the concept, it was noted that connectivity would be a difficult principle to apply in practice (ICAEW, 2013), and that it could be improved by further emphasising interconnections between capitals and stakeholders, and between strategy, materiality and capitals (FEE, 2013).

4.1.4 Consistency and Comparability and Other Suggested Principles

Some respondents to the discussion paper suggested that scope and boundary should be included in the guiding principles (ICAA, 2011; IFAC, 2011). It was further suggested that existing reporting Frameworks, including the IASB's Conceptual Framework for Financial Reporting and the Climate Disclosure Standards Board's Climate Change Reporting Frameworks, could be used to inform the development of additional principles for the Framework, such as timeliness and verifiability (ACCA, 2011; ICAA, 2011) and comparability and completeness (IFAC, 2011). There were also calls for greater emphasis to be placed on reliability (ACCA, 2011; FEE, 2011; ICAEW, 2011) and greater clarity in relation to materiality (AICPA, 2011; FEE, 2011; ICAS, 2011). Although consistency and comparability were subsequently included it was recognised as difficult to achieve given the emphasis on organisational strategy in the Framework (ACCA, 2013).

4.1.5 Materiality

The consultation draft specified that "an integrated report should provide concise information that is material to assessing the organization's ability to create value in the short, medium and long term" (IIRC, 2013a, p. 6). The document (IIRC, 2013, p. 21) subsequently defined 'materiality' as follows:

> A matter is material if, in the view of senior management and those charged with governance, it is of such relevance and importance that it could substantively influence

the assessments of the primary intended report users [providers of financial capital] with regard to the organization's ability to create value over the short, medium and long term.

Respondents though in general agreement, suggested that materiality should refer to providers of financial capital (ACCA, 2013; AICPA, 2013; CIMA, 2013; FEE, 2013; ICAS, 2013; IFAC, 2013). This however may lead to reports neglecting issues relating to the other capitals (e.g. social and environmental) (ACCA, 2013). This was of particular concern given the difficulty in quantifying judgments in respect of some of the capitals (ACCA, 2013). Others suggested that materiality should be determined from the perspective of key stakeholders and owners of all types of relevant capitals (IIA, 2013), and that the process of materiality determination should include involvement and engagement with intended users of the integrated report (ICAEW, 2013; IFAC, 2013).

Another key concern was how the measurement of materiality would interact with that of other organisational reporting e.g. financial reports (ICAEW, 2013; IFAC, 2013). There were varied opinions whether "there should not be a multiplicity of authoritative statements on what constitutes materiality in corporate reporting" (ICAA, 2013; IFAC, 2013; ICAEW, 2013, p. 8). Some respondents also suggested that there was a need to explicitly refer to the role of professional judgment in assessing materiality (CPA Australia, 2013; ICAA, 2013; IFAC, 2013).

4.1.6 Conciseness

Respondents were concerned with balancing conciseness with completeness (ACCA, 2013, 2013; ICAEW, 2013; ICAS, 2013) as disclosing all material issues may be overwhelming and lengthy (AICPA, 2013, p.6). It was suggested that "the requirement should be to disclose all material information (completeness), but to do so as concisely as possible" (ICAEW, 2013, p. 9).

4.1.7 Reliability and Completeness

There was broad agreement that robust internal reporting systems and related internal controls are important, as users are unable to assess the effectiveness of these systems relevant to IR themselves (ICAEW, 2013). Most respondents emphasised the need for assurance in demonstrating reliability of reporting systems and credibility of integrated reports to facilitate users' reliance on them for decision-making. It was proposed that organisations should have the flexibility to decide "how much of it [the integrated report] should be assured or exactly what form of assurance should be provided" (ICAEW, 2013, p. 4).

4.2 Business Model Reporting

Reporting on an organisation's business model and its ability to create and sustain value in the short, medium and long term was put forth by the IIRC as central themes for guiding IR in the discussion paper (IIRC, 2011). There was no universal authoritative definition of the term 'business model' as it "is often seen as the process by which an organization seeks to create and sustain value" (IIRC, 2011, p. 10).

Notwithstanding calls for greater clarity around the terms 'business model' and 'value-creation' and what form the description of a business model might take, there was general agreement that business model reporting was a suitable theme. Respondents believed that business model reporting would provide greater insights into organisations' long-term viability and sustainability (CIMA, 2011; FEE 2011). It was noted that the UK experience in reporting on business models might prove useful in informing this process, given existing requirements in the UK Corporate Governance Code and in The Companies Act 2006 (Strategic Report and Directors' Report) Regulations 2013. The FRC's Financial Reporting Lab was also suggested to be a potentially useful source for guidance (ICAEW, 2011).

Respondents generally agreed with the definition that a business model is "a chosen system of inputs, business activities, outputs and outcomes that aims to create value over the short, medium and long term" (IIRC, 2013a, p. 14). Nevertheless there were calls for greater context on how the elements (inputs, business activities, outputs and outcomes) interact to form value (CIMA, 2013; ICAA, 2013; IFAC, 2013). To recognise such interactions, the business model definition was subsequently revised to the organisation's "system of transforming inputs, through its business activities, into outputs and outcomes that aims to fulfill the organization's strategic purposes and create value over the short, medium and long term" (IIRC, 2013b, p. 25).

4.3 Multiple Capitals

In reporting on resources and relationships that contribute to organisational success, the Framework provides multiple capitals on which to report, while noting that the categorisations and definitions are not necessarily authoritative or universally applicable. These capitals are described in Table 3.

Respondents indicated that the use of multiple capitals was helpful in: prompting organisations to consider wider accountability obligations beyond merely financial and manufactured capitals; identifying key performance and risk indicators (ACCA, 2011, 2013; AICPA, 2011, 2013; ICAEW, 2011; IIA, 2011); and facilitating a better understanding of future prospects (IFAC, 2011).

Table 3 Multiple capitals and their definitions

Capital	Definition
Financial	The pool of funds that is: • available to the organization for use in the production of goods or the provision of services • obtained through financing, such as debt, equity or grants, or generated through operations or investments
Manufactured	Manufactured physical objects (as distinct from natural physical objects) that are available to the organization for use in the production of goods or the provision of services, including: • buildings • equipment • infrastructure (such as roads, ports, bridges and waste and water treatment plants)
Intellectual	Organizational knowledge-based intangibles, including: • intellectual property, such as patents, copyrights, software, rights and licenses • "organizational" capital such as tacit knowledge, systems, procedures and protocols
Human	People's competencies, capabilities and experience, and their motivations to innovate, including their: • alignment with and support for an organization's governance framework, risk management approach, and ethical values • ability to understand, develop and implement an organization's strategies • loyalties and motivations for improving processes, goods and services, including their ability to lead, manage and collaborate
Social and relationship	The institutions and the relationships established within and between communities, groups of stakeholders and other networks, and the ability to share information to enhance individual and collective well-being. Social and relationship capital includes: • shared norms, and common values and behaviours, • key stakeholder relationships, and the trust and willingness to engage that an organization has developed and strives to build and protect with external stakeholders • intangibles associated with the brand and reputation that an organization has developed • an organization's social license to operate
Natural	All renewable and non-renewable environmental resources and processes that provide goods and services that support the past, current or future prosperity of an organization. It includes: • air, water, land, minerals and forests • biodiversity and eco-system health

Source: International Integrated Reporting Council (IIRC) (2013), *The International <IR> Framework*

While some felt that the emphasis placed on financial capital as the primary capital was unjustified (ICAA, 2013; IFAC, 2013), others believed that investors were likely to be primarily interested in financial capital (ICAEW, 2011), and interested in non-financial capitals only to the extent that they (could) influence the organisations' financial position (ACCA, 2011).

Despite agreeing with the approach to capitals on a conceptual level, there was doubt about how this approach might be operationalized on a practical level (ICAA, 2013; IFAC, 2013). The practicality of reporting on the multiple capitals was perceived to be a challenge due to a number of issues such as reporting boundary and control of capitals, measurement issues, time horizon, terminology and categorisation. These are briefly discussed below.

4.3.1 Reporting Boundary and Control of Capitals

As some capitals might belong to stakeholders that include the organisation, or a more broadly defined society, there were concerns around the boundary of the organisation and the extent of ownership, legal rights and influence over capitals other than financial and manufactured capitals (ACCA, 2011; CPA Australia, 2011; ICAEW, 2011; IFAC, 2011).

AICPA (2013), supported by IFAC (2013), argued that with a multitude of potential approaches it is very challenging to report/value the change in capitals created for society. FEE (2013) also stated its concern about how capitals outside of organisations' span of control should be reported, and how far down the value chain organisations should go to report on these capitals.

4.3.2 Measurement Issues

There was concern that there was insufficient guidance on how to measure value in relation to these capitals, which would hinder efforts to create reliable and comparable reporting (ACCA, 2013; AICPA, 2013; CIMA, 2013; CPA Australia, 2011; FEE, 2011; ICAA, 2011, 2013; ICAS, 2013; IFAC, 2011, 2013). This problem is especially acute considering that some of the capitals are susceptible to misrepresentation due to their intangible nature (IIA, 2011). It was suggested that in order "to fully utilise the concept of multiple capitals then organisations will need to be able to convert non-financial value into discounted cash flows" (CIMA, 2011, p. 4), or that the IIRC should provide guidance in identifying appropriate high-quality indicators or measurement methods (AICPA, 2013).

The IIRC considered creating an online database of authoritative sources of indicators or measurement methods that preparers might refer to (IIRC, 2013a). Respondents were asked to suggest references to be included in such a database. Respondents were generally receptive to the idea of a database, believing that it would enhance consistency and comparability (ICAA, 2013) and provide useful guidance in identifying high-quality international frameworks for KPIs and non-financial disclosure elements (AICPA, 2013). However, some respondents believed that it would be better to let companies experiment and allow IR to mature before creating an authoritative list (AICPA, 2013; CPA Australia, 2013). It was also noted that IR was not "designed or intended as a basis for benchmarking" (CPA Australia, 2013, p. 2), and that ultimately the choice of indicators and measurement methods should be left to organisations' discretion, subject to their local regulatory and competitive environment (CIMA, 2013; CPA Australia, 2013). The ICAEW (2013, p. 6) suggested that the provision of an online database would be inappropriate as "if they are deemed to be authoritative, this implies that people should comply with them, which would put the IIRC into the inappropriate position of a sort of accreditation body for other bodies".

While this database of authoritative sources was not provided when the Framework was released, the IIRC has indicated that it is on their future work plan (IIRC, 2013c, p. 14). The GRI was the most cited reference with seven out of ten respondents referring to it, consistent with it being one of the two most quoted references in all responses to the consultation draft (the other was IASB).

4.3.3 Reporting Time Horizon

The issue of reporting time horizon received a substantial amount of attention by respondents, particularly around the reporting of long term prospects. The IIA (2011) suggested that reporting based on organisations' business model would overcome perceived deficiencies associated with the historical focus of financial reporting by providing a better indicator of future prospects. However, there was general concern that organisations might resist reporting forward-looking information, given its potentially sensitive and uncertain nature and the legal liabilities to which its reporting might give rise.

The ICAEW (2011) provided three reasons to support the focus on the immediate past in corporate reporting: to hold agents accountable; to provide an anchor for forecasting future performance; and to facilitate verification by a third party. However, there was recognition that while a longer term horizon is necessary for the impact of social and environmental factors, as these issues have intergenerational consequences (ACCA, 2011; IFAC, 2011), there is a need to incentivise market participants to consider the long-term perspective to provide the impetus for organisations to consider more fully and report on their longer-term impacts (ICAA, 2011; IFAC, 2011). As the FEE (2011, p. 7) suggested, "integrated reporting should be comprehensive and balanced... the right balance should be struck between reporting relevant information about the future and ensuring there is appropriate reporting on the past as a key feature of accountability".

4.3.4 Terminology, Categorisations and Prescriptiveness of Approach

The appropriateness of the term 'capitals' and the categorisation of the capitals were questioned by a number of respondents, as they argued that it was not commonly understood or embraced (AICPA, 2011; FEE, 2011), or overlapped with existing established classifications such that as used by economists (natural, man-made and social) without clear reconciliation between the two (IFAC, 2011).

A number of respondents advocated refinements or expansions to the capital categories. The ICAEW (2013) questioned the description of social and relationship capital, calling for greater emphasis on relationships. IFAC (2013, p. 8) called for an expansion of human capital to incorporate "setting the entity's vision, mission, outputs and outcomes". It was also suggested that intellectual capital should be the overarching category, with human capital, relationship capital and organisation capital as subcategories (AICPA, 2013; ICAS, 2013).

Although the IIRC has indicated that the categories are not authoritative, definitive (or mutually exclusive) or universally applicable, there remained concern that the language/expression in the consultation draft may be seen as overly-prescriptive (CPA Australia, 2013; ICAS, 2013). IFAC (2013) suggested that organisations should be given discretion in reporting on capitals.

4.4 Content Elements

The <IR> Framework provides eight content elements to be included in an integrated report, and poses a question in relation to each element for organisations to consider and respond to in preparing their reports. The content elements and their corresponding questions are provided below in Table 4. The Framework recognises that the content elements are "fundamentally linked to each other and are not mutually exclusive" (IIRC, 2013a, p. 24), and that the content of organisations' integrated reports may vary with their circumstances. Judgment is therefore required in applying the guiding principles to determine what information to include.

Respondents generally indicated that the contents elements were comprehensive and provided a sound foundation for report preparation (ACCA, 2011; CIMA, 2011; CPA Australia, 2011; ICAA, 2011; IIA, 2011). Some respondents noted

Table 4 Content elements in the International IR Framework and their corresponding questions

Content element	Question/s to address
Organizational overview and external environment	What does the organization do and what are the circumstances under which it operates?
Governance	How does the organization's governance structure support its ability to create value in the short, medium and long term?
Business model	What is the organization's business model?
Risks and opportunities	What are the specific risks and opportunities that affect the organization's ability to create value over the short, medium and long term and how is the organization dealing with them?
Strategy and resource allocation	Where does the organization want to go and how does it intend to get there?
Performance	To what extent has the organization achieved its strategic objectives for the period and what are its outcomes in terms of effects on the capitals?
Outlook	What challenges and uncertainties is the organization likely to encounter in pursuing its strategy and hat are the potential implications for its business model and future performance?
Basis of presentation	How does the organization determine what matters to include in the integrated report and how are such matters quantified or evaluated?

Source: International Integrated Reporting Council (IIRC) (2013), *The International <IR> Framework*

that the content elements were broadly aligned with elements from the IASB's Management Commentary Statement (ACCA, 2011), or AICPA's Enhanced Business Reporting Framework (AICPA, 2011). Other respondents requested greater guidance in respect of materiality and its application (ICAA, 2011), removing the undue emphasis on remuneration and 'future outlook',[4] since they would be captured in reporting on governance and other content elements respectively (AICPA, 2011).

Some respondents felt that the Framework was too specific and prescriptive in its approach to the content elements (ACCA, 2013; CPA Australia, 2013), and reiterated the need to keep the Framework principles-based (FEE, 2013). There was also continuing concern that the inclusion of 'future outlook' would hinder IR adoption, as it would entail reporting projections, forecasts and commercially sensitive information (AICPA, 2013; ICAA, 2013). It was suggested that greater emphasis should be placed on reporting information around risk (ICAA, 2013), and that the content element may be renamed 'risk management and internal control' (IFAC, 2013). Finally, it was suggested that the performance content element should be expanded to include the effect on capitals and that the interconnectivity of content elements should be made more explicit in reporting on future outlook (IFAC, 2013).

---

[4] The 'governance' content element was termed 'governance and remuneration' in the discussion paper, while 'outlook' was termed 'future outlook' in the discussion paper and consultation draft.

4.5 Focus on Providers of Financial Capital as Primary Audience

The IIRC's initial focus of the <IR> Framework would be reporting by larger companies to meet the needs of investors (IIRC, 2011). The subsequent consultation draft suggested "an integrated report should be prepared for providers of financial capital in order to support their financial capital allocations assessments" (IIRC, 2013a, p. 8).

With the exception of the ICAEW and the IIA, the focus on large companies was generally accepted as being practical given these organisations' wider "impact and reach" (CPA Australia, 2011, p. 5) and greater capacity to undertake the IR process (CIMA, 2011). FEE (2011) and AICPA (2011) suggested that the initial investor focus would provide preparers with greater ability to determine the relevance and materiality of information to report (AICPA, 2013; ICAEW, 2013). There was the belief that investors would be able to exert pressure on organisations to run more efficiently, profitably and with regards to all material risks (ACCA, 2011, p. 5). The initial investor focus could be a potential means to facilitate adoption of IR, as it would be less costly than reporting to a wider stakeholder base (FEE, 2011). Some respondents recognised that organisations need to be accountable to their wider stakeholders, with different information needs. A focus on investor information needs may also lead to neglect on environmental and social impacts (ACCA, 2011).

While a wider stakeholder perspective may "allow for a clearer link to sustainable stakeholder value generation in terms of achieving long-term sustainable organizational success as a public interest, or social, outcome for all stakeholders" (IFAC, 2011, p. 4), there was recognition that a trade-off between stakeholder needs and expectations might be necessary, and that transparency was needed in reporting how these trade-offs were managed (ICAA, 2011; IFAC, 2011). Investor centric focus does not align with the principle of responsiveness and stakeholder inclusiveness (in the discussion paper),[5] which promoted stakeholder engagement as a key process in determining materiality for ESG issues.

Notwithstanding these concerns, the IIRC moved towards crystallising the focus on providers of financial capital to inform financial capital allocation decision-making. This prompted concerns that the proposed approach privileged financial capital at the expense of the other capitals. It was noted that the allocation of all capitals may be material to all stakeholders, and that the emphasis on financial capital was contradictory to the objective of IR to enhance accountability and stewardship with respect to the broad base of capitals (ICAA, 2013; IFAC, 2013). The IIA (2013, p. 5) noted:

---

[5] The guiding principle of 'responsiveness and stakeholder inclusiveness' in the discussion paper was renamed 'stakeholder responsiveness' in the consultation draft and subsequently 'stakeholder relationships' in the Framework (refer to Table 4 above) 'to clarify that the integrated report should not attempt to satisfy the information needs of all stakeholders' (IIRC, 2013c, p34).

Reports on performance relative to financial capital are readily available but this is not so for other types of capitals. One of the strengths of <IR> is taking a broad view of capitals and going beyond the traditional focus on financial capital. The value of <IR> may not be optimized by focusing predominantly on providers of financial capital. If the IIRC wants to elevate the stature of non-financial capitals, it would make sense to acknowledge the users of reporting across all six capitals.

4.6 Support for IR

The IIRC (2011) noted that despite expanded corporate reporting, disclosure gaps persisted as critical interdependencies between financial reporting, governance and remuneration, management commentary and sustainability reporting were not explicitly considered and reported upon. It pointed out the increasing proportion of S&P 500 market value attributable to factors other than companies' physical and financial assets as being indicative of the fact that financial reporting does not sufficiently explain an organisation's business model and value.

Overall respondents to the discussion paper agreed that there was a need to improve corporate reporting to incorporate "future focused value creation activity" and greater insights into "interrelationships and interdependencies" (CPA Australia, 2011, p. 3). However, a number of respondents indicated that there remained work to be done in communicating the merits of IR (ICAEW, 2011) and called for "more compelling arguments to persuade businesses and investors of the benefits of integrated reporting" (FEE, 2011, p. 4) and for the IIRC to "demonstrate how it [IR] will work in real-life examples" (ICAS, 2011, p. 3). It was also noted that IR "should not strive necessarily to solve complexity in financial reporting in relation to the application of financial reporting standards" (IFAC, 2011, p. 2) and that IR might not capture all intangible items that contribute to the gap between the market value and balance sheet values of financial and tangible assets as indicated by the IIRC (ACCA, 2011). Some respondents also noted that there was limited basis for the IIRC's claims concerning the benefits of IR and called for a cost-benefit analysis to be performed to provide empirical evidence to support these claims (ICAA, 2011; IFAC, 2011).

Despite concerns about the IIRC's ability to obtain support across key regulators internationally (IIA, 2011) and cautioning that the development of IR had to be sufficiently flexible to accommodate jurisdictional and cultural differences (CPA Australia, 2011; FEE, 2011; ICAEW, 2011), respondents overwhelmingly supported a global approach to the development of IR.

Thus there is general support by the accounting and auditing profession of the inadequacies of current reporting, and the anticipated benefits regarding investors and wider stakeholders garnered in the potential for IR to improve decision-making and engagement with these parties and to contribute towards more resilient organisations that achieve long-term sustainable success (CIMA, 2011; IFAC, 2011).

However our analysis indicates a varied degree of uncertainty in IR application and there are concerns of the implications to the profession.

5 Discussion and Conclusion

5.1 The Evolution of IR Through a Regulatory Theory Lens

The evolution of the aims of the IIRC and Framework may be examined using three regulatory theories as stated earlier: public interest theory, private interest theory and regulatory capture theory.

At its inception, the aim of the IIRC was to "create a globally accepted framework for accounting for sustainability" by creating a framework that integrates financial and ESG information in corporate reporting "to meet the needs of the emerging, more sustainable, global economic model" (A4S & GRI, 2010, p. 1). Thus the regulatory theory best explaining the efforts to develop the Framework at the time was public interest theory. IR was an ambitious, idealistic project, intended to benefit the public and too recent an innovation to be captured by specific interest groups.

As time went on private interests appeared to increasingly take precedence over the public interest. The intended audience and focus changed from being *initially* on the needs of investors to firmly for "providers of financial capital in order to support their financial capital allocation assessments". Though IR was claimed to benefit wider stakeholders as they are also "focused on the creation of value in the short, medium and long term" (IIRC, 2013a, p. 8), the focus of the IIRC nevertheless shifted towards an emphasis on decision-usefulness, and away from stewardship and accountability.

From a private interest theory perspective, there is evidence of the profession attempting to influence the development of the Framework. This is unsurprising given its status as the incumbent service provider of corporate reporting and auditing/assurance, with significant financial gains to be made from "capturing the reporting and auditing and assurance requirements and ensuring that they are very closely aligned to those that already exist in relation to financial reporting" (Chalmers et al., 2012, p. 1016).

While there were varying degrees to which respondents espoused the need for change and an innovative approach, there was indeed a general tendency to refer to established accounting and assurance frameworks, particularly the IASB (especially in relation to verifiability/reliability/faithful representation) and IAASB Frameworks respectively. Further, professional bodies made explicit attempts to claim IR space. For instance, the IIA advocated for the role of internal audit and the chief audit executive, while CIMA argued for the need to improve internal management reporting systems and capabilities. Several other bodies, particularly FEE emphasised the role of auditors/assurance providers in relation to IR.

As such, the accounting profession appears to have had strong incentive to capture the Framework development process. However, regulatory capture theory may not best describe the profession's involvement in the process. While the Framework may arguably unduly emphasise reporting information amenable to quantification, the IIRC has kept a wide range of professions and interests groups on board, signing MOUs with the Carbon Disclosure Standards Board, the Sustainability Accounting Standards Board and the World Intellectual Capital Initiative, among others.

IR therefore remains a multidisciplinary effort, which limits the ability for the accounting profession to fully dominate and capture the Framework development process. Further, there were diverging views within the accounting profession regarding the focus and audience of IR, with several actually arguing for a wider stakeholder focus.[6]

Since its release, the uptake of IR by organisations internationally is still in its infancy, with mandatory requirements for South African listed companies. In Brazil there is a report or explain requirement for sustainability *or* integrated reports. The acceptance of the Framework is therefore dependent on the profession and the buy-in from domestic regulators and organisations in various jurisdictions. Future research may examine these efforts in local/regional areas. In examining the interactions between the IIRC and domestic organisations and regulators, it is likely that regulatory theory will continue to provide insights in understanding these processes and their outcomes.

5.2 Integrated Thinking and Building the Capacity of the Profession

The novelty in the IR initiative lies not in its calls for extended reporting, but in the application of integrated thinking in outlining the connectivity of information across the organisation's activities and reporting. The Framework (IIRC, 2013b, p. 33) defines integrated thinking as:

> The active consideration by an organization of the relationships between its various operating and functional units and the capitals that the organization uses or affects. Integrated thinking leads to integrated decision-making and actions that consider the creation of value over the short, medium and long term.

Integrated thinking, if applied as intended, promises significant benefits for internal management and external reporting purposes. As the Framework states, "the more that integrated thinking is embedded into an organization's activities, the

---

[6] It is worth noting that there were significant similarities across respondents' submissions in some instances. In particular, ICAA's and IFAC's responses to the consultation draft were substantially similar. A similarity check conducted on http://textmatch.eu/ between both submissions returned a similarity score of 75 %.

more naturally will the connectivity of information flow into management reporting, analysis and decision-making, and subsequently into the integrated report" (IIRC, 2013b, p. 16).

However, to achieve the intention of IR, significant challenges lie in the breaking down of silos and in transforming traditional mindsets and current approaches of the profession. Some respondents recognised the importance and centrality of integrated thinking to IR in their submissions, for example, with the IIA (2013, p. 5) stating that "this [integrated thinking] is the key differentiator and benefit of issuing an integrated report." Throughout the analysis of the responses, different concerns indicate the segmented views of reporting and the tradition of compartmentalised views of accounting. Examples of these are reflected in the views relating to the principles, the reporting boundaries, the content elements and the capitals and so on.

To apply and cultivate integrated thinking, accountants will need to reconceive their roles within the organisation and understand how their roles relate to those of other functions. It will no longer be sufficient to be a subject matter expert. It will be necessary to be able to communicate across functional departments as well as with stakeholders external to the organisation. The successful accountant in the era of IR will be one who does not conceive himself or herself as a financial accountant, or a management accountant. Those who succeed in the profession will likely be effective communicators comfortable with quantitative and qualitative information, both historical and prospective. They will also be innovative in how they engage and communicate with stakeholders.

In examining the professional bodies' submission, however, there was limited discussion around how the profession needed to adapt to apply and achieve integrated thinking. Despite the need to build capacity within organisations to facilitate IR being raised by several respondents to the IIRC's discussion paper in particular, respondents appeared all too willing to suggest that accountants and assurance providers (both internal and external) possessed the necessary competencies to improve internal systems and external reporting. In terms of building capacity within organisations, it was suggested that substantial resources would need to be invested into developing and integrating information systems and reporting processes, as well in building staff capacity, for example, by hiring personnel with specific expertise in non-financial areas (ACCA, 2011; AICPA, 2011; CIMA, 2011; CPA Australia, 2011; ICAA, 2011; IFAC, 2011; IIA, 2011). With only a few exceptions, consideration of the profession's ability to work with others within (and outside of) the organisation was lacking in the submissions, as was discussion around the need to build capacity in the profession through training and education in respondents' submissions.

Despite the accounting and auditing profession having an advantage in claiming the IR space as the incumbent corporate reporting and assurance service providers, it is imperative that the profession considers the need to evolve and adapt in order to ensure that it maintains its relevance and value in serving the public interest. This is pertinent, as questions are already being raised as to whether the accounting profession possesses the appropriate skills to undertake assurance services in relation to ESG information such as greenhouse gas emissions (Martinov-Bennie, Frost, & Soh, 2012; Simnett, Nugent, & Huggins, 2009). As such, the training and education of the current and future generation of accountants and assurance providers (both external and internal) warrant specific attention.

There is a need to reconsider the profession's approach to training and education, through professional accreditation requirements and ongoing continuing professional development initiatives for the current generation of accountants, as well as through tertiary and wider educational requirements for future generations. Accountants and assurance providers will need to develop their critical thinking and analysis skills in identifying interconnections, as well as their communication skills in working with a wide range of stakeholders and in undertaking their reporting and assurance roles effectively.

The antiquated accounting education curriculum in the evolving corporate reporting environment has also been discussed by Owen (2013), noting that it has traditionally focused on the transactional cycles, while failing to recognise, measure, or value the tactical and strategic levels of the business, nor adopting a longer term view of sustainability. He consequently suggests that aligning the accounting curricula with IR principles will require a fundamental paradigm shift from a transactional approach to an approach requiring accounting students to think about business more holistically. This will require students to be able to synthesise quantitative and qualitative information while acknowledging the effects that an organisation's position, performance and prospects have on business, society, the environment, and other stakeholders (Owen, 2013).

Conclusion
There is a strong drive to improve corporate reporting as a result of growing stakeholder demands and regulatory initiatives. IR, with its focus on integrated thinking and connectivity of information, has strong potential to provide these improvements in serving the public interest. However, it will take a shift in the mindsets of organisations, professionals and external stakeholders (including investors) for it to achieve its full potential. With the wide berth provided to reporting organisations using the Framework and its principles-based approach permitting departures from requirements, it is possible that organisations will go down the path of mimetic or isomorphic convergence in their (external) IR undertakings without substantive changes in their operations internally.

To prevent this from happening, those with a vested interest in the success of IR need to ensure that integrated thinking is widely encouraged and applied. A market-led approach driven by external stakeholders who are interested in the holistic performance of a company will require external stakeholders to adopt integrated thinking in exerting pressure on organisations to internally manage and externally disclose information. This will

(continued)

require a new mindset within and external to organisations, where transparency and wider accountability is the norm, and where it is no longer sufficient to only consider accountability to financial investors, but to wider stakeholders more explicitly.

The accounting and auditing profession also needs to deeply consider how it might work towards developing capacity in the profession, particularly in cultivating integrated thinking in the current and future generation of accountants and assurance providers. Relying on the present technical focus on financial reporting and auditing will limit the profession's ability to contribute to the development of integrated reporting and run the risk of the profession becoming increasingly irrelevant. Developments in IR present an opportune time for the profession to take up this challenge to reinvent itself and re-examine its value proposition in meeting the needs of organisations and society. As a starting point, a critical review of and research examining the education and development of future accountants is urgently needed.

Appendix

List of key professional bodies

Acronym	Professional body	Headquartered	Membership
ACCA	The Association of Chartered Certified Accountants	UK	>162,000 in 173 countries
AICPA	American Institute of Certified Public Accountants	US	>394,000 in 128 countries
CIMA	Chartered Institute of Management Accountants	UK	>100,000 in 177 countries
CPA Australia	Certified Practising Accountants Australia	Australia	>150,000 in 121 countries
ICAA[a]	The Institute of Chartered Accountants in Australia	Australia	>61,000 globally
ICAEW	The Institute of Chartered Accountants in England and Wales	UK	>142,000 globally
ICAS	The Institute of Chartered Accountants of Scotland	UK	>20,000 globally
IIA	Institute of Internal Auditors	US	>180,000 globally
IFAC	The International Federation of Accountants	US	179 members and associates in 130 countries and jurisdictions, representing approximately 2.5 million accountants
FEE	Fédération des Experts-comptables Européens—Federation of European Accountants	Belgium	47 member institutes from 36 - European countries with combined membership of >800,000

[a]Amalgated with the New Zealand Institute of Chartered Accountants in July 2014, forming Chartered Accountants Australia and New Zealand with > 100,000 members globally

References

Bakker, P. (2013). Accountants Will Save the World. *Harvard Business Review Blog Network*. Retrieved 21 June, 2014, from http://blogs.hbr.org/2013/03/accountants-will-save-the-worl/

Chalmers, K., Godfrey, J. M., & Lynch, B. (2012). Regulatory theory insights into the past, present and future of general purpose water accounting standard setting. *Acounting, Auditing & Accountability Journal, 25*(6), 1001–1024. doi:10.1108/09513571211250224.

Dey, C., & Burns, J. (2010). Integrated reporting at Novo Nordisk. In A. Hopwood, J. Unerman, & J. Fries (Eds.), *Accounting for sustainability: Practical insights* (pp. 215–232). London: Earthscan.

Ernst & Young (EY), & Boston College Center for Corporate Citizenship (BCCCC). (2014). *Value of sustainability reporting: A study by EY and Boston College Center for Corporate Citizenship*. Ernst & Young and Boston College Center for Corporate Citizenship.

European Commission. (2014). *Improving corporate governance: Europe's largest companies will have to be more transparent about how they operate*. Retrieved 15 April, 2014, from http://europa.eu/rapid/press-release_STATEMENT-14-124_en.htm

Fries, J., McCulloch, K., & Webster, W. (2010). The Prince's Accounting for Sustainability Project: Creating 21st century decision-making and reporting systems to respond to 21st century challenges and opportunities. In A. Hopwood, J. Unerman, & J. Fries (Eds.), *Accounting for sustainability: Practical insights* (pp. 29–45). London: Earthscan.

Global Reporting Initiative (GRI). (2013a). *G4 sustainability reporting guidelines*. Amsterdam: GRI.

Global Reporting Initiative (GRI). (2013b). *Report or explain: A smart EU policy approach to non-financial information disclosure: Global Reporting Initiative (GRI)*.

Gray, R. (2010). Is accounting for sustainability actually accounting for sustainability . . . and how would we know? An exploration of narratives of organisations and the planet. *Accounting, Organizations and Society, 35*(1), 47–62. doi:10.1016/j.aos.2009.04.006.

International Integrated Reporting Committee (IIRC). (2011). *Towards Integrated Reporting: Communicating Value in the 21st Century*. International Integrated Reporting Committee.

International Integrated Reporting Council (IIRC). (2013a). *Consultation draft of the international <IR> framework*. International Integrated Reporting Council

International Integrated Reporting Council (IIRC). (2013b). *The International <IR> Framework*. International Integrated Reporting Council (IIRC).

International Integrated Reporting Council (IIRC). (2013c). *Summary of significant issues*. International Integrated Reporting Council

International Integrated Reporting Council, & International Federation of Accountants. (2012). *Memorandum of understanding: IIRC and IFAC- 07 September 2012*. Retrieved 21 May, 2014, from https://www.ifac.org/sites/default/files/uploads/Comms/IFAC-and-IIRC-MoU-September-2012.pdf

KPMG. (2013). *The KPMG survey of corporate responsibility reporting 2013*. KPMG.

KPMG, United Nations Environment Programme (UNEP), Global Reporting Initiative (GRI), & Unit for Corporate Governance in Africa (UCGA). (2013). Carrots and sticks—sustainability reporting policies worldwide—today's best practices, tomorrow's trends (2013 ed.): KPMG Advisory N.V., United Nations Environment Programme, Global Reporting Initiative, Unit for Corporate Governance in Africa.

Martinov-Bennie, N., Frost, G., & Soh, D. S. B. (2012). Assurance on sustainability reporting: State of play and future directions. In S. Jones & J. Ratnatunga (Eds.), *Contemporary issues in sustainability accounting, assurance and reporting* (pp. 267–283). Bradford: Emerald Group Publishing Limited.

Mori Junior, R., Best, P., & Cotter, J. (2014). Sustainability reporting and assurance: A historical analysis on a world-wide phenomenon. *Journal of Business Ethics, 120*(1), 1–11. doi:10.1007/s10551-013-1637-y.

Ngwakwe, C. C. (2012). Rethinking the accounting stance on sustainable development. *Sustainable Development, 20*(1), 28–41. doi:10.1002/sd.462.

Owen, G. (2013). Integrated reporting: A review of developments and their implications for the accounting curriculum. *Accounting Education: An International Journal, 22*(4), 340–356. doi:10.1080/09639284.2013.817798.

Simnett, R., Nugent, M., & Huggins, A. L. (2009). Developing an international assurance standard on greenhouse gas statements. *Accounting Horizons, 23*(4), 347–363. doi:10.2308/acch.2009.23.4.347.

The Prince's Accounting for Sustainability Project (A4S), & Global Reporting Initiative (GRI). (2010). Formation of the International Integrated Reporting Committee (IIRC) Retrieved 15 April, 2014, from http://www.theiirc.org/wp-content/uploads/2011/03/Press-Release1.pdf

Submissions Examined and Analysed

American Institute of Certified Public Accountants (AICPA). (2011). [Submission concerning] *IIRC Discussion paper: Towards Integrated Reporting: Communicating Value in the 21st Century*. Retrieved 22 May, 2014, from http://www.theiirc.org/wp-content/uploads/2012/02/AICPA-USA.pdf

American Institute of Certified Public Accountants (AICPA). (2013). [Submission concerning] *IIRC Consultation Draft of the International Integrated Reporting Framework*. Retrieved 22 May, 2014, from http://www.theiirc.org/wp-content/uploads/2013/08/120_AICPA.pdf

Chartered Institute of Management Accountants (CIMA). (2011). [Submission concerning] *IIRC Discussion paper: Towards Integrated Reporting: Communicating Value in the 21st Century*. Retrieved 20 January, 2014, from http://www.theiirc.org/wp-content/uploads/2012/02/CIMA-UK.pdf

Chartered Institute of Management Accountants (CIMA). (2013). [Submission concerning] *IIRC Consultation Draft of the International Integrated Reporting Framework*. Retrieved 20 January, 2014, from http://www.theiirc.org/wp-content/uploads/2013/08/180_CIMA1.pdf

CPA Australia. (2011). [Submission concerning] *IIRC Discussion paper: Towards integrated reporting: Communicating value in the 21st century*. Retrieved 20 January, 2013, from http://www.theiirc.org/wp-content/uploads/2012/02/CPA-Australia-Australia.pdf

CPA Australia. (2013). [Submission concerning] *IIRC consultation draft of the International Integrated Reporting Framework*. Retrieved 20 January, 2013, from http://www.theiirc.org/wp-content/uploads/2013/08/094_CPA-Australia.pdf

Federation of European Accountants (FEE). (2011). [Submission concerning] *IIRC Discussion paper: Towards Integrated Reporting: Communicating Value in the 21st Century*. Retrieved 22 May, 2014, from http://www.theiirc.org/wp-content/uploads/2012/02/FEE-Belgium.pdf

Federation of European Accountants (FEE). (2013). [Submission concerning] *IIRC Consultation Draft of the International Integrated Reporting Framework*. Retrieved 22 May, 2014, from http://www.theiirc.org/wp-content/uploads/2013/08/301_Federation-of-European-Accountants-combined.pdf

The Association of Chartered Certified Accountants (ACCA). (2011). [Submission concerning] *IIRC Discussion paper: Towards Integrated Reporting: Communicating Value in the 21st Century*. Retrieved 20 January, 2014, from http://www.theiirc.org/wp-content/uploads/2012/02/ACCA-UK.pdf

The Association of Chartered Certified Accountants (ACCA). (2013). [Submission concerning] *IIRC Consultation Draft of the International Integrated Reporting Framework*. Retrieved 20 January, 2014, from http://www.theiirc.org/wp-content/uploads/2013/08/184_ACCA.pdf

The Institute of Chartered Accountants Australia (ICAA). (2011). [Submission concerning] *IIRC Discussion paper: Towards Integrated Reporting: Communicating Value in the 21st Century*. Retrieved 20 January, 2014, from http://www.theiirc.org/wp-content/uploads/2012/02/Institute-of-Charterd-Accountants-1-Australia.pdf

The Institute of Chartered Accountants Australia (ICAA). (2013). [Submission concerning] *IIRC Consultation Draft of the International Integrated Reporting Framework.* Retrieved 20 January, 2014, from http://www.theiirc.org/wp-content/uploads/2013/08/146_Institute-of-Chartered-Accountants-Australia.pdf

The Institute of Chartered Accountants in England and Wales (ICAEW). (2011). [Submission concerning] *IIRC Discussion paper: Towards Integrated Reporting: Communicating Value in the 21st Century.* Retrieved 22 May, 2014, from http://www.theiirc.org/wp-content/uploads/2012/02/ICAEW-UK.pdf

The Institute of Chartered Accountants in England and Wales (ICAEW). (2013). [Submission concerning] *IIRC Consultation Draft of the International Integrated Reporting Framework.* Retrieved 22 May, 2014, from http://www.theiirc.org/wp-content/uploads/2013/08/342_ICAEW.pdf

The Institute of Chartered Accountants in Scotland (ICAS). (2011). [Submission concerning] *IIRC Discussion paper: Towards Integrated Reporting: Communicating Value in the 21st Century.* Retrieved 22 May, 2014, from http://www.theiirc.org/wp-content/uploads/2012/03/ICAS-UK1.pdf

The Institute of Chartered Accountants in Scotland (ICAS). (2013). [Submission concerning] *IIRC Consultation Draft of the International Integrated Reporting Framework.* Retrieved 22 May, 2014, from http://www.theiirc.org/wp-content/uploads/2013/08/183_ICAS.pdf

The Institute of Internal Auditors (IIA). (2011). [Submission concerning] *IIRC Discussion paper: Towards Integrated Reporting: Communicating Value in the 21st Century.* Retrieved 20 January, 2013, from http://www.theiirc.org/wp-content/uploads/2012/02/IIA-2-USA.pdf

The Institute of Internal Auditors (IIA). (2013). [Submission concerning] *IIRC Consultation Draft of the International Integrated Reporting Framework.* Retrieved 20 January, 2013, from http://www.theiirc.org/wp-content/uploads/2013/08/114_Institute-of-Internal-Auditors.pdf

The International Federation of Accountants (IFAC). (2011). [Submission concerning] *IIRC Discussion paper: Towards Integrated Reporting: Communicating Value in the 21st Century.* Retrieved 22 May, 2014, from http://www.theiirc.org/wp-content/uploads/2012/02/PAIB-USA.pdf

The International Federation of Accountants (IFAC). (2013). [Submission concerning] *IIRC Consultation Draft of the International Integrated Reporting Framework.* Retrieved 22 May, 2014, from http://www.theiirc.org/wp-content/uploads/2013/08/219_IFAC.pdf

United States Accounting Firms Respond to COSO Advice on Social Audit, Sustainability Risk and Financial Reporting

Katherine Kinkela

1 Introduction

The Big 4 United States Accounting firms (Ernst and Young LLP; KPMG LLP, Deloitte LLP and PwC LLP) have developed their sustainability and social audit practice units in a substantial way over the past few years. The Big 4 Accounting firms were motivated to build these new practice areas based in large part on new guidance by COSO, the Committee of Sponsoring Organizations. In the United States, corporate stakeholders perceive added value when corporations demonstrate that they will commit to sustainability goals on an environmental or social level. The measurement and disclosure of these sustainability goals is not required to be reported under a standard format or required to be included with the filing of United States financial statements. There is a movement in the American community to demonstrate how sustainability goals have become a part of the fabric of the strategic goals of the corporation, and to follow this extended commitment to financial reporting of sustainability goals within an organization. One leader in this effort of increasing reporting of sustainability goals and adding disclosure metrics of sustainability goals to financial reporting is COSO, the Committee of Sponsoring Organizations, an organization that has gained strength in the years following the passage of the Sarbanes-Oxley Act of 2002. COSO released a white paper on sustainability risks and goals in 2013, and United States firms have responded by adjusting their outlook and practice based on this new guidance. We will examine the impact of this COSO paper on sustainability risk on corporate views of sustainability goals, and the related increase in emphasis of this practice area in the Big 4 firms.

Financial reporting and managerial accounting practices in the United States reflect the Generally Accepted Accounting Principles (GAAP), but United States

K. Kinkela (✉)
Iona College, 1111 Midland Avenue, Bronxville, New York 10708, USA
e-mail: Kkbuslaw@hotmail.com

GAAP does not have specific requirements regarding implementation of sustainability or social goals within a corporation. While typical disclosures within the financial statements indicate legally required compliance with environmental and other concerns, social and sustainability plans and audits are typically not disclosed. A recent paper by the Committee of Sponsoring Organizations (COSO), an American organization comprised of Accounting industry associations, advocates integrating sustainability goals and social audit practices as a part of corporate strategy. In addition, COSO encourages corporations to assess sustainability risk as an ongoing process in business planning, and monitor sustainability programs through social audit.

This chapter examines the white paper on sustainability risk written by COSO in 2013. The white paper seeks to advise corporations of practical ways that corporations may include sustainability goals as a part of the risk assessment process that is ongoing within a corporation. In addition, the white paper addresses benefits to the corporation, and seeks to demonstrate how moving sustainability goals into the overall strategy planning and risk assessment functions of a corporation can also achieve other synergistic efficiencies within the corporation. We will also examine the way that the sustainability practices within the Big 4 accounting firms have expanded in light of the new developments.

2 Background on COSO and Enterprise Risk Management

The Committee of Sponsoring Organizations of the Treadway Commission (COSO) is dedicated to providing thought leadership through the development of comprehensive frameworks and guidance on enterprise risk management, internal control, and fraud deterrence, designed to improve organizational performance and governance and to reduce the extent of fraud in organizations. COSO is a private-sector initiative jointly sponsored and funded by five organizations: American Accounting Association (AAA), American Institute of Certified Public Accountants (AICPA), Financial Executives International (FEI), The Institute of Management Accountants (IMA) and The Institute of Internal Auditors (IIA).

COSO was expanded as a result of the Sarbanes Oxley Act of 2002 (SOX). With so many American corporations looking for guidance on SOX compliance, COSO prepared a series of studies on how to implement better internal control systems. COSO advocates the use of Enterprise Risk Management (ERM), a systematic and global approach to setting strategies and assessing risks within the corporate environment. The ERM process includes an eight step process; assessing internal environment, objective setting, risk identification, risk assessment, risk response, objective setting, information and communication, and monitoring.

COSO's work on sustainability includes a recent white paper, entitled "Demystifying Sustainability Risk: Integrating the Triple Bottom Line into an Enterprise Risk Management System." (COSO White Paper, 2013) The concept is that corporations should include sustainability as a part of the ERM planning process and goals on sustainability should be included as part of the goal setting and audit process.

2.1 The Triple Bottom Line, ERM and Sustainability

The COSO White Paper starts with a discussion of how sustainability goals add value to the organization, by adding intangible value that is perceived by stakeholders and stockholders. This intangible value has been referred to as a "triple bottom line". The COSO White Paper frames the issue in this way:

> **Intangibles Identify an Organization's True Value**
> The confluence of risks and opportunities associated with environmental, social and economic performance has made sustainability a strategic priority for companies as part of their overall business strategy. Measuring an organization's environmental, social and economic performance is often referred to as the "triple bottom line" (COSO White Paper, 2013).

2.1.1 Integrating Sustainability Goals Within the ERM Process

The white paper takes the eight steps of the Enterprise Risk Management ERM process and describes how to incorporate sustainability and social goals into the ongoing ERM process.

2.1.2 ERM Step One: Internal Environment

The first step in the ERM process is an examination of the internal environment of the corporation, understanding the resources strengths and weaknesses of the current corporation. When an understanding of the internal environment is achieved, the corporation can identify risk tolerances and risk appetite, and specifically look at opportunities and risks associated with social and sustainability goals. Risk tolerances and risk appetite are set by the board of directors of the corporation; management must understand the risk appetites that are set by the board of directors and must adhere to these risk appetites when engaging in the everyday operations of the corporation. COSO indicates:

> The internal environment reflects the tone of an organization and how it considers and manages risk. It sets the stage for what is defined in the corporate risk appetite, as well as related activities and decisions. Internal environment considerations should not simply be a summary of the status quo. Rather, it is an opportunity to proactively align and drive the organization. The internal environment should be the actualization of leadership vision and strategic aspirations (COSO White Paper, 2013).

The COSO White Paper notes that formalizing the risk appetite process allows the board of directors and management a unique framework to discuss issues regarding operational and strategic risk in greater detail than simply stating vague guidelines. This stage of formalizing risk appetites is an ideal time for the board of directors and management to discuss the integration of social and sustainability goals within the organization, to discuss feasibility, risk and benefit, and to consider alternative solutions for achieving social and sustainability goals. Formalizing risk appetites can identify and solidify the corporation's commitment to achieving social and sustainability goals, by finding workable solutions to strategic challenges and identifying potential future issues in the overall processes of the corporation.

The vision and mission statements of a corporation may provide insight into the recommended decision making process, however, taking time to create a clear understanding of the risk appetites of the board of directors can provide substantial benefits in the operation and management decision making processes.

> Although many organizations have an internalized set of assumptions that reflect the values and guidelines they use for their decision making, few have taken the step of defining their risk appetite. Formalizing the fundamental assumptions and preferences in the form of a risk appetite drives better alignment of risk and establishes a clear foundation for formulating practical risk tolerances (COSO White Paper, 2013).

For social and sustainability goals, a well-developed set of risk appetite guidelines can convey to management, the board of directors intent to prioritize strategies and processes that support and advance social and sustainability goals. The board of directors can also use the opportunity of the process of setting and explaining risk appetites to management, to fully explore stakeholder desires in terms of social and sustainability goals. A discussion of expectations and priorities in strategy for the board of directors, the management and stakeholders can help to clear up the direction necessary in the future decision making processes on all levels.

> When formulating or reviewing the enterprise-wide risk appetite, organizations should also establish their sustainability risk boundaries. For example, a basic scenario analysis which tests the acceptability of various sustainability impacts to the organization can help set the tone for what sustainability risks the organization should or should not accept. Other approaches, such as comparing stakeholder expectations to current sustainability strategies and exposures, can help set the management tone by indicating the weighting applied to various considerations and potential impacts (COSO White Paper, 2013).

In a strategic plan, social and sustainability goals must be designed to incorporate many levels of an organization, and to integrate social and sustainability goals at each level and function of the organization as a part of the primary strategy development of the individual parts of the organization.

> Organizations should also evaluate whether business sustainability should have its own strategy or be a part of the larger picture. We advocate that sustainability should be an embedded consideration in all organizational strategies and tactics rather than a stand-alone initiative. However, each company's decision on this aspect will weigh heavily on the internal tone of its ERM efforts as it pertains to sustainability. Ideally, this should occur when an organization creates or updates the organizational strategy and related tactical initiatives. This aligns initiatives and work steps which, in turn, helps mitigate risk and

reduce costs. For those organizations that only update their overall strategy on a periodic basis (e.g., every 5 years), it may be prudent to develop a sustainability strategy with the intent of integrating it into the overall organizational strategy during the next period of strategy update and renewal (COSO White Paper, 2013).

The overall goal of introducing the concepts of social and sustainability goals, at this early stage in the strategy and risk assessment process, is to create a holistic vision of social and sustainability goals as a part of the overall corporate strategy. At this early stage, the corporation emphasizes its commitment to social and sustainability goals, and it encourages all participants in the strategy making and risk assessment processes to take responsibility for these social and sustainability goals from the outset of the planning process. This ownership makes it easier to follow through with the same objectives throughout the whole operational process.

> This requires considerable coordination to ensure that the sustainability strategy is not developed in isolation and then simply "tacked on" to the overall strategy (COSO White Paper, 2013).

The authors of the white paper also noted the importance of an examination of the external factors, the opportunities and threats in the external environment when formulating strategy and risk appetite. External environmental factors are important to success of operational strategies.

> In addition to thinking about sustainability in the context of the internal environment, organizations may also wish to consider the external environment. Although not explicitly called out in this area of the COSO ERM Framework, external scanning is essential to truly connect a company's internal environment to the world in which it operates. This is especially important relative to sustainability to accommodate a full range of business models and more fully account for the interaction and interdependencies of internal and external forces (COSO White Paper, 2013).

2.1.3 ERM Step Two: Objective Setting

The Second ERM step is objective setting; objective setting is critical to the measurement of desired outcomes, including social and sustainability audit outcomes. Objective setting must be informed by the considerations set out in the first step, internal environment. The COSO White Paper does not address the objective setting section of the ERM in great detail.

> All ERM programs need to start with the basis of organizational objectives as the backdrop for risk considerations and management activities. This doesn't change when considering sustainability objectives. Incorporating sustainability considerations broadens the range of possible risks that can impact organizational objectives. It can also serve to align potential exposures with the risk appetite and highlight risks associated with chosen strategies and pursuits (COSO White Paper, 2013).

2.1.4 ERM Step Three: Event (Risk) Identification

Risk Identification is the third aspect of the enterprise risk management cycle. Risk Identification is the process of choosing the risks with the highest impact to the operational system so that the corporation can be studied. The white paper suggests that all risks that have been analyzed by the corporation in the past should be reconsidered with a specific view towards the impact of implementing additional social and sustainability goals. It is important to analyze whether the additional steps required in the overall process to achieve social and sustainability goals create additional risk. At this stage, alternatives of different processes and methods to achieve social and sustainability goals can be explored. This additional risk presented by adding social and sustainability initiatives should be analyzed and compared to the risk appetites in the first part of analysis. The analysis and identification of risk based events is also significant for different levels and functions of the corporation and this impact should also be discussed.

> Sustainability should be top-of-mind when considering risk identification as a whole, but particularly when comparing sustainability risks and opportunities against the full spectrum of a company's risk universe and specific profile. At this level, sustainability can pose a higher-level impact, which subsequently defines how the organization evaluates the risks and opportunities (COSO White Paper, 2013).

Social and Sustainability issues can also provide an important reason to reprioritize examination of objectives and resource deployment. Re-examining original designations can be beneficial to overall success of newly created objectives.

> Organizations need to evaluate all risk exposures relative to potential sustainability issues, as well as how those sustainability issues may impact other risks present within the organization. Organizations can then prioritize the issues within traditional considerations of impact and probability (COSO White Paper, 2013).

Risk identification should be a systematic process to determine materiality and priority, and sustainability should be incorporated in the levels of measurement of risk and impact. The idea is to make the measurement process as useful as possible by putting as much information about social and sustainability goals as possible.

> Most risk identification scales include three to five impact dimensions, which are graduated from low (minimal) impact to high (catastrophic) impact. Organizations can integrate sustainability impacts into this scale to expand awareness and prioritize risks. For example, sustainability can be a component of identifying operational risk objectives by considering the type and level of effects sustainability events could present (COSO White Paper, 2013).

Operational evaluation and integration with social and sustainability goals should be examined and refined over time.

> To gain a comprehensive view of the potential, possible and likely sustainability threats and challenges to an organization's objectives, organizations should bring together both sustainability subject matter experts as well as the operational and strategic business content experts. Sustainability knowledge experts can identify and articulate interdependencies, unintended consequences and non-intuitive impacts stemming from social, environmental

and economic considerations that often do not come to light in a traditional approach (COSO White Paper, 2013).

2.1.5 ERM Step Four: Risk Assessment

Once Risk Identification is complete, the corporation should seek to assess the probability and impact of the risk on the overall process; this is Risk Assessment. Risk Assessment requires examining the risks identified in the prior examination and determining what the likelihood of occurrence of the risk is and what the impact of the risk will be if it occurs.

> Most organizations include a risk root cause and sensitivity analysis to understand the drivers and pathways of organizational risks. Because of the changing nature of company value perceptions, sustainability also provides an increased ability to further analyze risk by enabling a range of potential value impairment estimates tied to the changing perceptions of an organization. For example, by tracking reputational impacts linked to sustainability missteps (yours or another company's), an organization can build a database that enables correlations and scenario modeling relative to stock impacts, top line revenue impairments and even market dynamics. This is an area that is rapidly developing and provides a valuable dimension to risk assessments (COSO White Paper, 2013).

Connecting social and sustainability goals with associated risks is critical so that materiality of risks can be determined. Connecting risks to other operational objectives and risks can be beneficial to the overall process.

> However, it is important to note that sustainability discussions related to materiality can become complex very quickly. Often, there are a number of engaged stakeholders who want to influence which risks the organization should prioritize. In addition, it can be hard for organizations to accurately measure the impact a risk has on its sustainability initiatives. For example, an organization that treats the community in which it operates, or its employees, poorly, could expose itself to operations, financial and reputation risks (COSO White Paper, 2013).

A complete risk assessment also considers the extended effects of the identified risks, as an additional indication of materiality.

> Because sustainability concerns extend beyond financial impacts, organizations would do well to also evaluate directional impacts. These may include the eventual impact actions or activities that do not present themselves as a discrete event, such as ignoring an emerging stakeholder group—the risk that those stakeholders gain influence over consumer sentiment and ultimately brand value (COSO White Paper, 2013).

2.1.6 ERM Step Five: Risk Response

Once the risk is identified and risk assessment is completed by understanding the probability of occurrence and the potential for damage as evaluated, Risk Response strategy must be formulated by the management. Risk response is an analysis of potential solutions to the problems that might be generated by the risk. Risk response must consider social and sustainability issues.

As noted earlier, risk responses should be tied to the drivers of risk and anchored in what is an acceptable range of solutions. Sustainability factors that form the core of an organization's values can help frame what will or won't serve as an acceptable risk response, and why (COSO White Paper, 2013).

Considering the impact of social and sustainability goals can be important to deciding on the nature and importance of risk responses; the more socially responsible solutions may have the greatest long term benefit even where the initial cost may be greater to the corporation. Choosing appropriate risk responses can also be important to the public perception of the corporation. Crisis management is important to the stakeholders of the corporation.

For example, if a key sustainability precept is protecting cultural history, artifacts or sites where it operates, then risk responses likely include production capacity issues, limitations on facility footprint or building height. Such self-imposed risk responses can significantly impact facility design, but can also provide positive impacts on how the market views the organization (COSO White Paper, 2013).

Proper communication with management is critical at this step. As decision makers, management must take a global and holistic view of the issues.

In addition to specific action planning, organizations should consider these factors when designing business cases or making investment decisions. For example, as an extension of the ERM process, all business cases may incorporate a section, or suite of questions that probe the potential sustainability impacts of the investment. Accordingly, a well-designed set of leading questions can enable management to identify and address potentially overlooked linkages and unintended consequences (COSO White Paper, 2013).

2.1.7 ERM Step Six: Control Activities

The sixth element of the enterprise risk management process is the creation of control activities. Creating effective internal controls is a collaborative effort between the board and management. Controls should be created as a timely indicator of the success of processes and in addition the study of the results of the controls can indicate the emergence of additional risk factors that might potentially be material threats to operations.

Sustainability resources, the controller's office, operations and other relevant stakeholders can work closely together to develop policies and procedures that effectively execute risk responses. It is also important that the sustainability function collaborate with a wide range of stakeholders who thoroughly understand the risks and opportunities being addressed. Control activities should not be defined in a vacuum. Once internal controls are identified and implemented, they require continuous measurement, monitoring and evaluation to ensure effectiveness (COSO White Paper, 2013).

The Internal Audit process in existence prior to setting social and sustainability goals should be reviewed and revised to add reviews and controls relevant to the social and sustainability process. This reevaluation should make the overall control process stronger.

Internal audit and other control monitoring functions within an organization (e.g., legal, compliance or safety) can also perform audits to evaluate the effectiveness of sustainability practices, communication protocols and reporting initiatives. These audits enable the organization to obtain an independent analysis of the design and operating effectiveness of sustainability initiatives. They can also provide valuable recommendations to improve initiatives or activities based on emerging trends within and outside the industry (COSO White Paper, 2013).

2.1.8 ERM Step Seven: Information and Communication

Once the results of the evaluation process are over, the results must be communicated to the proper decision makers within the organization. Communication is necessary so that timely implementation of changes may be completed. Reputation management goals are closely connected with the communication of the information gained through the value process.

Information and communication are critical factors for managing risks and opportunities, particularly those associated with sustainability. We have already discussed the importance of communicating clearly and truthfully to avoid reputation risks. This same rule applies when communicating sustainability performance to investors and analysts through sustainability reporting (COSO White Paper, 2013).

The triple bottom line is connected with reputation management for a corporation. Accountability of corporate board and management on sustainability issues, through the triple bottom line or similar measurement formats is expected and important to corporate stakeholders. Stakeholders form a community and feel personally about the importance of corporate social and sustainability goals and objectives. Accountability about incorporation of sustainability practices is important to stakeholders.

Stakeholders within the sustainability ecosystem expect organizations to not only share their successes, but also their failures or areas of improvement. This expectation creates an element of reputational risk in the short term. However, in the long term, this risk is often outweighed by the benefits. These benefits include: better measurement of the organization's triple bottom line performance, greater stakeholder trust, improved risk management and increased operational efficiency (COSO White Paper, 2013).

COSO has advised corporations about the benefits of identifying Key Performance Indicators (KPI), as a part of the overall risk assessment process. KPI are the optimal factors to measure to determine if performance goals have been successfully met. KPI make sure that the relevant items are being measured so that consistent and continuous improvements can be made. COSO has provided guidance on using these KPI as the basis for evaluating risk and strategic goals. KPI are the critical factors in evaluating operational performance.

Many of these benefits are derived from the internal processes and controls organizations put in place to help them collect, store and analyze financial and non-financial key performance indicators (KPI). Obtaining real-time, quality data on such issues as GHG emissions, water use and supply chain activities can help organizations enhance decision making, while reducing risks and enhancing opportunities (COSO White Paper, 2013).

Transparency in operations is a major consideration for corporations. As sustainability goals and reporting become the norm, stakeholders will demand more accountability from corporations on social and sustainability goals and related timelines. Stakeholders will want to understand the social and sustainability goals, and the level of achievement reached, in order to determine the overall commitment of the corporation and its board and management to achieve social and sustainability goals.

> Choosing *not* to report on sustainability, by contrast, can increase reputation risks or limit opportunities. Organizations that do not release sustainability information may appear less transparent than competitors that do, and come across as laggards even if they are not. Furthermore, those that report incompletely, or with insufficient rigor, may find that if reporting becomes mandatory and standards are tightened, glaring discrepancies might appear between past reports and newer ones (COSO White Paper, 2013).

Overall, a constant and consistent sustainability analysis, with its examination of long-term benefits and challenges to the corporation, is an integral part of the risk management analysis. Timely analysis of sustainability goals can provide significant insight into improvement of business practices.

> Internally, sustainability reporting is critical to decision making. It validates risk response effectiveness and overall sustainability performance. It can also identify changes to the risk environment, upon which business units can take action, and it can reflect changes to the organization's overall risk profile (COSO White Paper, 2013).

2.1.9 ERM Step Eight: Monitoring

The true essence of the social audit is contained within the monitoring component of the ERM process. In this monitoring segment of the ERM process, we see whether social and sustainability goals have been added correctly within the process, whether social and sustainability goals have been prioritized and the overall success of achieving social and sustainability goals within the overall operational process. Social audit objectives are confirmed in the monitoring process as the fulfillment of objectives are measured and evaluated.

> To ensure that an organization is achieving its objectives, staying within its risk tolerance threshold and satisfying stakeholders, it should constantly monitor and evaluate the sustainability activities it undertakes. Questions organizations should be asking as part of their measurement, monitoring and evaluation activities include:
>
> - Are activities or processes aligned to the corporate strategy?
> - Are they being executed in such a way to enable the business to better achieve its strategic objectives?
> - Are activities adding value in terms of risk awareness and understanding?
> - Are they agile enough to respond to changes in the risk environment as issues arise?
>
> (COSO White Paper, 2013)

The format of monitoring and the social audit implemented within the corporation will vary, and the process of social audit will be tailored and streamlined to fit the individual corporation's needs. Several useful formats used in the evaluation process are the balanced scorecard, and a dashboard approach. The balanced scorecard provides financial and nonfinancial measures of success, analyzing financial, customer, operational and employee goals.

> One approach organizations use to keep track of how well they are doing in their sustainability objective is the use of balanced scorecards. Using key risk indicators, organizations can plan, measure and monitor their sustainability risk management at each level of the organization. Management can then communicate this information using executive dashboards to senior executives and the board (COSO White Paper, 2013).

The usefulness of information provided in Social Audits depends on the timeliness of Social Audit information. Information must be provided to coincide with the times that processes are reviewed and strategic changes will be made. Management must also have confidence in the usefulness of Social Audit information and have a commitment to use of Social Audit report in critical decision-making processes.

> In the end, the effectiveness of monitoring approaches lies in the timeliness, integrity and transparency of the results, as well as what is done with the results to manage sustainability initiatives and mitigate the corresponding risks. Having a scorecard alone doesn't alleviate management's responsibilities for monitoring sustainability performance. Rather, the scorecard should enable management to make decisions on how to improve performance and achieve a competitive advantage in the marketplace (COSO White Paper, 2013).

3 White Paper, Social Audit Practices and Competitive Advantage

The white paper closes with a reiteration of the practical benefits of using the triple bottom line and social auditing practices as a part of corporate strategy. "Organizations that choose to embed sustainability into a COSO-based risk management program can achieve the following competitive advantages:" (COSO White Paper, 2013)

First, the white paper shows a holistic view of the corporation reveals a strong connection between sustainability and strategy.

- Alignment of sustainability risk appetite to the organization's corporate strategy and the new world view of company value. Having a holistic view of sustainability risk that looks across the entire enterprise enables organizations to do a better job of anticipating and responding to issues as they arise (COSO White Paper, 2013).

Sustainability and Social Audits allow corporations a better understanding of the global environment in which they operate. This added level of review that a social audit provides improves operational performance because it allows familiar issues to be viewed in a new way.

- Expanded visibility and insights relative to the complexity of today's business environment. Embedding sustainability into an organization's ERM framework enables the sustainability function to gain valuable insights regarding the sustainability risks the organization faces and the materiality of those risks. These are insights the sustainability function can then share with management and the board so that they have a clear understanding of the sustainability risks relative to the complexity of the business environment (COSO White Paper, 2013).

When corporations embrace sustainability, the corporation demonstrates that they find value in intangible and nonfinancial goals; and that the decision makers within the corporation understand the connection between sustainability goals and strategic success.

- Stronger linkage of company values and non-financial impacts to the organization's risk management program. Identifying sustainability risks and opportunities can be challenging. However, organizations that understand how to link them to their value drivers are better able to understand the impacts on the business in non-financial ways (COSO White Paper, 2013).

Using a "sustainability lens" is an additional level of review, and this additional level of review can provide definite benefits. The additional level of review that sustainability lens provides helps make strategy and operations more effective, comprehensive and innovative. Management must also incorporate a long-term approach for sustainability goals, and this long-term consideration can benefit other comprehensive program goals.

The implementation of this "sustainability lens" can also be a benefit as an aspect of reputation management, as stakeholders perceive a more aware and effective management team, a management that is in tune with social and sustainability needs.

- Better ability to manage strategic and operational performance. Organizations can create competitive advantage by managing sustainability risk to improve business performance, spur innovation and boost bottom- line results. Companies that conceive their products or services through a sustainability lens will attract funding from external investors and boost stakeholder confidence. Sustainability as part of the value proposition is also becoming as relevant to market capitalization as innovation or R&D (COSO White Paper, 2013).

Finally, the Social Audit practices connected with sustainability help corporations to deploy capital in the most efficient way to achieve sustainability and systematic goals. The corporation can examine the benefits and multiple efficiencies achieved with effective capital deployment.

- Improved deployment of capital. Organizations that have used the COSO ERM Framework to embed sustainability risk management practices have better opportunities to allocate capital more effectively—in ways that maximize capital efficiency or that send the right messages to stakeholders based on the organization's corporate values and strategy, but in all ways enable the organization to reach its sustainability and, more importantly, its corporate objectives (COSO White Paper, 2013).

Corporations should strive for transparency in reporting and full disclosure should include social audit and sustainability goals. Demand for transparency in social and sustainability programs by stakeholder groups makes sustainability increasingly important for attention. Social and sustainability goals can be aligned with existing corporate policies. Incorporating the social goals with corporate policies is an important step to complete integration of the social and sustainability goals within the business process.

Overall, the white paper provides a good starting point for analysis and discussion of incorporating social and sustainability goals into the fabric of the organizational strategy of the corporation. In the future, COSO could address the process of objective setting in greater detail. Objectives should be tailored to include both the operational and related social and sustainability goals, and additional COSO guidance in this area would be beneficial.

American corporations have an opportunity to embrace sustainability fully by incorporating social audit into their Enterprise Risk Management systems. Although the current norm for strategic planning and financial reporting for American Corporations is to report on sustainability issues (other than required legal environmental compliance) separately from the annual financial reports of a corporation, integration of social audit practices and results into the financial reporting process will be a more holistic view of the operational and strategic success of a corporation. In addition, the process of examining and crafting sustainability goals will benefit the corporation by reviewing and refining existing processes. COSO's work in the white paper provides an excellent starting point for corporate management to integrate these principles.

4 Accounting Firms Responses to COSO Guidance

Each of the big 4 United States Accounting Firms has incorporated the COSO White paper guidance into their sustainability practices; each firm also has been monitoring the sustainability marketplace in financial reporting.

4.1 Ernst and Young LLP

4.1.1 Sustainability Survey

Ernst and Young prepared a 2013 sustainability survey. The Ernst and Young survey examined how companies are responding to a wide range of internal and external forces related to environmental sustainability risks and how well companies are prepared to address them. The key results that were determined as a result of the sustainability survey were six trends (E&Y Sustainability Survey, 2013).

The six trends in Fig. 1 indicate that most corporations have not yet identified sustainability goals as a part of the overall business strategy. An Executive's involvement is key to getting sustainability issues prioritized as a part of financial reporting (E&Y Sustainability Survey).

> Six trends emerged:
>
> 1. The "tone from the top" is key to heightened awareness and preparedness for sustainability risks.
> 2. Governments and multilateral institutions aren't playing a key role in corporate sustainability agendas.
> 3. Sustainability concerns now include increased risk and proximity of natural resource shortages.
> 4. Corporate risk response is not well paired to the scale of sustainability challenges.
> 5. Integrated reporting is slow to take hold.
> 6. Inquiries from investors and shareholders are on the rise.

Fig. 1 Six trends

4.1.2 Executive Involvement

In addition, the Ernst and Young publications made the following observations on financial reporting and the connection with the involvement with the commitment of upper level financial executives.

> "Companies that have a greater level of engagement from the CEO and the board have much closer alignment between what they voluntarily disclose (such as CDP and DJSI) and what they are mandated to disclose (such as 10-K filings). When the CEO and the board are involved, there is much greater alignment in risk identification and disclosure. While 22 % of surveyed companies indicated total alignment on both mandated and voluntary sustainability disclosures, 36 % acknowledge "total alignment," indicating both a fully engaged board and CEO. Heightened CEO and CFO attention to sustainability reflects the gradual ascent of sustainability issues within the corporate risk register. C-suite involvement also underlines the growth of corporate sustainability as a strategic differentiator" (E&Y Tone from the Top, 2014).

At this stage, according to Ernst and Young the majority of companies have not prioritized sustainability issues as a part of strategic planning.

4.1.3 E and Y Vision 2020 Program

The internal sustainability commitment for E&Y as a corporation is a campaign with multiple facets, including a companywide initiative for sustainability. "Recently, the global EY organization launched Vision 2020. A global initiative, Vision 2020 details our purpose, ambition, strategy and positioning for building a better working world in four distinct categories:"

Clients. Through timely and transparent information, we provide help to build trust and confidence in the capital markets and in economies across the world. Through our professional services, we help our clients improve and grow, resulting in higher living standards and more opportunities for growing local economies. And through our Strategic Growth Markets practice, we recognize and help entrepreneurs, who are the key to economic health.

People. We are committed to a highest-performing teaming culture with great people who develop into future leaders; inclusive, borderless teams; and people who live our values. Our people support one another in pursuit of their personal best, and they possess an unwavering commitment to diversity and inclusiveness.

Communities. Our people support the wider marketplace and communities. Globally, we work with organizations such as the World Economic Forum (WEF). Locally, our people give their time, skills and knowledge to the communities in which we live and work.

Environment. We recognize that the biggest positive environmental impact we can have is by supporting our clients in their goals to operate more sustainably. We help our clients improve their environmental performance, lower costs, manage risk and increase transparency. We also recognize our own environmental responsibility.

As such, we continue to challenge ourselves to work in a more environmentally responsible manner and find new ways to reduce our carbon footprint by engaging with our stakeholders.

Although we have launched Vision 2020, we are only at the beginning of our journey. We have a lot of hard work ahead to execute on our Vision 2020 objectives and fulfill our commitment.

In a world filled with uncertainty, our role in the years to come is to support our clients, our people and our communities in making the right decisions. EY is working toward a better working world every day, and we are committed to doing all we can to achieve it" (E&Y Vision 2020).

The Ernst and Young sustainability practices incorporated in the Vision 2020 program demonstrate the internal commitment of the firm to sustainability principles (E&Y Americas Sustainability Report, 2013).

5 PwC

5.1 PwC Strategic Sustainability Practice

PricewaterhouseCoopers (PwC) has a Strategic Sustainability practice that integrates the COSO white paper principles. The PwC description of the practice notes:

As sustainability moves up the boardroom agenda, it is increasingly being integrated into corporate level strategic planning. Management now needs to balance increased regulation, protecting the brand and ensuring stable supply chains with seeking opportunity for enhanced performance and using the sustainability agenda for strategic advantage (PwC Strategic Sustainability).

PwC consulting offers an integrated approach to different levels and aspects of sustainability planning:

"Developing and integrating a detailed sustainability vision into your long-term strategic plan in a way that creates lasting value whilst also building public trust is a common challenge for all types of organisations.

We can help you to:

Identify your issues and goals to determine where the pressures are likely to be and raise awareness of what needs to happen to make your business more sustainable.

Prioritise these issues from both a sustainability and commercial point of view. This will help you recognise and better manage risk, improve efficiency, revenue potential, growth and other opportunities.

Map the short and long-term ambitions for your sustainability vision, assess the risks, and address any gaps in delivery.

Support the alignment and integration of your sustainability vision into your overall corporate strategy.

Develop and deliver a robust sustainability programme that includes prioritised initiatives, enablers, milestones, key performance indicators, and measurable targets" (PwC World Watch Sustainability News, 2014).

As noted, the PwC consulting covers long and short term priority goals integrating sustainability goals at all levels of strategic planning and implementation goals. There is also usage of mapping and KPIs to monitor effectiveness of sustainability initiatives.

5.2 Social Value v. Shareholder Value

Studies prepared by PwC also note that upper level executives are motivated by both shareholder value on the long and short term, and by the desire to leave a legacy of social values for the corporation (Preston, PwC Social and Shareholder Value, 2014).

6 Deloitte

6.1 The Era of Sustainability Reporting

Deloitte's sustainability practice is anchored by a study entitled "the Era of Sustainability Planning". Deloitte has combined the information gathering into an environmental, social and governance (ESG) component.

"During the past few decades, the primary drivers of business value have shifted significantly. Formerly, capital market performance most closely tracked an organization's tangible assets, but today's markets are more strongly correlated with intangible assets in the form of goodwill or brand equity, which can include research and development, brand, reputation, management of external social and environmental factors, and social license to operate. Many feel that traditional financial metrics may not effectively capture a company's long-term value creation potential, but rather serve as indicators of short-term performance. This shift in value drivers and a broader recognition of the importance of environmental, social, and governance (ESG) performance have been accelerated by market forces demanding greater transparency by companies. In addition, there has been an increase in initiatives to promote and, in some instances, enforce more structured ESG reporting" (Sullivan Deloitte Internal Audit, 2014).

The Deloitte ESG practice area is devoted to incorporating sustainability issues and metrics into long term corporate goals.

6.2 Deloitte Studies Shareholder Value and Sustainability

Deloitte also emphasizes the shareholder value that is perceived by identifying social goals.

> Sustainability has also made it to the top of shareholders' agendas. In 2013, the number of social and environmental policy proposals filed has grown to comprise the second largest proportion of shareholder-sponsored proposals (after proposals focusing on corporate governance). Many companies are responding to the proposals by publishing sustainability reports and shedding more light on their approach to social and environmental issues raised by shareholders (Sullivan Deloitte Internal Audit, 2014).

7 KPMG's Internal Sustainability Goals and Reporting

7.1 KPMG Corporate Incentives "Sustainable Firm of the Year"

KPMG's internal sustainability initiatives received an award for excellence. KPMG highlighted its efforts"

> In securing the repeat award for "Sustainable Firm of the Year", the KPMG Global Citizenship team demonstrated KPMG's global leadership across a range of corporate sustainability initiatives.
> Highlights include the Global Green Initiative for environmental sustainability, which has seen KPMG firms meet aggressive goals in reducing emissions and increasing energy efficiency, while KPMG's Climate Change and Sustainability Services (2014a) was recognized by the 2013 *Verdantix Global Sustainability Leaders Survey: Brands*, which scored KPMG highest among all organizations on brand preference for sustainability assurance services.

Other sustainability initiatives highlighted included the KPMG Global Development Initiative, which is focused on helping find sustainable solutions to global and local poverty issues. The IAB Awards judges took special note of KPMG's dedication to new corporate citizenship initiatives across its network, offering thought leadership, research projects as well as internal initiatives making the organizations increasingly sustainable worldwide (KPMG IAB Award, 2014b).

7.2 KPMG Provides Consulting Services on the Journey to a Sustainable Business Model

In advising clients how to integrate sustainable practices into their business strategy, KPMG advocated the following six part model in Fig. 2 to integrating sustainability into the strategic business plan:

> Phase 1: Analyze
> The journey to building long-term value begins with analysis of an organization's situation in relation to environmental and social trends.
>
> Phase 2: Plan
> Goals and milestones must be fully integrated with the broader business strategy.
>
> Phase 3: Implement
> An organization needs a number of fundamentals in place in order to implement the strategy effectively.
>
> Phase 4: Monitor
> In order to monitor performance effectively, organizations need to have the right data-gathering processes, analysis tools and methodologies.
>
> Phase 5: Report and assure
> Reporting on social and environmental performance is now standard practice in business wherever in the world you may operate.
>
> Phase 6: Evaluate
> Regular evaluation is essential for companies taking the journey to a more sustainable business model.(KPMG Sustainable Business Model)

As you can see this model incorporates sustainability into the business planning model. It also insures that the actions are monitored and evaluated as a part of the overall process.

Fig. 2 Six phase model

Conclusions

The Big 4 Accounting firms in the United States have identified sustainability reporting as a priority. The growth of the sustainability practices reflects the guidance of the COSO working paper and this change in view will solidify sustainability planning practices in internal control and social audit in the United States in future years.

References

COSO White Paper. (2013). *Defining sustainability risk*. Ernst and Young LLP
Ernst and Young. (2013). *Six growing trends in corporate sustainability*.
Ernst and Young. (2014). *The tone from the top for sustainability reporting*.
Hickox J. R., KPMG LLP. (2014). *Working with you to create long term value*.
KPMG. (2014a). *Climate Change and Sustainability Services*, The journey to a sustainable business model, global center for excellence for climate change and sustainability.
KPMG. (2014b). Press release: *KPMG captures two International Accounting Bulletin Awards*.
Leisha, J., Starbuck, S., Ernst and Young LLP. (2013). *E&Y Americas Sustainability Report*. Ernst and Young LLP.
Preston, M., PricewaterhouseCoopers. (2014). *What motivates your CEO over the long term: Shareholder or social value?*
PricewaterhouseCoopers. (2014). *World watch sustainability news*.
Sullivan, K., Alan, F., Anna, N., & Mark, M. (2014). *Sustainability and non-financial reporting: The role of internal audit*. NY: Deloitte and Touche LLP.

Social Audit Regulation Within the NGO Sector: Practices of NGOs Operating in Bangladesh and Indonesia

Vien Chu and Belinda Luke

1 Introduction

NGOs are essential actors in delivering aid programs and their purpose in this context is primarily to address the needs of beneficiaries. Accordingly, NGO accountability to beneficiaries in the form of efficient and effective programs and operations is essential. This is particularly so, given that funds are entrusted by donors to NGOs on beneficiaries' behalf (Agyemang, Awumbila, Unerman, & O'Dwyer, 2009; Najam, 1996). Various researchers have noted the need for strengthening NGO accountability to beneficiaries (O'Dwyer & Unerman, 2008) in order to better meet beneficiaries' needs. Specifically, benefits identified from developing NGO accountability to beneficiaries include increasing the poor's sense of ownership of poverty alleviation projects, enhancing their self-esteem and confidence, and eliminating the risk of program fraud and inefficiencies (Mango, 2010).

Social audit is identified as one mechanism to strengthen NGO accountability to beneficiaries. Specifically, this process involves identification of and dialogue with salient stakeholders, developing performance indicators or benchmarks, evaluation of performance enabling continuous improvement, and public disclosure of findings (Ebrahim, 2003a). However, its application in practice is very limited (Agyemang et al., 2009). The value of social audit lies in it representing both a participation and evaluation process. However, conducting social audits involves both time and cost, particularly through the participation of multiple stakeholders (Agyemang et al., 2009; Assad & Goddard, 2010). Further, given the lack of prevailing regulation, the conduct of social audits remains largely voluntary. This results in self-selection bias, limiting the effectiveness and usefulness of this mechanism (Ebrahim, 2003a, 2003b).

V. Chu (✉) • B. Luke
School of Accountancy, QUT Business School, Queensland University of Technology, 2 George St, Qld 4001 Brisbane, Australia
e-mail: thithanhvien.chu@qut.edu.au

© Springer International Publishing Switzerland 2015
M.M. Rahim, S.O. Idowu (eds.), *Social Audit Regulation*, CSR, Sustainability, Ethics & Governance, DOI 10.1007/978-3-319-15838-9_5

NGOs engaging in microenterprise development programs (MED NGOs) have a strong focus on poverty alleviation through social and economic development. Specifically, MED involves working with and training the poor to develop their own microenterprises as a means to progress out of poverty. However, the participation of various actors (e.g. beneficiaries, companies, local governments) is central to link microenterprises with public sector resources and support, and private sector customers and suppliers in order to develop sustainable social and economic outcomes. While participation is central within MED, a similar participatory approach in the context of social audit is also essential. In particular, this approach has the potential to increase program effectiveness, enhancing MED NGOs' accountability to beneficiaries (Bhatt, 1997; LeRoux, 2009; Schmitz, Raggo, & Vijfeijken, 2011). However, the nature of social audit practices within these NGOs has received only limited attention in the literature.

Accordingly, the objective of this chapter is to identify the prevalence, scale, and scope of social audits in MED NGOs. From this process, the benefits and constraints of conducting social audits are also investigated. The next section reviews literature on social audits as one accountability mechanism for NGOs—its characteristics and limitations. The research method section follows. The findings and discussion sections are then presented, reflecting on social audit practices of MED NGOs, suggesting a better distinction between social audits and other mechanisms of NGO accountability is required, and outlining a pragmatic approach to social audit to expand the scope of this practice.

2 Social Audits as an Accountability Mechanism

Within the NGO sector, Ebrahim (2003a) identifies five mechanisms for accountability, being reports and disclosure statements, performance assessments and evaluations, participation, self-regulation, and social audit. While tools such as reports and disclosure statements, performance assessment and evaluation are relatively well-developed in practice (serving upward accountability to donors), participation, self-regulation, and social audit (addressing accountability to a broader range of stakeholders) are noticeably less developed (Agyemang et al., 2009; Ebrahim, 2003a). As noted by various researchers (Jäger & Rothe, 2013; Mason, Kirkbride, & Bryde, 2007; Sinclair, 1995), however, accountability is perhaps most appropriately viewed as a web involving multiple dimensions and multiple actors. Thus, accountability mechanisms which extend to and incorporate a range of stakeholders are essential for a more holistic recognition of accountability.

Of the three less developed accountability mechanisms identified by Ebrahim (2003a)—participation, self-regulation, and social audit—social audit is somewhat unique in that it represents both a tool and process, combining participation and transparency. Participation focuses on meaningful involvement of various stakeholders to evaluate NGOs' operations, reinforcing collaboration and co-ordination between NGOs and their stakeholders (Jordan & Tuijl, 2006; O'Dwyer & Unerman, 2010). Specifically, this process takes into account stakeholders' (including beneficiaries') views of an organisation's goals and operations, in particular the impacts of an NGO's activities on beneficiaries' lives (Agyemang et al., 2009). Transparency evaluates the level at which stakeholders can access information on organisational procedures, structures, and assessment processes on a timely basis (Hammer & Lloyd, 2011). Through the social audit process, an organisation is able to evaluate its performance according to the expectations of stakeholders including the communities in which they operate, learn from feedback and disclose findings publicly, promoting transparency (Deegan, 2002). This process helps NGO achieve effective long-term program outcomes, strengthening their social responsibility and ethical behavior (Dawson, 1998; Mason et al., 2007; Owen, Swift, & Hunt, 2001). It also assits in reinforcing an NGO's organisational legitimacy.

While NGO legitimacy is in part derived from compliance with legal requirements, in the context of social audit, legal and regulatory frameworks in both developed and developing countries are largely silent. Thus, typically NGOs' wider legitimacy is morally derived from their social mission and performance (Slim, 2002). Intangible sources of legitimacy such as trust, integrity and reputation (Slim, 2002) are invaluable. However, how NGOs develop trust and demonstrate integrity is somewhat discretionary. While social audits are one such mechanism, the literature suggests this mechanism is not widely adopted in practice, in part due to lack of formal regulation.

As noted by Owen, Swift, Humphrey, and Bowerman (2000), the term social audit relates more to 'taking a pulse' of the organisation's operations and effectiveness based on feedback from various stakeholders, rather than attesting to verifiable standards. Indeed, one of the limitations of social audit (considered further below) is the absence of such standards. Hence, social audits, while not legally required, can be a valuable resource—not only for an NGO's reputation and legitimacy, but also for the learning it generates, and the operational benefits which ensue. As such, social audit practice is typically dependent on self-regulation (Gugerty, 2008; Gugerty, Sidel, & Bies, 2010), whereby the adoption of social audit rests on individual NGOs or donors recognising or identifying the need to engage in this activity.

In theory, social audit represents a tool for strategic planning and organisational learning. In practice, however, social audit has not been widely used due to several organisational level challenges (Agyemang et al., 2009; Dawson, 1998). These challenges include time and cost constraints (particularly if the audit is externally verified), as well as lack of agreed processes (e.g. systematic approach, use of appropriate indicators for performance evaluation) and experienced staff to conduct these audits (Owen et al., 2000). However, given NGOs engage in a wide range of activities, development of universal performance indicators or benchmarks for comparison or evaluation within the NGO sector remains a significant challenge (Dawson, 1998). Performance assessments typically involving quantitative measures are provided mainly to donors or organisational managers for decision making. Incorporating qualitative or non-economic outcomes generated from NGOs' projects is often considered more complex (Jäger & Rothe, 2013). Other constraints and challenges of social audit include privileging the voices of some (more powerful) stakeholders over others, and encouraging those less powerful (e.g. beneficiaries who are often vulnerable and dependent on NGO assistance) to voice concerns (Kang, Anderson, & Finnegan, 2012). While these challenges are not insurmountable, the need for a pragmatic approach underpinned by the legitimacy of the process is highlighted, such that the social audit process is useful in serving the function of stakeholder accountability, rather than simply stakeholder management (Owen et al., 2000).

3 Social Audit in an MED Context

Unlike other types of NGOs (e.g. emergency help, healthcare service or education), NGOs engaging in MED have a strong focus on the social and economic development of poor communities by helping the poor to engage in income-generating activities, as a way of progressing out of poverty (Strier, 2010). In this context, a participatory approach to accountability has been promoted, such that beneficiaries are actively involved in various aspects of NGOs' poverty alleviation programs. Specifically training and networking processes allow the poor to develop the skills and experience required to establish micro or small businesses (under NGOs' guidance) and continue operating these businesses once NGOs' support ceases, rather than being passive recipients of aid (Brown & Moore, 2001; Choudhury, Hossain, & Solaiman, 2008).

Importantly, however, effective MED programs rely on the participation of various stakeholders (Islam & Morgan, 2011; Peredo & Chrisman, 2006), consistent with the notion of socialising accountability (Roberts, 2009). Engagement with multiple stakeholders recognises the limited resources of NGOs, and the importance of connecting microenterprises established by the poor with both the public and private sector (Janvry & Sadoulet, 2009; Jones, Kashlak & Jones, 2004; Karnani, 2007). Such connections enhance program effectiveness by developing long-term working relationships with these groups which can continue once NGOs' support ceases (Jones et al., 2004). Specifically, public sector support—both at the national and local level—is important to ensure participation in MED programs is encouraged and promoted, available public sector resources are identified and utilised, and the proposed business activity is supported (Karnani, 2007). Similarly, private sector buy-in is important to ensure connections with suppliers and buyers are made, and microenterprises' goods are tailored to market demands, as the poor learn to operate sustainable (long-term) businesses (Jones et al., 2004). Thus, a collective approach is important for effective and sustainable (long-term) poverty alleviation program outcomes.

Given effective MED relies on the participation of multiple stakeholders, it is perhaps not surprising that social audit has particular relevance to this form of aid program. From a financial perspective, social audit allows donors to hold NGOs responsible for effective use of funds received through social audit reports and reporting systems (Deegan, 2002). From an operational perspective, however, the value of social audit extends to participation of beneficiaries and those in the local community (both public and private sector actors) to provide feedback and input and voice and concerns through a participatory process. However, as noted by Keystone (2006), few (26 %) donors expressly request that NGOs involve beneficiaries in developing performance indicators, and fewer still (5 %) showed an interest in discussing beneficiary feedback with NGOs. While creating an institutional culture among MED NGOs that encourages and values beneficiaries' participation in social audits remains challenging (Kang et al., 2012), conducting social audits within the NGO sector often rests on the individual interests and priorities of donors or NGOs themselves. This effectively results in self-selection bias, limiting the effectiveness and usefulness of social audits more broadly as a control and evaluation mechanism (Ebrahim, 2003b). As such, competent and effective NGOs have strong incentives to conduct social audits and participate in self-regulation, whereas struggling NGOs may have little to gain from revealing their underperformance and ineffectiveness (Burger, 2012; Ebrahim, 2003b).

While the notion of socialising accountability reflects a sense of shared responsibility, social audit represents an opportunity for shared communication and understanding on what is perceived as effective, and what could be improved. Ultimately, however, without regulation requiring social audit in some form, the use of this mechanism rests on NGOs or powerful stakeholders identifying the need and allocating the resources for this task. Yet, given the power imbalances within accountability relationships between NGOs and other stakeholders (Assad & Goddard, 2010; Ebrahim, 2003a), it is unlikely beneficiaries or poor communities would be in a position to require or request social audits. However, from the perspective of more powerful stakeholders (e.g. donors), and consistent with the notion that with power comes responsibility (Keystone, 2006; Kilby, 2006), arguably, donors could require this process be undertaken on a regular (e.g. annual) basis, if NGOs do not initiate it voluntarily.

The focus on a participatory approach within MED NGOs has strong potential to increase the effectiveness of social audit in this context, yet the nature of this mechanism has received only limited attention in the literature. As such, it is necessary to understand what forms of social audits exist in practice, and their scale and scope. Before examining this issue, the next sections present an overview of the context and research method employed in this study.

4 Contextual Background: Bangladesh and Indonesia

Bangladesh and Indonesia were selected for this study as they are well known for their poverty reduction needs and activities. In 2012, total development aid provided by the Organisation for Economic Co-operation and Development [OECD] to Bangladesh and Indonesia was US$2,252 million and US$7,076 million respectively (OECD, 2014a, 2014b). Economic development initiatives (such as MED) are considered a central approach to poverty alleviation in both countries. However, despite the large number of poverty alleviation programs being established in both countries, success has been limited (Deen, 2010; Islam & Morgan, 2011). Bangladesh in particular remains one of the world's poorest countries. In 2012, nearly 43 % of its population were classified as extremely poor (living on less than $1.25 a day) (AusAID, 2013a). Further, the unstable social and political environment of Bangladesh presents additional challenges for both the poor and NGOs trying to assist them (AusAID, 2013a; Islam & Morgan, 2011).

Unlike Bangladesh, Indonesia has a much higher Gross Domestic Product (GDP) per capita ($3,557 in 2012 compared to $1,679 for Bangladesh). However, economic growth has not benefited Indonesia's population consistently, with poverty remaining a challenge. In 2012, more than 120 million (approximately 48 %) of Indonesia's population lived on less than $2 a day, 44 million of which (18.7 %) were living on less than $1.25 a day (AusAID, 2013b).

Notably, public sector corruption in both countries has been identified as an issue. While Indonesia was ranked 114 among 177 countries, Bangladesh was considered to be one of the most corrupt countries (ranked 136)[1] (Transparency International, 2014). These high levels of inequality and corruption often result in low levels of trust among local communities when engaging with the public sector. This effectively weakens their voices, affecting the quality of dialogue with local communities (Ghuman & Singh, 2013; Warhurst, 2005). In particular, it likely hinders the quality of social audit processes aiming to consider local communities' views, feedback, or complaints (Ahmad, 2008; O'Dwyer, 2005).

A high level of poverty and a large number of economic development programs implemented in these two countries presents a valuable context for the examination of MED programs and the use of accountability mechanisms such as social audits. Essentially, different social and economic contexts influence NGOs' operations and approaches to achieving project outcomes and developing successful accountability mechanisms (Gibelman & Gelman, 2004; Islam & Morgan, 2011; Jordan & Tuijl, 2006). Accordingly, by examining NGO activity with respect to MED in these countries (which have made modest progress in alleviating poverty, with at times limited success), this study will provide valuable insights and understandings into social audit as an accountability mechanism and its barriers within NGOs' practice.

5 Research Method

Given little information exists regarding social audit practices within the NGO sector, MED NGOs in particular, this study adopted an exploratory process and explanatory approach (Cavana, Delahaye, & Sekaran, 2001). Within this approach, this study was undertaken on the basis that "reality exists only in the context of mental framework" (Guba, 1990, p. 25). Therefore, realities are multiple, they exist in people's minds and are constructed based on individuals' social experiences (Creswell, 2009; Guba, 1990). Interaction with participants allows the researcher to adopt an interpretivist approach uncovering the realities constructed and held by participants, within the local and specific contexts that have given them meaning

---

[1] 1 representing lowest corruption; 177 representing highest corruption.

(Guba, 1990; Liamputtong, 2009). As such, this study adopted a qualitative approach, involving semi-structured in-depth interviews with NGO senior executives and beneficiaries in two developing countries where MED programs operate: Bangladesh and Indonesia. In addition, documentary analysis of publicly available data relating to the participating NGOs was conducted.

The latest directory of NGOs operating in each country is listed on the website of Directory of Development Organisations (2011)[2]. Based on the list of local and international NGOs mainly focusing on MED detailed in this directory, there were 57 such organisations operating in Bangladesh and 31 in Indonesia as at 2011. Interview invitations were sent to all MED NGOs detailed in the list and senior executives of 20 NGOs (12 in Bangladesh and 8 in Indonesia, including local and international NGOs) accepted the invitations. The most recent annual report and other publicly available documents of the participating MED NGOs were reviewed to gain an understanding of their operations. Interviews of approximately one and a half hours each were then conducted with one to five senior executives of each NGO. In addition, one NGO operating as a donor ('donor NGO') providing funds to the Indonesian Government (which then funded NGO projects) was interviewed. With the support of the participating NGOs, ten interviews (six in Bangladesh and four in Indonesia) were also conducted with individuals and groups of beneficiaries from four NGOs (two in Bangladesh, two in Indonesia), in order to gain an understanding of beneficiaries' perceptions of the NGOs' projects and accountability mechanisms. Interviews with beneficiaries lasted approximately 30 minutes. A summary of the interview participants is detailed in Table 1.

As noted in Table 1, of the 20 interviews conducted with NGOs, seven interviews were conducted with more than one executive (as a group interview) at NGOs' request. In total, the interviews with NGOs involved 34 NGO staff (23 in Bangladesh, 11 in Indonesia), and were conducted in the cities Dhaka, Bangladesh and Jakarta, Indonesia, where the NGOs' offices were located. Interviews with beneficiaries were conducted in rural areas (up to 100 km from the main cities of Dhaka and Jakarta), where the NGOs' projects were based. Interview sites, both metropolitan and rural, provided the researcher with the opportunity to observe differences in the social and economic conditions of both countries (e.g. infrastructure, living conditions).

All interviews were conducted by the lead researcher in English. Interviews with beneficiaries were arranged with the support of the relevant NGO staff and a local interpreter (independent of the NGOs). Interviews were audio-recorded (with permission) and transcribed. NVivo was then used to assist with data analysis, allowing data to be deconstructed but also reviewed as a whole.

Thematic analysis was conducted to analyse the interview data. The coding process involved four stages, as suggested by Boyatzis (1998). As a first step, the structure of the interview protocol was used to identify general themes or nodes in Nvivo. These themes included forms of social audits, frequency of conducting social audits, who instigates them and how they are conducted, and the benefits

---

[2] Most recent directory available at the time of data collection.

Table 1 Summary of interview participants

NGOs			Number of executives interviewed	Beneficiaries	
No.	Local	International		Number of interviews	Number of beneficiaries interviewed
Bangladesh					
1		✔	1	3	3 (2 groups[a], 1 individual)
2		✔	2		
3	✔		1		
4	✔		5	3	3 (individuals)
5	✔		1		
6	✔		1		
7	✔		5		
8		✔	1		
9		✔	1		
10		✔	1		
11	✔		2		
12	✔		2		
∑	7	5	23	6	6
Indonesia					
1		✔	2		
2		✔	1		
3	✔		3	2	3 (a group of 2, 1 individual)
4		✔	1		
5		✔	1		
6		✔	1		
7		✔	1	2	2 (individuals)
8	✔		1		
∑	2	6	11		
Total					
20	9	11	34	4	5

[a]While these groups each involved approximately 15–20 people, 1 beneficiary in each group predominantly spoke on behalf of the group

and constraints of the process. Next, these themes were coded systematically for each transcript to maintain consistency. Codes were then refined to capture the essence of the data. Through deductive and inductive analysis, themes identified in the literature were refined and new themes emerged with respect to social audits employed in practice (e.g. variations in their form, scope and purpose). In the fourth stage, the themes were interpreted to identify the underlying meaning of the data. The findings were compared with publicly available data (e.g. annual reports, and other publicly available documents) and observation during the interview process, enabling triangulation. Comparison of the data sources provided clarification on the findings and helped to avoid bias (Guba, 1990), ultimately increasing reliability (Berg, 2009). The next section presents the findings, incorporating excerpts from interviews which were supported by secondary data, but provided significantly richer detail.

6 Findings

6.1 Prevalence and Forms of Social Audit

While social audit is a relatively well established term within the private sector (Bauer & Fenn, 1972), within the NGOs investigated, only two NGOs (both operating in Indonesia) acknowledged and understood the term 'social audit'. The large majority (executives from 16 of the NGOs interviewed) expressed an understanding of this practice in more general terms such as monitoring, evaluation, or review processes that could be conducted by NGO staff or a third party. These three terms were neither completely distinct nor used consistently by interviewees. 'Monitoring' was commonly referred to as a process often conducted during the implementation of a project, and its results were intended to help NGOs learn from their current practices and improve their project design for better outcomes. Both evaluation and review processes were generally conducted at the end of projects. While evaluation was mainly used to collect data regarding the outcomes or effectiveness of the particular projects, review processes often involved a more comprehensive approach, encompassing multiple projects or the NGO's whole operations. For example, an executive of one NGO used the term 360-degree review process referring to reviewing its whole operations involving various projects, operational areas and stakeholders.

Of the 16 NGOs which adopted monitoring, evaluation or review processes (rather than social audit specifically), typically, they considered these processes as a form of social audit, but interpreted and applied them in different ways, with an emphasis on impact and final evaluation.

> That is [an] impact evaluation obviously, final evaluation (Senior Executive NGO 4, Bangladesh, 2013).
> ...it's the monitoring part, we call it monitoring and evaluation. We never call it social audit (Senior Executive NGO 2, Bangladesh, 2013).
> Yeah, like a social impact assessment (Senior Executive NGO 6, Indonesia, 2013).

The remaining two NGOs (both operating in Bangladesh) did not conduct any form of review and considered their social audit practice to be 'under-developed' or not their key focus. Executives from both NGOs acknowledged it as a systematic process, distinct from financial audit, indicating some level of awareness regarding what a social audit represents.

> No, we have no systematic social audit...but we have a very strong process in financial auditing. But we are thinking now about the social audit (Senior Executive NGO 11, Bangladesh, 2013).
> Not yet, not yet...there is a system. The system is an initiative for social review. But that is not so strong (Senior Executive NGO 5, Bangladesh, 2013).

Table 2 Forms of social audit adopted

Forms of social audit	NGO Bangladesh	Indonesia	Total
Social audit	0	2	2
Monitoring or evaluation	6	4	10
Review	4	2	6
None	2	0	2
Total	12	8	20

Further, the two NGOs that acknowledged using the term social audit were international and well-known with established operations. Thus, very few of the NGOs conducted social audits, however monitoring, evaluation and review process were more common. The NGO executives' responses in relation to whether they adopt social audits (or variations thereof) are summarised in Table 2.

6.2 Frequency of Conducting Social Audit

Findings indicate that NGOs had different timeframes for conducting social audits, monitoring, evaluation, or reviews, depending on a project's design or requirements, or the NGO's own purposes. With respect to project design or project requirements (typically required by and agreed with donors), executives from eight of the 20 NGOs indicated that their NGOs conduct social audits either throughout the project's duration for projects which extend over several (e.g. 3–5) years, or when the projects finish for short-term projects (e.g. 1–3 years).[3] Thus, from a practitioner perspective, emphasis was placed on impact, often at the end of a project.

> If it's a 5 year project, we'd normally have a mid-term evaluation, then end-project evaluation. And if it's a short 3 year project, we have an end-of-year evaluation (Senior Executive NGO 3, Bangladesh, 2013).
>
> Yeah, if it is a 1 year project, they come only once at the end of the project. If it is a 3 year project, maybe they come in the middle of the project and also at the end of the project. It's called Mid-term Evaluation and the Final Evaluation (Senior Executive NGO 12, Bangladesh, 2013).
>
> It depends on the project design or budget that is given (Senior Executive NGO 1, Indonesia, 2013).

Further, executives from 11 of the 20 NGOs (seven in Bangladesh and four in Indonesia) noted that their social audit process (or variations thereof) was conducted periodically. Typically, this was every 6 months, annually, or longer

---

[3] There was, however variation in terms of what was considered a 'short-term' project (e.g. 1 v. 3 years).

intervals (up to 5 years); with annually (eight NGOs) being the most common time period.

> ...we are now doing once every year in every community (Senior Executive NGO 3, Bangladesh, 2013).
> Evaluation activities, we do have 3–5 years, including from external [evaluation]...We always do like that, every 3 years or 5 years (Senior Executive NGO 1, Indonesia, 2013).

6.3 Conduct of Social Audits

Social audits (or variations thereof) were conducted by NGOs themselves, third parties nominated either by donors or NGOs, or government bodies. A summary of NGOs conducting social audits (or variations thereof) is presented in Table 3.

For NGOs conducting social audits as an internal audit process, the process and timeframes were often flexible, based on NGOs' internal policies or initiatives (rather than imposed by donors or government).

> Internally, we have the audit sector, our audit people are frequently visiting the area, and they're providing feedback to me (Senior Executive NGO 4, Bangladesh, 2013).
> [Conducting social audits] is [pretty] much every year. I mean we do our own internal [review]...we have this annual review process (Senior Executive NGO 8, Indonesia, 2013).

NGOs also nominated third parties to conduct social audits at the end of MED programs or during long-term projects.

> ...every 5 years, we do evaluation using external consultant to look at the progress of our activities (Senior Executive NGO 7, Indonesia, 2013).
> ...we agreed [to] appoint this independent auditor to audit our work...actually [we have] two [social audit] mechanisms, internal mechanism and external mechanism...For external audit we always do it at the end [of] the programs (Senior Executive NGO 5, Indonesia, 2013).

Typically, both internal social audits and external (third party) social audits were required by donors who wanted an independent view of the projects they funded in terms of progress achieved, and ongoing work required.

> ...evaluation by a third party, not the project team, third party, they'll do [the social audit] on behalf of donor (Senior Executive NGO 2, Bangladesh, 2013).

Table 3 Actors conducting social audits

Actors	NGO Bangladesh	Indonesia	Total
NGO itself	9	5	14
Third party	6	5	11
Government organisation	1	0	1

N.B. Categories are not mutually exclusive. Some NGOs had more than one actor conducting audits (or variations thereof)

> If they want to go in-depth, then they hire one evaluator and they evaluate the program and they submit a report. As an example, like [one particular donor], after 5 years, they are doing some evaluations to find out what benefit we did in the area and what more requirements [are] needed (Senior Executive NGO 1, Bangladesh, 2013).

One NGO operating in Bangladesh noted that social audits were also conducted by a government body, with an emphasis on the NGO's transparency and accountability.

> And MRA, Micro-credit Regulatory Authority people is also coming, visiting our office, and they're going to the field, so they're checking...transparency is maintained, accountability is maintained (Senior Executive NGO 4, Bangladesh, 2013).

Thus findings suggest that within the NGOs investigated, the conduct of social audits (or variations thereof) was commonly an internal process. Importantly, however, both internal and external social audits were typically undertaken due to donor requirements within the broader project requirements as a condition of donor funding.

6.4 Social Audit Processes

Among the NGOs interviewed, social audits (or variations thereof) were commonly conducted using different methods, including focus group discussions, surveys or questionnaires, and case studies. These methods were used individually or collectively, depending on the NGOs' purposes. The number of NGOs using these methods is summarised in Table 4.

As detailed in Table 4, NGOs mainly used focus group discussions as a method for conducting social audits. Under each of the methods, facilitating the participation of various actors involved in and benefiting from NGOs' projects, particularly beneficiaries, was considered important.

> We use a methodology called focus group discussion (FGD). That is one and a half to 2 hour discussion with selected people and with representatives [e.g. community leaders, local government], and through that we collect this kind of information [on what these people think about the NGO's projects]. And that is reflected in our annual planning (Senior Executive NGO 6, Bangladesh, 2013).
>
> ...the consultants will visit our programs and they will meet the beneficiaries, visit the villages, conduct interviews directly [with various] people (Senior Executive NGO 5, Indonesia, 2013).

Participants in this process included local partners (e.g. civil and government organisations and local private sector businesses) within the communities.

> We sit together, we facilitate [review activities] so that they [actors within communities] can review [our operations] by themselves...where they are, what they need more from their [co-operation]. Did the [co-operation] work properly or not? So we get some answers from them, sometimes good, sometimes bad (Senior Executive NGO 7, Bangladesh, 2013).
>
> [At the] end of the project, around two and a half years back, when we did the monitoring we set up questions, we interviewed some of our producers, suppliers, raw material suppliers, company staff... (Senior Executive NGO 2, Indonesia, 2013).

Table 4 Methods for conducting social audit

Methods	NGO Bangladesh	Indonesia	Total
Focus group discussions	9	7	16
Surveys or questionnaires	2	0	2
Case studies	0	1	1

N.B. Categories are not mutually exclusive. Some NGOs adopted more than one method when conducting social audits

> We do FGD with the communities, we do FGD with the partners—local organisations—their staff, we talk with the government and other local governments to understand what went well, what are their suggestions, what they think will be better for their own communities. So, this kind of thing we always do for learning and for improvement (Senior Executive NGO 10, Bangladesh, 2013).

Facilitating the participation of the above actors, particularly beneficiaries, was identified as a central approach to social audits (or variations thereof), by all NGO executives interviewed. One of the main reasons for this was to ensure sustainable (long-term) outcomes were developed for poverty alleviation programs, such that poor communities (with the support of local partners) had the capacity to continue operating microenterprises once NGOs' projects finished.

> We want the continuation of the project [such] that the community people will be in [the] driving seat, take the decisions, continue the work of the project after [us] phasing out of the project. If it is [managed], then okay. Somehow, it is satisfactory level. All the time, there are unsatisfactory [issues identified], and from the dissatisfaction our innovation will increase, our innovative work will increase. If there is no dissatisfaction, there will not be any good work [changes and improvement implemented] (Senior Executive NGO 4, Bangladesh, 2013).

However, the benefits (e.g. sustainable project outcomes) of social audits depended on the way that NGOs utilised the results of this process, and the beneficiaries' perceptions towards this process. These issues are discussed in the next sections.

6.5 Social Audits as Performance Measures

While donors imposed reporting requirements on NGOs regarding economic and other quantitative measures (e.g. number of beneficiaries reached, amount of money spent), in the context of social performance reports and evaluations, indicators or measures of social performance were voluntarily identified and reported by NGOs rather than required or specified by donors.

> The one that we have to do as mandatory is more on the income and then in sanitation, sanitation coverage...so ours is more focused on that...With social issues [the] main

approach of the projects should have [appropriate measures] but it's not a mandatory [part] of what we report on (Senior Executive NGO 6, Indonesia, 2013).

Interestingly, the donor NGO operating in Indonesia viewed social audit as equivalent to monitoring and evaluation and noted this was tailored to individual projects and NGOs.

> ...we have what we call monitoring/evaluation, and usually each project will develop their own evaluation system (Donor NGO, Indonesia, 2013).

Economic and other quantitative measures in the social audit reports prepared internally by NGOs were compared with evaluation or review reports prepared by government bodies (e.g. economic, agricultural departments).

> ...the report produced by the district [local government] management office, they get by talking with the communities (not through the NGO)...we can check and recheck [compare performance measures used between] the government and NGO. So the [government] people will have their evaluation [report on] the NGO but the NGO will also have their own evaluation [report] to give (Donor NGO, Indonesia, 2013).

6.6 Benefits of Conducting Social Audits

NGOs identified various benefits of conducting social audits, including sharing information with beneficiaries and actors participating in NGOs' projects, and learning from the feedback.

> Usually we use the participatory approach to collect and share information regarding this project (Senior Executive NGO 12, Bangladesh, 2013).
>
> The benefit of course we will get what we call learning. Learning and also critiques and also warning; we call it warning about [what] we have done for one phase, like 3 years or 5 years, so I think that's the benefit. We will get a lot of information in regards to how we do our work...people say, we need to change, what we do in the last 5 years is not really targeting our goals, so we need to change the other projects, for example (Senior Executive NGO 7, Indonesia, 2013).
>
> ...to learn about how the projects make changes in term of the outcomes, the objectives, and addressing the problem faced by the communities (Senior Executive NGO 5, Indonesia, 2013).

One NGO also noted that actions needed to be taken based on the social audit reports. If not, the process would be an ineffective tool.

> ...[based] on the review, there is scope to [act], to rectify, or to reduce. If it is done only for the theoretical words then both [parties will] decide they don't feel interested. After the review, if anybody makes some comments, [partners] should take action...[beneficiaries would] give up if they find that nothing is happening after that. So it doesn't work (Senior Executive NGO 7, Bangladesh, 2013).

To effectively address issues identified from social audits, NGOs used the results from this process to compare with their initial objectives or benchmark performance indicators.

...We have to have some baseline [initial indicators], and after that we compare all the baseline things to final things (Senior Executive NGO 2, Bangladesh, 2013).
...to come in and look at what we were doing, how we were working, whether we were achieving our objectives. So that kind of external evaluation...we found that very useful (Senior Executive NGO 8, Indonesia, 2013).

The results from this process were also used to improve NGO staff capabilities.

...and then we have our [staff] meeting, where we look at what's come from the villages, and we talk amongst ourselves about things that have worked, things that can be improved in our work (Senior Executive NGO 8, Indonesia, 2013).

NGOs also emphasised using social audit findings to modify the designs of current or future projects for more effective outcomes.

Through this user survey, we know where we are. That last [project] we did, we conducted four, five user surveys—last 5 years, five user surveys. From the first user survey we found that our design needed to be modified. We did [this] (Senior Executive NGO 8, Bangladesh, 2013).
...on the basis of [social audit findings] we put together a plan for the next year. So there's an internal review, an internal annual review process, which is quite rigorous (Senior Executive NGO 8, Indonesia, 2013).

Through the process of social audit, NGOs identified opportunities to be closer to beneficiaries, listen to their needs, feedback, or complaints.

...you are more people-friendly and people-oriented, you are reaching your targets [the poor]. We are here for the benefit of the poor. So if you are more people-friendly, you are doing more good to the people (Senior Executive NGO 6, Bangladesh, 2013).

Once the social audit results were reviewed and verified by NGOs' field staff, findings were then discussed with communities to identify required actions, and report back to NGO managers. This process helped NGOs learn from beneficiaries' feedback and complaints, potentially improving current practices and achieving more effective program outcomes.

There are some audit reports [auditors] are submitting to me, we're sending it to the field level, and field level people are asked to [review and reply to] it. With that, they'll find out if it's [correct], or has been written [incorrectly] by the audit people. They're collecting [comments] from people, and they're sending again to us, then we're sitting with the people [who give complaints] and the audit report with the audit people. Then we're minimising [problems] (Senior Executive NGO 4, Bangladesh, 2013).

For NGOs focusing on poverty alleviation through economic development, reporting in a way that reflected NGOs' performance in quantitative and qualitative terms was perceived as both challenging and rewarding, improving NGOs' reporting tools.

Actually benefit [of qualitative data] is more...it makes our reporting system [more useful] where we can write the proper stories of quantitative information and qualitative information. We can include all the [relevant] things, the impacts [with] our own eyes (Senior Executive NGO 2, Bangladesh, 2013).

Importantly, the independence of external auditors in the social audit process was valued by NGOs and helped them access unbiased information on NGOs' operations. These reports were used to compare project data collected internally by NGO staff.

> That's why, sometimes we have to be more [connected] to the audit sector, because the audit sector is independent. So, in both ways, it helps us. [Internal] program people are providing some sort of information, audit people are providing some sort of information. So, being a manager, we're matching both information (Senior Executive NGO 4, Bangladesh, 2013).

Hence external social audit findings provided not only independent views but also an important tool for comparability at a management level. However, despite of the benefits derived from social audits, NGOs also faced constraints illuminated by this process. These constraints related to their own operations, beneficiaries, or local contexts, and are considered below.

6.7 Constraints of Conducting Social Audits

6.7.1 NGO-related constraints

Despite NGOs acknowledging the benefits gained from conducting social audits, they also indicated some major constraints. From an internal perspective, conducting social audits was perceived as costly and time-consuming.

> When it comes to the socio-economic data, yes, definitely that's always a challenge, because it becomes very costly to try and do that for every single household. So, there's no way we can get that level of detail (Senior Executive NGO 2, Indonesia, 2013).
>
> Another thing is time-consuming. [We] sometimes hired the monitoring team, evaluation team, and also we're implementing the project, so sometimes we have to do some implementing and also do the monitoring, side by side. So this is time-consuming, time is the main factor. Sometimes, [it is] difficult to complete within the time period (Senior Executive NGO 2, Bangladesh, 2013).

In some cases, the NGO staff implementing programs were also data collectors for review and evaluation processes, leading to potential conflict of interest and data distortion or bias.

> The program facilitators also visit the field, they're also taking some time from the field level. But sometimes program facilitators are also reluctant [to collect data] because they're implementers (Senior Executive NGO 4, Bangladesh, 2013).

While concerns about lack of independence were identified regarding social audits conducted by internal auditors, limitations were also acknowledged regarding social audits conducted by external auditors. In particular, external auditors often did not fully understand NGOs' activities, or were not able to provide the same depth or level of understanding of some internal auditors due to time and resource constraints. This situation limited the benefits of the social audit process, particularly in terms of identifying areas for improvement.

They [external auditors] need to clarify again and again due to [being] outsider evaluators. So, this is our very big constraint (Senior Executive NGO 12, Bangladesh, 2013).

I think it's difficult to find [external auditors] who really understand this work (Senior Executive NGO 7, Indonesia, 2013).

...so we're not entirely sure, you know, what is the impact at every level in the value chain of a particular activity or initiative that we have supported, because [external] monitoring does not go deeply enough into the [value] chain (Senior Executive NGO 2, Bangladesh, 2013).

Another challenge within NGOs was that when areas requiring improvement were identified, NGOs sometimes lacked the capacity to implement changes.

So in the design of projects, there is a need for improvements, but when looking back there [were] things that we couldn't have known more thoroughly [due to lack of capacity] (Senior Executive NGO 1, Indonesia, 2013).

6.7.2 Beneficiary-related constraints

From a beneficiary-related perspective, one of the main challenges was their perception of and participation in the social audit process. Given beneficiaries typically perceived that they had a weak negotiating position or wanted to please NGOs, they were often not willing to express any criticism (constructive or otherwise) in their feedback.

There is some distorted data. Sometimes they try to please us. So [what] they would say are not really their real thoughts; to share it, rather they try to say things which they think would please us (Senior Executive NGO 6, Bangladesh, 2013).

In some cases, beneficiaries wanted to receive or retain benefits from the projects, and tailored their responses accordingly, at times dishonestly.

Sometimes people are getting benefits, but they are saying we are not getting any [benefit]. Because if they are saying they are getting any [benefit], they have to pay the installments. There is a non-paying culture in this subcontinent. Not all the customers, like 5-10 %, or 20 % [of beneficiaries] are like this, and when you ask people are you happy...they will say, well we are happy with [activities] but we have some problems [that they need more time to repay investment loans or need more financial assistance] (Senior Executive NGO 8, Bangladesh, 2013).

Further, beneficiaries were often uneducated and lacked confidence to share information and provide feedback during social audits.

Then also sometimes, maybe this has been due to years of exclusion, years of isolation, years of not being respected of their opinions...some are reluctant to give feedback (Senior Executive NGO 3, Bangladesh, 2013).

...when you are talking to people who are essentially rejected, discriminated against, stigmatised, judged by society at large...even if you try hard to not do that, it impacts their ability to trust anybody else, right; to listen, to come up with ideas; their self-esteem as well (Senior Executive NGO 4, Indonesia, 2013).

Given beneficiaries were poor, they were typically pre-occupied with basic livelihood activities (e.g. family commitments and earning extra income to feed their families). As such, sitting with NGOs for lengthy review processes was considered a burden and less of a priority.

> Mostly I think because of the time. Because you know the community, they're very busy with not just their livelihood activity but also because of the family. If we asked them to come and sit together for 3 days or 4 days, it's really difficult. They don't have a lot of time like that (Senior Executive NGO 7, Indonesia , 2013).

This constraint was observed by the researcher during the interviews conducted with beneficiaries. Specifically, when asked about social audits, none of the beneficiaries seemed to show enthusiasm for this process. For them, social audits seemed more of an administrative process done for the benefit of NGOs, rather than a process for improvement, ultimately benefiting beneficiaries.

> Yes, some staff visited...they checked IGA [Income Generating Activities] every week and took photos (Beneficiary 4, Bangladesh, 2013).

6.7.3 Context-related constraints

Compared to Bangladesh, Indonesia's social and economic setting appeared to be better developed.[4] Such development plays an important role in the success and constraints of poverty alleviation programs and social audits. In particular, differing levels of participation by beneficiaries in the social audit process were evident. For example, executives of several NGOs operating in Indonesia indicated that, within an economic environment where the poor have more opportunities to participate in economic mainstream activities, there were often conflicts of interest between individual beneficiaries with working groups. This resulted in people (beneficiaries) being less motivated to participate in groups. This situation however, was less prevalent in Bangladesh where opportunities for the poor were limited, and participation by the poor in MED programs was typically more positive, resulting in less conflict within groups.

> Manage the conflicts in a group. Manage the self-interests in a group. Manage the self-motive. So these are the some [boundaries] in development process (Senior Executive NGO 5, Indonesia, 2013).
>
> I found that in Indonesia to make them solid as a group is difficult. This is our challenge (Senior Executive NGO 7, Indonesia, 2013).

However, compared with Indonesia, Bangladesh's persistent political instability and less developed regulatory environment (with a high level of corruption) (Islam & Morgan, 2011) often limited the effectiveness of the social audit process.

---

[4] Based on observation when visiting areas in each country (e.g. infrastructure (roads, facilities), street life, visible signs of poverty, such as people regularly searching through public bins), and consistent with social and economic data for the relevant countries (AusAID, 2013a, 2013b).

Sometimes here in Bangladesh, lots of bad political situations we are facing. In that case we are facing the problem [of people in public sector organisations asking NGOs for money for doing things (e.g. to participate in social audits)] where we cannot complete our monitoring or evaluation on time (Senior Executive NGO 2, Bangladesh, 2013).

But the government officers go to these offices [of microenterprises] and seek grants and bribes and [we] suffer many problems...and sometimes you know, [they are like] the burglars (Senior Executive NGO 5, Bangladesh, 2013).

...for any informal microfinance, [we] have some sort of different types of risk. Risk means corruption...(Senior Executive NGO 11, Bangladesh, 2013).

6.8 Beneficiaries' Perspectives of Social Audits

Amongst the 10 interviews with beneficiaries, in seven interviews it was noted that beneficiaries either participated in monitoring, evaluation, or review process; or were aware of this process within their NGO. Further, beneficiaries primarily referred to it as an annual process in their NGO.

We had discussion about how to perform certain things, how much was the loan amount, and how much was the savings (Beneficiary 1, Bangladesh, 2013).

...yeah, it was just about yesterday that we had an evaluation (Beneficiary 2, Indonesia, 2013).

The remaining three beneficiaries interviewed (all based in Bangladesh, all from the same NGO), however, did not seem to be familiar with or have an understanding of the terms social audit, social review or evaluation process; reinforcing that social audit is not a common practice, or does not involve beneficiaries more broadly in terms of participation or communication. This raises concerns regarding how NGOs conduct such evaluations, monitoring or review processes, and the extent to which beneficiaries are involved in or informed about this process. Table 5 summarises the responses of beneficiaries compared to NGO executives regarding forms of social audit adopted by the participating NGOs.

For NGOs, social audits (or variations thereof) provided opportunities to engage in dialogue with beneficiaries on programs' progress and impact. However, as noted by one Indonesian beneficiary, the meetings they were required to attend were perceived to be time-consuming and of little benefit.

Because they have to congregate in meetings out there...in farmer groups, in women entrepreneur groups. It's a problem for them to congregate with others [for so long] (Beneficiary 3, Indonesia, 2013).

Further, the depth and value of the dialogue was at time questionable, as this beneficiary indicated that NGOs came and gave groups questionnaires rather than having in-depth interviews with beneficiaries in order to gain a deeper understanding of beneficiaries' situations, issues and feedback.

...no-one ever asked this [e.g. one on one interview], instead [auditors used] a questionnaire (Beneficiary 3, Indonesia, 2013).

Table 5 Forms of social audit: comparison of NGOs' and beneficiaries' accounts

Forms of social audit	NGO			Beneficiary		
	Bangladesh	Indonesia	Total	Bangladesh	Indonesia	Total
Social audit	0	2	**2**	0	0	**0**
Monitoring or Evaluation	6	4	**10**	3	4	**7**
Review	4	2	**6**	0	0	**0**
None	2	0	**2**	3	0	**3**
Total	**12**	**8**	**20**	**6**	**4**	**10**

Thus, findings indicate that formal social audits are not widely used within the MED NGOs investigated; rather NGO executives refer to monitoring, evaluation or review processes conducted by both internal and external parties. This practice remains unregulated and is based largely on donors' requirements or NGOs' own initiatives. Through these processes, NGOs focused strongly on the participation of different actors involved in and benefitting from NGOs' projects. These actors often included beneficiaries, local private sector businesses, and local government organisations. The social audit practices examined reveal both benefits and limitations. Reflections on how social audit is adopted, and compares with the literature on social audit regulation, is discussed in the next section.

7 Discussion

7.1 Distinction of Social Audits and Other Mechanisms for NGO Accountability

Returning to Sects. 6.1 and 6.2, findings suggest that social audit is not a well-established term among MED NGOs investigated. More commonly, these NGOs refer to monitoring, evaluation, or reviews as tools and processes to engage in dialogue with their stakeholders, particularly beneficiaries. These forms of social audits are conducted throughout the MED projects, when projects finish, or across multiple projects or activities of NGOs' operations in order to assess effectiveness and consider opportunities for learning and change. The finding is similar with Mason et al. (2007) who suggest that the routines of social audit practice provide a process for achieving ongoing accountability. However, the practices of social audits identified from the findings are far less rigorous than the expectations of a social audit detailed in the literature. Specifically, within the NGOs examined, the conduct of social audits mainly rested on donors' requirements or to a lesser extent, NGOs' own interests or initiatives. However, none of the NGOs investigated published social audit reports on their website, suggesting transparency is limited.

Rather, these reports were mainly used by donors or NGOs themselves. Further, conducting social audits at the end of projects (particularly short-term 1–3 year projects) and learning from social audit findings and results once projects had been completed, reveals results were of limited benefit for the projects reviewed.

The lack of distinction between social audit and monitoring, review, or evaluation, made by NGO executives suggests a blurring of boundaries, which are assumed by others in the literature. Ebrahim (2003a) for example, details five accountability mechanisms with social audits distinct from performance assessments and evaluations and participation, with performance assessments and evaluations argued mainly as being used for NGOs' internal accountability or accountability to donors. In the context of this study, forms of social audit (while blurred in practice) represented important tools for broader accountability to a range of stakeholders. However, reporting of social audits remained mainly to NGOs and donors. Specifically, this process was used for monitoring, evaluation and assessment involving participation of stakeholders through communicating, sharing and learning from NGO stakeholders' feedback; in particular beneficiaries.

7.2 Participatory Approach to Social Audits

As mentioned in Sect. 2, NGOs (characterised as self-governing, non-profit seeking, charitable organisations) typically rely on funding from donors (Martens, 2002; Vakil, 1997). Thus it is not surprising that the conduct of social audits is often based on donors' requirements (agreed within the project design). This process provides the opportunity to review the outcomes and impacts of NGOs' projects, such that NGOs' responsibility and accountability to a range of stakeholders (e.g. beneficiaries, donors, local communities) is addressed (Ebrahim, 2003a). As such, the participatory approach embedded in social audits facilitates participation of not only beneficiaries, but also the private sector and local governments. Their involvement and feedback through social audits (or variations thereof) helps NGOs to achieve more effective program outcomes so that beneficiaries, local private sector business, and local public sector organisations can continue working together after NGO projects finish.

Whilst focus group discussions were identified from the findings as the most popular method for conducting social audits, this finding is slightly different to the existing literature, which suggests questionnaire is the most common method (Owen et al., 2001). However, as noted in the findings regarding beneficiary-related constraints, beneficiaries are often not interested in participating in evaluation or assessments as they perceive these tools to be time-consuming and primarily for NGOs' interests. As such, the findings of this study suggest that when social audits involve engaging in dialogue with stakeholders (including beneficiaries), facilitators should emphasise the importance of open and honest communication, constructive criticism and suggestions, and the importance of the process for both NGOs and stakeholders, including beneficiaries. This approach of sharing and learning from stakeholders' feedback and suggestions is also preferred by beneficiaries, as they are more interested in the social audit approach where NGOs engage in dialogue (i.e. two-way exchange), listen and acknowledge their needs and constraints, to help them progress out of poverty.

7.3 A Pragmatic Approach to Social Audits

Returning to the central characteristics of social audit as identified by Ebrahim (2003a), and based on our understanding of how social audit is currently being adopted in practice, we suggest a modified, pragmatic approach may be most appropriate to promote the increased use of this tool within MED NGOs. The five characteristics of social audit identified in the literature and the modified pragmatic approach to social audits based on our finding are presented in Table 6.

First, identification of and dialogue with stakeholders is important based on both the literature and the findings. Further, this process should encompass a range of stakeholders from different sectors, and with different power bases, emphasising the importance of two-way exchange such that stakeholders (including beneficiaries) understand the importance of giving open and honest feedback. With respect to development of performance indicators or benchmarks, given the very early development stage of the social audit process in the context of the MED NGOs examined, arguably individual benchmarks and indicators relevant to individual programs is an important first step. While the concept of universal benchmarks has been raised in the literature (Dawson, 1998; Owen et al., 2000), we argue a tailored approach has more relevance to individual programs, given that NGOs' operations cover a wide range of activities. Thus, universal benchmarks will have less relevance and risk decreased learning, by potentially overlooking what is most important to individual programs. Continuous improvement, while an important goal, is perhaps somewhat subjective. Thus, we propose a more immediate objective of 'organisational learning' through a shared or collective responsibility which shifts the emphasis from accountability as a more traditional concept of being held responsible for results, to collective or socialising accountability to develop more effective (long-term) outcomes. Importantly, this objective should be acknowledged and supported by both beneficiaries and donors, to encourage leaning from failure rather than focusing solely on successful outcomes and incidences. Last, while the literature promotes public disclosure of social audit reports, given that none of the NGOs investigated provided social audit reports on their websites, and that many seemed to focus on reporting internally or upwards to donors, we propose stakeholder disclosure rather than public disclosure is a more realistic development aim.

While the proposed steps are a more simplistic approach to social audit (and do not extend to consideration of external verification, for example) they represent an important and practical development, which would help NGOs adopt a more

Table 6 A pragmatic approach to social audit

Ebrahim (2003a)	A modified pragmatic approach
Stakeholder identification	Involve a wide range of stakeholders, particularly powerless stakeholders (e.g. beneficiaries)
Stakeholder dialogue	Emphasise two-way exchange or dialogue
Development of indicators or benchmarks	Focus on individual benchmarks and indicators relevant to individual programs
Continuous improvement	Emphasise shared or collective responsibility for learning; developing more effective (long-term) outcomes
Public disclosure	Focus on stakeholder disclosure rather than public disclosure

formalised and accessible process. This is particularly important given the early development stages of the sector, its limited adoption of social audit, and lack of an inclusive stakeholder focus (i.e. specifically including beneficiaries in both obtaining feedback and communicating results) (Kang et al., 2012). Further, once adoption of social audit becomes a more widespread, institutionalised practice (Owen et al., 2000) involving a range of stakeholder input and broader stakeholder feedback, attention can then turn to standardising this process, sharing indicators and developing benchmarks, external verification, and communicating findings more broadly to the public at large.

From an internal NGO perspective, an appreciation of the need for feedback and organisational learning should be promoted. This is consistent with Mills and Friesen's (1992) notion of the learning organisation (transferring learning, commitment to knowledge development, and openness to the outside world) being important for commercial organisations to survive and progress. While organisational goals may differ between commercial and non-commercial organisations (such as NGOs), the value of learning remains. In 2000, Owen et al. argued it was too early to provide formal regulation for social audits, contending that to strengthen accountability and transparency within the NGO sector, social audits should first be considered as a norm required within NGOs through self-regulation or donor expectations. Our findings in 2014 continue to support this view, given that monitoring and evaluation have become accepted practice, but have not necessarily progressed to formalised social audits. Thus, the challenge (and opportunity) presented to NGOs (and their stakeholders) is to create a culture which encourages and values evaluation, and recognises mechanisms such as social audit as a valuable approach to achieve this.

Social audits offer opportunities for learning and improving, however the process needs to take into account obstacles that can affect their effectiveness. Along with elements identified in the literature (e.g. cost, time, lack of agreed processes), the findings of this study suggest additional issues related to NGOs, beneficiaries, and the local context. NGOs' projects are often perceived as charitable by beneficiaries. As such, their feedback on NGOs' projects may be compromised or presented in a way that aims to ensure benefits continue to be received. In addition, changing beneficiaries' perception of social audits is also important, such that they understand and appreciate this process is essential not only for NGOs' purposes, but also for beneficiaries to benefit from more effective NGO projects. Regarding the local context, both the social and political environment are important influences (Gibelman & Gelman, 2004; Islam & Morgan, 2011; Jordan & Tuijl, 2006). Unlike Indonesia, in Bangladesh—a country struggling with a high level of poverty—the poor have fewer opportunities to participate in economic mainstream activities. Thus, working in groups and supporting each other to gain benefits from NGOs' projects and beneficiaries' collective actions is common, potentially increasing beneficiaries' participation and contributing to effective program outcomes. However, the effectiveness of social audits also depends upon the participation of different actors, particularly local government and the private sector. As such, an unstable political environment and high levels of corruption (e.g. Bangladesh) can adversely affect the effectiveness of social audits.

Conclusion
While the findings presented in this chapter are limited by the relatively small sample from two countries, engaging with both NGO executives and beneficiaries provides valuable insights into social audit within two developing countries working to alleviate poverty. Specifically the benefits and limitations of this mechanism for strengthening NGO accountability have been highlighted, particularly for beneficiaries. Through these findings, NGOs' social audit practices highlight the need to develop more systematic approaches to social audits (e.g. reporting to stakeholders) to distinguish social audits from performance assessment and evaluation and participation. As social audits emphasise the important role of engaging in dialogue with stakeholders, a participatory approach involving meaningful dialogue (e.g. focus group discussions) is essential. In addition, examination of current practice also suggests a more pragmatic approach to social audit. This approach encourages dialogue with various actors benefiting from and involved in NGOs' projects as a first step, and tailored performance indicators or benchmarks (before considering the development of universal benchmarks for the NGO sector). It also emphasises stakeholder disclosure, rather than public disclosure. Further, in developing social audit processes, challenges need to be considered such as honesty and openness of beneficiaries, and the local context where NGOs operate. As social audit regulations within the NGO sector remain limited, the findings provide stakeholders and regulators with valuable guidance for better understanding the value of social audit as a mechanism to strengthen accountability of the NGO sector, particularly accountability to beneficiaries.

References

Agyemang, G., Awumbila, M., Unerman, J., & O'Dwyer, B. (2009). *NGO accountability and aid delivery (Research report 110)*. London: The Association of Chartered Certified Accountants.

Ahmad, R. (2008). Governance, social accountability and the civil society. *Journal of Administration & Governance, 3*(1), 10–21.

Assad, M. J., & Goddard, A. R. (2010). Stakeholder salience and accounting practices in Tanzanian NGOs. *The International Journal of Public Sector Management, 23*(3), 276–299.

AusAID. (2013a). *Bangladesh*. Retrieved April 2, 2013, from http://www.ausaid.gov.au/countries/southasia/bangladesh/Pages/bangladesh-statistics.aspx

AusAID. (2013b). *Indonesia*. Retrieved April 2, 2013, from http://www.ausaid.gov.au/countries/eastasia/indonesia/Pages/home.aspx

Bauer, R. A., & Fenn, D. H. (1972). *The corporate social audit*. New York: Russell Sage.

Berg, B. L. (2009). *Qualitative research methods for the social science*. (2nd ed.). Boston: Allyn & Bacon.

Bhatt, N. (1997). Microenterprise development and the entrepreneurial poor: Including the excluded? *Public Administration & Development (1986-1998), 17*(4), 371–386.

Boyatzis, R. E. (1998). *Thematic analysis and code development, transforming qualitative information*. Thousand Oaks: Sage Publications.

Brown, L. D., & Moore, M. H. (2001). Accountability, strategy, and international nongovernmental organizations. *Nonprofit and Voluntary Sector Quarterly, 30*(3), 569–587.

Burger, R. (2012). Reconsidering the case for enhancing accountability via regulation. *Voluntas: International Journal of Voluntary and Nonprofit Organizations, 23*(1), 85–108.

Cavana, R. Y., Delahaye, B. L., & Sekaran, U. (2001). *Applied business research: Qualitative and quantitative methods*. Milton, QLD: Wiley.

Choudhury, M. A., Hossain, M. S., & Solaiman, M. (2008). A well-being model of small-scale microenterprise development to alleviate poverty. *The International Journal of Sociology and Social Policy, 28*(11/12), 485–501.

Creswell, J. W. (2009). *Research design qualitative, quantitative, and mixed methods approaches* (3rd ed.). Los Angeles: Sage.

Dawson, E. (1998). The relevance of social audit for Oxfam GB. *Journal of Business Ethics, 17*(13), 1457–1469.

Deegan, C. (2002). The legitimising effect of social and environmental disclosures—A theoretical foundation. *Accounting, Auditing & Accountability Journal, 15*(3), 282–311.

Deen, T. (2010). *Is global poverty reduction a political myth?* Retrieved July 15, 2011, from http://ipsnews.net/news.asp?idnews=52142.

Directory of Development Organisations. (2011). *Asia and the Middle East*. Retrieved July 2, 2012, from http://www.devdir.org/asia_middle_east.htm

Ebrahim, A. (2003a). Accountability in practice: Mechanisms for NGOs. *World Development, 31*(5), 813–829.

Ebrahim, A. (2003b). Making sense of accountability: Conceptual perspectives for northern and southern nonprofits. *Nonprofit Management and Leadership, 14*(2), 191–212.

Ghuman, B. S., & Singh, R. (2013). Decentralization and delivery of public services in Asia. *Policy and Society, 32*(1), 7–21.

Gibelman, M., & Gelman, S. R. (2004). A loss of credibility: Patterns of wrong doing among Nongovernmental Organizations. *Voluntas, 15*(4), 355–381.

Guba, E. G. (1990). *The paradigm dialog*. London: Sage Publication, Inc.

Gugerty, M. K. (2008). The effectiveness of NGO self-regulation: Theory and evidence from Africa. *Public Administration and Development, 28*(2), 105–118.

Gugerty, M. K., Sidel, M., & Bies, A. L. (2010). Introduction to minisymposium: Nonprofit self-regulation in comparative perspective-Themes and Debates. *Nonprofit and Voluntary Sector Quarterly, 39*(6), 1027–1038.

Hammer, M., & Lloyd, R. (2011). *Pathways to accountability II, The 2011 revised Global Accountability Framework. Report on the stakeholder consultation and the new indicator framework*. London: One World Trust.

Islam, M. R., & Morgan, W. J. (2011). Non-governmental organizations in Bangladesh: Their contribution to social capital development and community empowerment. *Oxford University Press and Community Development Journal, 47*(3), 369–385.

Jäger, U. P., & Rothe, M. D. (2013). Multidimensional assessment of poverty alleviation in a developing country: A case study on economic interventions. *Nonprofit Management and Leadership, 23*(4), 511–528.

Janvry, A. D., & Sadoulet, E. (2009). Agricultural growth and poverty reduction: Additional evidence. *The World Bank Research Observer, 25*(1), 1–20.

Jones, R. M., Kashlak, R., & Jones, A. M. (2004). Knowledge flows and economic development through microenterprise collaboration in third-sector communities. *New England Journal of Entrepreneurship, 7*(1), 39–48.

Jordan, L., & Tuijl, P. V. (2006). *NGO accountability, politics, principles & innovations*. London: Earthscan.

Kang, J., Anderson, S. G., & Finnegan, D. (2012). The evaluation practices of US international NGOs. *Development in Practice, 22*(3), 317–333.

Karnani, A. (2007). The mirage of marketing to the bottom of the pyramid: How the private sector can help alleviate poverty. *California Management Review, 49*(4), 90–111.

Keystone. (2006). *Downward accountability to 'beneficiaries': NGO and donor perspectives*. Keystone: Accountability for social change.

Kilby, P. (2006). Accountability for empowerment: Dilemmas facing non-governmental organizations. *World Development, 34*(6), 951–963.

LeRoux, K. (2009). Managing stakeholder demands. *Administration & Society, 41*(2), 158–184.

Liamputtong, P. (2009). *Qualitative research methods* (3rd ed.). Australia & New Zealand: Oxford University Press.

Mango. (2010). *Accountability to beneficiaries checklist. How accountable is your organisation to its beneficiaries?* Version 2. Retrieved November 20, 2012, from http://www.mango.org.uk/Pool/G-Accountability-to-beneficiaries-Checklist.pdf

Martens, K. (2002). Mission impossible? Defining nongovernmental organisations. *Voluntas: International Journal of Voluntary and Nonprofit Organisations, 13*(3), 271–285.

Mason, C., Kirkbride, J., & Bryde, D. (2007). From stakeholders to institutions: The changing face of social enterprise governance theory. *Management Decision, 45*(2), 284–301.

Mills, D. Q., & Friesen, B. (1992). The learning organisation. *European Management Journal, 10*(2), 146–166.

Najam, A. (1996). NGO accountability: A conceptual framework. *Development Policy Review, 14*(4), 339–354.

O'Dwyer, B., & Unerman, J. (2008). The paradox of greater NGO accountability: A case study of Amnesty Ireland. *Accounting, Organizations and Society, 33*, 801–824.

O'Dwyer, B., & Unerman, J. (2010). Enhancing the role of accountability in promoting the rights of beneficiaries of development NGOs. *Accounting and Business Research, 40*(5), 451–471.

O'Dwyer, B. (2005). The construction of a social account: A case study in an overseas aid agency. *Accounting, Organizations and Society, 30*(3), 279–296.

OECD. (2014a). *Bangladesh*. Retrieved April 11, 2014, from http://www.oecd.org/dac/stats/documentupload/BGD.JPG

OECD. (2014b). *Indonesia*. Retrieved April 11, 2014, from http://www.oecd.org/dac/stats/documentupload/IDN.JPG

Owen, D. L., Swift, T., Humphrey, C., & Bowerman, M. (2000). The new social audits: Accountability, managerial capture or the agenda of social champions? *European Accounting Review, 9*(1), 81–98.

Owen, D. L., Swift, T., & Hunt, K. (2001). Questioning the role of stakeholder engagement in social and ethical accounting, auditing and reporting. *Accounting Forum, 25*(3), 264.

Peredo, A. M., & Chrisman, J. J. (2006). Toward a theory of community-based enterprise. *Academy of Management Review, 31*(2), 309–328.

Roberts, J. (2009). No one is perfect: The limits of transparency and an ethic for 'intelligent' accountability. *Accounting, Organizations and Society, 34*(8), 957–970.

Schmitz, H. P., Raggo, P., & Vijfeijken, T. B. (2011). Accountability of transnational NGOs: Aspirations vs. practice. *Nonprofit and Voluntary Sector Quarterly, 41*(6), 1175–1194.

Sinclair, A. (1995). The chameleon of accountability: Forms and discourses. *Accounting, Organizations and Society, 20*(2/3), 219–237.

Slim, H. (2002). *By what authority? The legitimacy and accountability of non-governmental organisations*. Paper presented at the The International Council on Human Rights Policy International Meeting on Global Trends and Human Rights—Before and after September 11, Geneva.

Strier, R. (2010). Women, poverty, and the microenterprise: Context and discourse. *Gender, Work and Organization, 17*(2), 195–218.

Transparency International. (2014). *Corruption perceptions Index 2013*. Retrieved April 11, 2014, from http://cpi.transparency.org/cpi2013/results/

Vakil, A. C. (1997). Confronting the classification problem: Toward a taxonomy of NGOs. *World Development, 25*(12), 2057–2070.

Warhurst, A. (2005). Future roles of business in society: The expanding boundaries of corporate responsibility and a compelling case for partnership. *Futures, 37*(2–3), 151–168.

Social Audit for Raising CSR Performance of Banking Corporations in Bangladesh

Md. Tarikul Islam

1 Introduction

Corporations' commitment to their social responsibilities has long been a global concern with reference to sustainability, transparency and fair practice, which is of great concern to stakeholders. Corporations are expected to be responsible to the society they operate within and do business in a socially responsible manner. To some extent, nowadays, corporations do so by participating in various social activities and causes in the hopes that this will strengthen their accountability; this is commonly known as Corporate Social Responsibility (CSR). There is no clear picture yet about the social impact that these involvements are creating around the globe. Moreover, there is speculation about the intention behind and the strategic aim of corporations' involvement in CSR activities. This causes a widening gap between corporations and stakeholders, which is damaging for both parties.

Social audit, a tool used to check performance of organizations against their commitments, can be an option to bring relevant stakeholders in a single platform where communication of multiple parties will bring more transparency. The concept of social audits has been around for a long time, however it has only fairly recently been put into practice. India is pioneer in social auditing and received much attention for its implementation. Some companies in Europe have their own social audit systems in which they check their suppliers in terms of pro society business practice (Björkman & Wong, 2013). Various researchers have concluded that social audits are usefulness for monitoring if well designed (Boyd, Spekman, Kamauff, & Werhane, 2007), and can be used to create transparency in the system. Social audits can be helpful in the management of organizations too (Owen, Swift, Humphrey, & Bowerman, 2000). The audit process involves relevant stakeholders

---

M.T. Islam (✉)
QUT Business School, Queensland University of Technology, Australia
e-mail: imtarikul@gmail.com

and is usually well accepted by them (Locke, Qin, & Brause, 2007). It should be noted that social audits usually recommend change by pointing to non compliance, however it does not automatically bring change (Pruett, Merk, Zeldenrust, & de Haan, 2005); instead it redirects the attention of organizations toward social welfare from mere wealth maximization.

Wealth maximization (Friedman, 2007) is one of the major aims of an organization, but how the organization performs this task changes over time (Dodd, 1932; McElhaney, 2009). Among many changes, a prominent one is the focus on society to create shared values (Rangan, Chase, & Karim, 2012) and the enhancement of those values in the society (Cochran, 2007). Organizations try to give stakeholders the feeling 'that we were here for you', 'we are here for you', and 'we will be here for you'. There are lots of debates why firms do this, but the majority of the views are that the focus is on the future (Isaksson, Kiessling, & Harvey, 2014). Some researchers argue that with the intense competition in the market (Burke & Logsdon, 1996), this is merely a selling strategy (Cotten & Lasprogata, 2012; Rangan et al., 2012) to create a positive image for more stakeholders about the firm. Another argument is that firms become socially responsible as they realize the fact that without benefiting society they cannot benefit themselves (Hiller, 2013). Nowadays there are firms aligning themselves with eco themes and producing eco friendly products i.e. green products in their opinions. Irrespective of the argument sustaining the debate, it is clear that there is a shift in thinking toward the welfare of society.

This shift has affected the governance mechanism of organizations. Those who are currently governing believe that investing in social welfare has at least two benefits, if not more. Firstly, it fosters social development and secondly it improves the firms' positioning in the market. However, there is not yet proof that a firm's investment in CSR activities has a positive relationship with their financial return (Burke & Logsdon, 1996). Still there are arguments that because of the unwillingness of the people running these organizations to involve, firms are staying away from social activities. Policymakers like directors and top level executives in the firms say that their duty is to work on behalf of the owners of the firms i.e. shareholders, and shareholders want their return on investment at the maximum level. This attitude of directors and top management stops organizations becoming actively engaged with the society other than via their regular activities like production, marketing, financing etc. But the inherent nature of business activities and the 'social contract' based on which business organizations operate require business organizations to be in contact with the stakeholders i.e. society. Business organizations depend on the society for some of their basic requirements, for example, human resources, financial resources, information resources, raw materials and many more. On the other hand, stakeholders benefit from organizations because of job creation, new product offers, innovative solutions and many others. Therefore, it is a mutual relationship and both parties have responsibilities to each other.

Business organizations are especially criticized for their misuse and exploitation of natural resources, which has ultimately brought about various disasters such as rising sea levels, temperature increase, ozone layer depletion, different diseases.

Similar scenarios can be found for social resources too; there are many organizations that employ child labour, do not offer appropriate working conditions, do not train the employees for the job, do not pay salaries regularly. Therefore, these kinds of organizations neither care for their internal stakeholders nor for the external ones. This invites pressure from consumer groups or associations, regulators, competitors in the market for these firms to behave in a socially responsible way. Corporations, therefore, either willingly or unwillingly, become involved in social programs in various ways. The question remains: do corporations do enough considering the damage they cause to society? Also in the absence of appropriate CSR disclosures (Belal & Cooper, 2011; Guthrie & Parker, 1989), it is practically impossible to know how firms design their CSR programs and what they do once they are in the field. When CSR is the subject matter, organizations usually disclose qualitative information about it. Also, there is not yet any uniform measurement instrument to measure the actual achievement of CSR programs. In addition to that, as CSR is voluntary in most counties and corporations have the freedom to design their CSR policies (Alam, Hoque, & Hosen, 2010), the dimensions of CSR are very much diversified which makes it difficult to compare across regions.

One of the possible ways of solving these problems is to regulate CSR with a uniform law. This would create one single system to be followed and make it easy to measure and compare CSR practices. This has its own problems as it might demotivate organizations from going above and beyond (Rahim, 2011), and it would be an interruption to the free market economy system. Stakeholders can argue that if CSR is made mandatory, it is then a law rather than a responsibility. However, with the current scenario where environmental degradation is sometimes directly related to corporations' activities, market regulators have started rethinking the system. India is working its way to regulating CSR to some extent (Van Zile, 2011). Indonesia has already made it mandatory (Utama, 2007). In Bangladesh, CSR disclosure has been mandatory for banks (Islam, Islam, & Ahmed, 2013).

This chapter focuses on the CSR practices of the banking sector in Bangladesh to explore if social audits are a potential too for ensuring social accountability. It begins with a brief introduction to the banking sector in Bangladesh followed by a discussion on the major CSR activities in this sector. The next two segments focus on the sustainability issues of CSR activities from a societal perspective and regulators' viewpoints on banks' CSR programs respectively. In the final segment of the chapter, social auditing has been introduced to the Bangladesh scenario to analyse its viability.

2 Overview of the Banking Sector in Bangladesh

The banking sector is the key player in the financial system of Bangladesh, which is still in the developing stage. The financial system has three sectors—formal, semi formal, and informal. In the formal sector, there are banks and other non-bank financial institutions that are regulated and controlled by the Bangladesh Bank, the central bank of the country. Although the Bangladesh Bank creates the regulations, the semi formal sector is not fully controlled by the, this sector is under the control of some other entities. The informal sector has private intermediaries and is fully unregulated. Therefore, banks fall into the formal sector under the guidance and regulation of the central bank of the country.

The major function of the banking sector is the financial intermediation and the level of intermediation has been increasing over time since the independence of the country in 1971. Immediately after independence, there were six nationalized commercial banks, two state owned specialized banks, and three foreign banks. As a new country, Bangladesh needed to accumulate savings in a more organized way and therefore started planning to expand the banking sector. As a result, during the 80s the banking sector expanded and new banks were opened. One of the major aims of this was to extend banking services into rural areas so that money being saved by rural households could come to banks, one of the formal channels of monetary flows. However, over time mismanagement (Nguyen, Islam, & Ali, 2012), corruption, and loss by banks have arisen and in the 1990s the government began reforming this sector.

Currently there are 56 banks in the sector (Islam, Islam, & Ahmed, 2013). Among them 52 are scheduled banks and the remaining are non-scheduled banks. Scheduled banks are those, which operate under the Company Act 1991 (amended in 2003) while non-scheduled banks have been introduced with a specific purpose. Out of the 52 scheduled banks, four are either fully or majorly owned by the government. Similarly with the specialized banks, four are either fully or majorly owned by the government. The number of private commercial banks is 35, of which 28 offer conventional banking services and seven offer Islami Sharia-based banking. The remaining nine banks are foreign commercial banks. The four non-scheduled banks are Ansar VDP Unnayan Bank, Karmashangosthan Bank, Probashi Kollyan Bank, and Jubilee Bank. Figure 1 is a representation of the banking sector in Bangladesh.

Banking sector experts in Bangladesh have another way of classifying the banks in the banking sector based on when the bank began operating. Banks that started their operation within the period of 1982–1988 are called first generation private banks. Second generation private banks started their operation in between 1992 and 1996. Third generation private banks started their operation after 1998.

Recent major changes within the banking sector occurred in 2013 when the Bangladesh Bank permitted five new banks to operate in the country. These five banks, namely Union Bank Limited, Modhumoti Bank Limited, the Farmers Bank Limited, Meghna Bank Limited, and South Bangla Agriculture and Commerce Bank Limited, are already up and running. This is the fourth time that the Bangladesh Bank has allowed this. In 1983 it happened for the first time, then in 1995 followed by a third time in 2001 (Islam et al., 2013).

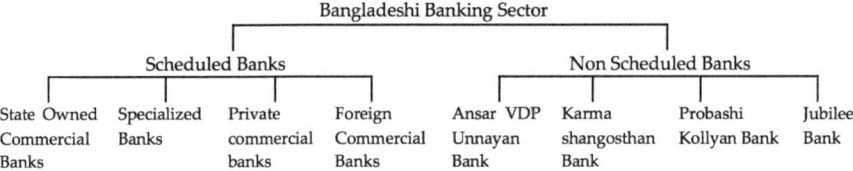

Fig. 1 Banking sector of Bangladesh (Islam et al., 2013)

3 Corporate Social Responsibility Practices in Bangladesh

CSR practices are comparatively new in Bangladesh; large scale operations started in 2007. Since then the financial sector has been the largest CSR contributor, however, pharmaceutical and telecom companies also contribute significantly. The issue of CSR is gaining increasing attention from both academics and corporations. In brief, CSR moves companies from legal compliance to the ethical or philanthropic standpoint where they go beyond their legal requirements and invest part of their profits in social welfare. To some extent, it is an extra expense for the firms as many researchers argue that there are returns on investment in the various forms of CSR. However, starting from Milton Friedman who says that maximizing profits (Friedman, 1962) through operating within the legal boundaries in a competitive market is CSR, to Carroll's approach in (1999) that businesses become ethical and philanthropic after being profitable, there are lots of debates on the CSR. Hanlon agreed (Hanlon, 2008) that CSR should not influence the firm's profitability in a negative way, while Banerjee's (Banerjee, 2008) opinion was that CSR is rather narrow and firms should focus on broader perspectives of stakeholders' interests.

Talking about interests, areas like community/society, employee, environment, and customer/consumer arise. However, among them, society and environment are of most importance because of the sustainability issues. When stakeholders become more aware of these issues they want to know that organizational activities have minimum impacts on these issues. On the other hand, historical events, such as the Exxon Valdez oil spill in Alaska (Patten, 1992) and the Union Carbide gas leak in India (Blacconiere & Patten, 1994), show that a firm's activities can have significant impacts on the environment. Therefore, social and environmental disclosure, as well as philosophical discussions on social and environmental accounting dominated the financial research agenda during the 1990s (Belal, 1999).

In 2000, Quazi and O'Brien came up with a modified version (Quazi & O'Brien, 2000) of Carroll's four factors' model. In the original model, Carroll suggested that the firms try to create a balance between their economic, social, legal and discretionary responsibilities (Carroll, 1979). Economic responsibilities are profit making functions and the basis of all social obligations. Legal responsibilities are mandatory and ethical responsibilities go beyond legal compliance. Discretionary responsibilities are those that are neither mandatory nor expected by the stakeholders; companies, as a result of their altruistic principles, uphold these. In the model of

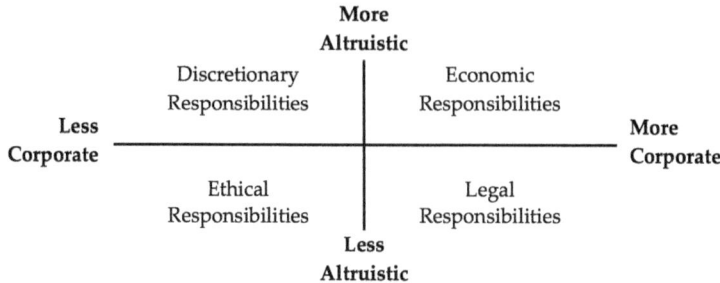

Fig. 2 Social responsibility categories: adapted from Carroll (1979), Quazi and O'Brien (2000)

Quazi and O'Brien, they portrayed activities of a firm as either altruistic or corporate (Fig. 2).

Irrespective of the debates, CSR issues continue to draw attention mainly in the developed world. Jamali and Mirshak (2007) concluded that although there is no question on CSR in developed countries, it is still in its infancy in developing countries where it faces questions on its worthiness. Khan (1985) said that because of the socio economic factors and behavioural patterns, the list of expectations are very different between developed and developing nations, and also consumers and stakeholders in developing countries demand less regarding CSR. Gray et al. (Gray, Kouhy, & Lavers, 1995) said that firms' CSR activities are different across regions because of area-specific demands. Also in course of times, new categories emerge and the concept of CSR changes in terms of definition and scope. Belal in 2001 believed that CSR is usually believed to be a developed country phenomenon (Belal, 2001) and it has been overlooked in developing countries followed by few researches on the issue.

CSR represents firms' overall commitment to the welfare of the community and environment in which they operate (Rana, Kalam, & Halimuzzaman, 2012), and over time it has become a powerful tool in gaining competitive advantage (Ferdous & Moniruzzaman, 2013) in the market. It plays a role in decision making (Cochran, 2007) and with the growing demand of it firms in the market changed their attitude from classical "profit maximizing approach" to "social responsibly" approach (Almona, 2005). Although researchers and different experts or expert bodies define CSR in different ways, the overall theme is similar. As per the European European Commission (2011), CSR is the responsibility of an organization for its impacts on society. World Bank defines CSR as "the commitment of business to contribute to sustainable economic development, working with employees, their families, the local community and society at large to improve quality of life, in ways that are both good for business and good for development" (Petkoski & Twose, 2003). The World Business Council for Sustainable Development in 1999 defined CSR as the continuous contribution of a firm toward its stakeholders, which includes the workforce and their families, and society at large (Development, 1999). Fontaine described CSR as a form of corporate self regulation integrated into the business model itself (Fontaine, 2013) and Ahmed (2013) concluded that CSR is not just philanthropic, but rather creates a bond between employees and the firm and its stakeholders (Ahmed, 2013). Many research studies have devoted efforts to measure firms' performance on CSR (Waddock & Graves, 1997; Zappi, 2007) or the impact of CSR on society. In 1979, Abbott and Monsen (1979) worked with Fortune 500 companies and developed a social disclosure scale by focusing on six areas namely environment, products, equal opportunities, personnel, community involvement, and other disclosures. There is an ongoing debate about how CSR influences a firm. Du et al. concluded that CSR creates favourable stakeholder attitudes (Du, Bhattacharya, & Sen, 2010), while Henderson (2001) believed the opposite, saying that CSR is a distraction from the main objective of profit maximization.

The history of CSR initiatives in Bangladesh is not very long and it actually only officially started in 2008 with a Bangladesh Bank directive on CSR. However, before the directive there were already CSR initiatives. In Bangladesh, CSR is voluntary but organizations have to declare their CSR policies and show their CSR expenditures in annual reports. Banks and other financial institutions play a major role in CSR in the Bangladesh market. However, other companies, especially telecommunication firms are quite active regarding CSR. Though donation in various forms was the major form of initial CSR activity, it has been expanded to include many more. The following table (Table 1) presents a list of the major forms.

This list in Table 1 is not exclusive; there are some other forms that are nontraditional. As CSR is still an evolving concept (Alam et al., 2010) in Bangladesh, it's very difficult for firms to work within a fixed framework, and they also have to balance CSR with other ethical obligations (Azim, Ahmed, &

Table 1 Major forms of CSR activities in Bangladesh

Category	Item	Sector performing
Humanitarian and Disaster	Donations, gifts, and money to disaster relief fund	Banks, telecommunication
Education	Offering scholarships, giving education materials, donating computers to educational institutions, offering educational and skill development training programs, developing online schools	Banks, telecommunication, pharmaceuticals
Health	Blood donation camps, health check up camps, donation to the treatment of different diseases, pure drinking water projects, and motherhood and infant care projects	Banks, telecommunication
Sports	Sponsoring various sports events	Banks, telecommunication
Arts and Culture	Sponsoring cultural events, helping in local cultural development	Banks, telecommunication
Environment	Providing solar panels, environmental awareness campaigns, beautification of cities	Banks, telecommunication, private universities, hospitals

Source: Annual reports and websites of banks in Bangladesh

D'Netto, 2011). In reality, in a small economy like Bangladesh economic growth (Ali, 1994) gets more priority than social causes and therefore CSR has not been of great importance. However pressures from various bodies like multinational buyers (Belal, 2001), civil society, and consumer groups have made the government aware of CSR, and it started taking action in this regard. Initially the government campaigned against the manufacturers who used to adulterate consumer products. The good thing is that the involvement of firms in CSR initiatives increases gradually in the Bangladesh market.

4 The Banking Sector's CSR Activities in Bangladesh

Banks contribute the highest amount in CSR expenditure in Bangladesh, and the amount has increased since 2008. But because of the lack of appropriate CSR reporting and communication, stakeholders are not fully aware of the facts and figures related to CSR (Belal & Cooper, 2011). Therefore, there is a lack of transparency about the banks' CSR policies and reasons behind why some invest more or less in a particular sector. Banks sometimes invest in some sectors that are not their field of operation. They ignore or invest lowest amounts in areas such as the environment, which deserves more attention in a country like Bangladesh. It raises questions about banks' the commitment to a sustainable society.

However, history shows that banks and other financial institutions lead CSR initiatives in Bangladesh by having the largest share of CSR investment in the market. In the last couples of years, the investment has increased and CSR is now a competitive strategy for banks. The Dutch Bangla Bank was a pioneer in this case, putting highest amount of money into CSR initiatives. There were no explicit policies on CSR in Bangladesh prior to 2008 when the Bangladesh Bank, the central bank of the nation, issued a directive namely 'Mainstreaming Corporate Social Responsibility (CSR) in the Banks and Financial Institutions in Bangladesh'. In the directive, the Bangladesh Bank asked banks to formulate their own CSR policies with an annual outlay for CSR programs included in their mainstream banking activities instead of short-term social contributions such as providing grants, aids and donations. Since then banks' CSR spending has steadily increased. The expenditure increased to Tk. 3,046.69 million in 2012 compared to that of Tk. 226.4 million in 2007 (Bank, 2012). There was a big jump in the expenditure in 2010 when CSR expenditure increased by 320 %, i.e. from Tk. 553.8 in 2009 million to Tk. 2,329.8 million (Bank, 2013). After the regulation in 2008, CSR expenditure has increased by 641 % till 2012. Local private commercial banks play a major role in CSR expenditure; they claim about 85.06 % of the banking sector's total expenditure. Eight state-owned banks are next and contribute 10.35 % to the total CSR expenditure. Foreign banks, usually used by the high income people in society spend the lowest in CSR, i.e. 4.59 % (Bank, 2013) of the banking sector's total investment.

Various fields where banks invest their money as part of CSR include but are not limited to education, humanitarian and disaster relief, health, art and culture, sports, and environment. The education sector, in most cases, receives the highest amount of contributions. Banks offer scholarships to students, provide educational materials to needy students, offer training to students, provide computers to educational institutions, donate funds to develop infrastructure, and invest in the development of online schools. However, major investment goes with the scholarships. In 2012, banks invested 32.3 % of their total CSR money into education. In the same year, humanitarian and disaster relief received the second highest priority claiming about 25.9 % of the banks' total CSR funds. This makes sense, as Bangladesh has had a number of natural disasters in the last couple of years. The whole country came forward to help the victims and banks got involved too. Examples of humanitarian and disaster relief support are: donations to the Prime Minister's Relief Fund, help in the rehabilitation of affected people, donations directly to the affected people, and arranging food for the victims. The health sector had the third highest amount of contributions at 14.3 %. The health standard in Bangladesh is very poor and many people suffer from various diseases. Also when there are natural disasters, some diseases like cholera, fever, and headaches rapidly increase. Therefore, it is one of the top areas where banks or other entities can invest under their CSR programs to gain attention. The environment received the least at 4.6 %. This area should be of much greater concern to corporations.

CSR data for the period 2008–2012 (presented in Table 2) presents a similar pattern of expenditure; with environment the lowest priority and education the top. Since 2007, contributions to the education sector have greatly increased. Compared to Tk. 14.3 million in 2007, the total investment in the sector in 2012 increased to Tk. 983.69 million, representing a 410 % increase. Contributions to the health sector began increasing in 2010 and at aggregate level crosses humanitarian sector.

Dutch Bangla Bank Limited (DBBL), a joint venture between Bangladesh and the Netherlands, topped all the banks in the Bangladeshi banking sector by investing Tk. 527.7 million under their CSR program in 2012. One of the major characteristics of the bank's CSR program is that it invests heavily in the education sector and this is one of the reasons that keep education sector at top continuously in terms of CSR benefits. The Export Import Bank of Bangladesh Limited (EXIM) spent the second highest amount, Tk. 399 million, on CSR programs in 2012. Only two state owned banks, Janata Bank Limited and Agrani Bank Limited, reached a place in the top ten banks investing the most in CSR programs in 2012; they invested Tk. 137.6 million and 104.1 million respectively.

Table 2 clearly shows that the environment, the degradation of which is a major concern in Bangladesh, receives the lowest amount of attention. Almost all the banks invested a minimum amount of their CSR funds in this field. This does not make sense, as the environment should be of great concern to Bangladesh in the face of severe challenges from environmental degradation and climate change. Therefore it seems like banks ignore this national priority and place their focus elsewhere. One of the reasons behind this is that banks link their reputation and branding with their CSR investments. There is return on CSR investments in

Table 2 Contribution to CSR activities (in million Taka)

Sectors	2008	2009	2010	2011	2012	Total	%
Education	30.5	94.8	400.79	612.48	983.69	**2,122.26**	**24.9**
Health	112.1	245.5	689.07	520.42	435.43	2,002.52	23.5
Humanitarian and disaster	58.6	125.1	460.41	188.03	788.37	1,620.51	19.0
Others	158.9	86.9	125.58	198.73	301.81	871.92	10.2
Sports	49.8	1.2	265.23	359.07	183.85	859.15	10.1
Arts and culture	0.8	0.3	328.91	171.52	213.31	714.84	8.4
Environment	–	–	59.78	138.07	140.23	**338.08**	**4.0**
Total	410.70	553.80	2,329.80	2,188.33	3,046.69	8,529.28	100

Source: CSR review (2013)

various forms, for example increased customers loyalty, increased reputation or brand value, lower employee turnover, and less legal fees. In order to see results from their investments, banks do not choose the environment sector as feedback on these investments take longer; in other fields, such as education, health, and humanitarian there are relatively quick results. Therefore, even though banks are not experts in education, they invest heavily in this area. On the other hand the environmental sector is more complex and therefore don't receive much attention from the media. This discourages organisations to invest CSR funds in the environment, because without media coverage it is harder to promote this to the masses. If most people don't get the message, the investment does not benefit the banks. Therefore the media has a role to play and companies look for media attention to get in touch with their stakeholders so that they remain in their favour. Surprisingly, these same companies don't yet communicate enough about the whole scenario of their CSR to stakeholders.

Overall CSR reporting has actually improved around the world. According to the KPMG report, for the world's largest companies, reporting increased from 52 % in 2005 to 70 % in 2008 (KPMG, 2008). There are various discussions on how firms should disclose information about their social concerns. However, a format recommended by Global Reporting Initiative (GRI) has been widely recognised and many companies are adopting the format (Raman, 2006). In Bangladesh, the issue of CSR is comparatively new and the GRI format is still unfamiliar; therefore, the disclosure of CSR information is not market a level with the rest of the world. However, due to competitive pressures, the perceived ability of disclosure to enhance corporate reputation (Kabir, 2003) and pressure from stakeholders (Belal, 2001), companies now usually disclose some information in their annual reports in Bangladesh. Imam (2000), after studying 40 companies listed with Dhaka Stock Exchange (DSE), found that only 22.5 % of companies disclosed environment-related information in Bangladesh. In 2006, Hossain et al., found that 8.33 % of Bangladeshi companies disclosed social and environmental information (Hossain, Islam, & Andrew, 2006); their study included 107 non-finance companies in Bangladesh.

In 2008, the Bangladesh Bank made it mandatory for firms to disclose their CSR expenditure information and banks were required to submit a report on this to the central bank. Yet still the information that banks disclose is not thorough enough to include the details of CSR programs. There are many reasons why banks do not publish detailed information on CSR. Firstly, existing laws do not require this as CSR is voluntary (Belal, 2001). Banks sometimes publish CSR-related information in brief and many experts believe that they do it to develop their reputation in the market. They strategically select which information to promote to gain the maximum out of the communication. Secondly, banks do not want to let others know about their CSR activities in detail because they do not want to draw attention from parties in need, as they might not be able to fulfill the needs of all (Azim, Ahmed, & Islam, 2009). Thirdly, sometimes bank executives do not want to let the banks' shareholders know that they are spending their money on social causes, as there might be shareholders who are against it (Azim et al., 2009). Therefore, executives do not want to put their jobs in jeopardy.

5 Social Auditing as a Solution

The introduction of social auditing can be an ideal option to connect stakeholders with the CSR programs of banks toward an all inclusive CSR management. It would bring more transparency to the system and banks would then find themselves in the upper level of social accountability status. Stakeholders desire a firm to be more than an artificial living being in acts rather than mere in theory. They want to know in detail how the firms' activities affect them either directly or indirectly. Also there is a huge demand from the stakeholders that corporations stick to their commitments regarding inclusive growth, i.e. growth with the stakeholders. Therefore, the pressure for transparency and accountability is high. This is an opportunity and firms can be successful in this regard by involving stakeholders in the social auditing process.

5.1 Introducing Social Auditing

A social audit is a process of evaluating and reporting on a business's performance regarding its commitment to economic, legal, social and philanthropic responsibilities. Many different researchers have defined a social audit in, however, the working definition (Natale & Ford, 1994) came in 1990 from Gerald Vinten (1990). He defined a social audit as this:

> A review to ensure that an organisation gives due consideration to its wider and social responsibilities to those both directly and indirectly affected by its decisions, and that a balance is achieved in its corporate planning between these aspects and the more traditional business-related objectives.

Here Vinten talks about incorporating social values into corporations' planning. Before Vinten, Buchholz defined a social audit as:

> The social audit is an attempt by an individual corporation to measure its performance in an area where it is making a social impact ... an attempt to identify, measure, evaluate, report, and monitor the effects a corporation is having on society that are not covered in the traditional financial reports (Buchholz, 1989).

In this definition, Buchholz focused on the total disclosure of corporations by recommending that firms should disclose those parts too, which were not included in the annual report. But Steiner and Steiner (1991) stated that this type of voluntary audit was not meaningful. They believed that there are two types of audits, mandatory one required by the government and voluntary. Voluntary audits are just that and corporations do nothing but describe what they have done in a particular time period. In 1986, Davis and Blomstrom defined the social audit as:

> A social audit is a systematic study and evaluation of an organization's social performance, as distinguished from its economic performance. It is concerned with possible influences on the social quality of life instead of the economic quality of life. The social audit leads to a social performance report for management and perhaps outsiders also (Davis & Bromstrom, 1975).

Though there are many different definitions, the major theme of defining a social audit is more or less similar across studies. The majority of the researchers concluded that corporations can benefit implementing social audits within their organizations. Through a social audit a firm can identify and measure its progress and challenges to share with stakeholders (who include, but not limited to employees, customers, investors, suppliers, community members, activists, the media, and regulators). This process can help a firm to increase its attractiveness to investors, improve relationships with stakeholders, identify potential liabilities, improve organizational effectiveness, and reduce the risk of misconduct and adverse publicity. A social audit looks at a company's record of charitable giving, energy consumption, voluntary involvement, work environment, employee salary packages, and transparency to evaluate how a firm influences the location it operates within. To do so, the auditor or the audit team collects data from both primary and secondary sources. While checking relevant documents is very common, audit team can collect data from the stakeholders through questioning or interviewing or any other mechanism to collect primary data. At the end, a report is prepared and is communicated with the relevant stakeholders.

5.2 History and Development of Social Auditing

The term social audit was first mentioned by Howard R. Bowen in 1953 in one of his articles 'social responsibilities of a businessman' (Sushmita, 2013). However, the theme of social audits can be linked to the academician, Theodore Kerps, who called on companies to accept responsibility toward citizens in the wake of the depression in 1940. It got a positive wave during the 60s in the USA and Europe when public repudiation of the Vietnam War triggered a movement to boycott the products and shares of some companies that were associated with the war. People wanted the firms to change their ethical standards and to be more accountable. Therefore some firms started publishing information on their social actions and objectives that eventually led to social audit. With the shift of ruling mechanisms toward democracy in various parts of the world the concept of a social audit was gradually recognised. After its transition to democracy in 1985, Guatemala started to disclose information on its military regimes. In 2000, a civil society group (GAM-CIIDH-Observatorio Ciudadano) started examining the documents and finally requested the court to start an investigation against military officials (UNDP, 2011). In line with this, Uganda and the Indian city of Bangalore introduced citizen report cards to measure the transparency in their health sectors; this was eventually successful (Vian, 2008) as a later review showed a decrease in corruption. Positive outcomes have been reported from Nepal too in case of their health sector (NHSSP, 2011).

There are three types of social audits: (a) first party audit in which an organization conducts a self evaluation using its own internal auditors, (b) second party audit where and organization audits its suppliers to its own codes of conduct or to an external standard, and (c) third party audit which happens when an organization is audited by an independent external organization.

In India, social auditing was introduced in 2005 and legalized in their system. They call it MGNREGA or Mahatma Gandhi National Rural Employment Guarantee Act. They developed this act, prepared trained workforce to run the program, and published reports on their activities for the period of 2006–2010. The project involves the government administration increasing transparency and getting people involved in the decision making process, however, there is mixed feedback on the project from stakeholders. Some believe that there is no transparency in the project itself and there is not yet an accurate or complete format to run the project. There is also the opinion that India does not yet have enough resources such as qualified or trained auditors to run the program. Afridi (2014) states that although social audits decreased administration-related complaints, it did not reduce corruption-related complaints.

However, the root of social auditing in India can be traced from the activities of Mazdoor Kisan Shakti Sangathan (MKSS) in Rajasthan province. This format has been in practice since 1994 following five stages:

- Gathering information: Citizen auditors collect information from government agencies; they collect the documents relevant to the scope of the audit.
- Collating information: Collected information is structured in an easily understandable format for the next level action. The format is designed in such a way that all the relevant stakeholders can understand the process.
- Sharing information: Structured information is shared with the stakeholders so that they can prepare for public hearings.

- The public hearing: This is the most important step where stakeholders discuss government records and actual scenarios. Public officials are asked questions and provide their opinions. A panel, with socially well reputed and well accepted persons is formed to administer the process.
- Follow-up to the public hearing: After the public hearing a report is prepared and communicated with the relevant stakeholders, which include senior government officials, media, community.

With the concept of MKSS, there can be two ways to conduct a social audit; one is government-led and the other citizen-led. In government-led social audits the government leads the process and it is easy to get access to the documents or information. On the contrary, in citizen-led audits, a community committee leads the process and there are usually some difficulties in gathering information or relevant documents. However citizen-led social audit teams usually have more information on community activities and in most cases are more accepted than the government-led social audit teams. On the other hand, government-led audit teams face less resistance from firms' employees compared to citizen-led audit teams (IBP, 2012). In Kenya, it has been found that the collection of information is a very difficult issue (IBP, 2012); social audit teams have been known to wait for more than 6 months following their request to gain access to information. Sometimes there is no response from concerned authorities, let alone access.

Since 1997 there has been a standard, namely SA 8000, in the market, developed by New York based Social Accountability International (SAI) to accommodate customers' concerns on the production of the products they purchase or consume. The standard works with nine core elements namely child labour, forced labour, health and safety, freedom of association and right to collective bargaining, discrimination, discipline, working hours, compensation, and management systems. As per the SAI webpage, they are working in 65 countries across 65 industrial sectors with more than 3,000 facilities certified for the SA 8000 standard. The institute has a wing called Social Accountability Accreditation Services (SAAS). Any organization wishing to be certified has to apply for certification and following an inspection from SAAS they are granted a certificate if applicable. This is a voluntary certification but it represents a lot about a firms' performance on social and environmental issues related to sustainability.

To ensure that social audits work properly, political will to implement it, access of common people to information, social audit infrastructure at affordable cost, and political will to punish the culprits are mandatory. In regards to donors' projects or projects funded by donors, it is not difficult to conduct a social audit. But when about it involves the government or corporations, there has to be a legal framework to supplement social auditing. For example, if a social audit reveals that a firm is guilty of damaging the environment, there have to be laws to take appropriate action(s) on that issue. Andhra Pradesh in India is an example of this; the Pradesh government did the following:

- Initiated a social audit scheme and the state cabinet passed social audit rules and rights to information,

- Set up a dedicated social audit unit, Andhra Pradesh Society for Social Accountability & Transparency (APSSAAT), and
- Trained and recruited state, district, and village level social audit personnel.

Based on the findings from the initial pilot social audit, APSSAT fired three technical assistants and 34 field officers for their involvement in malpractice and corruption. Two First Information Reports (FIR) have been lodged with the police against officials involved in malpractice and 59,786 equivalent Indian currency has been recovered from those involved in corruption (CGG, 2009).

5.3 Social Audits in the Context of the Banking Sector of Bangladesh

Social audits, with the support of government or relevant government offices, can be introduced into the banking sector of Bangladesh to increase social accountability of firms, greater transparency regarding the CSR initiatives of banks, and a safer environment. The following segments discuss the challenges with banking sector CSR and the place for social auditing in this sector.

5.3.1 Lack of Transparency and low Stakeholder Participation in the Banking Sector CSR

CSR programs carried out by banks in Bangladesh do not have the stakeholders' orientation in reality and there is a lack of transparency along with minimum stakeholder participation. Over time, banks' CSR initiatives have continuously expanded in scope in terms of financing and sectors. As a result, the number of people or clients benefiting from banks' CSR programs has increased over time. But stakeholders are speculative about the impact of these CSR activities, decision making process of CSR investors, accuracy of the banks' CSR claims, and the justification of current CSR programs. As mentioned in Sect. 4, banks do not engage much with environment-related CSR programs, even though stakeholders expect this. This means that banks ignore one of the three core pillars of CSR. Usually the environment is at the centre of CSR campaigns and sustainability; therefore the long term success of CSR is not attainable without a major focus on environment. This raises the question on how banks choose their CSR programs. Banks do not communicate the reasons behind their choices and various researches conclude that CSR reporting in Bangladesh is far from the global standard. Also, when banks communicate about their CSR programs, they share either qualitative or subjective information, which is difficult to analyse or compare; the main objective of their communication is to gain media attention. In addition, banks, in general, strategically select activities for their CSR programs that best suit their interests. They do not consider stakeholders' needs and the form in which stakeholders might be served better. Stakeholders do not know how banks decide on their CSR programs and they are not part of the decision making process.

The accuracy of CSR reporting in their annual report is also questionable. There is no audit process for that particular expenditure and banks simply report on their expenditure without further details or explanations. Although they are not required to do so by law, from the stakeholders' perspective this minimal information creates uncertainty. Also, as there is a 'tax reduction' benefit based on the CSR expenditure, this amount has to be cross-checked to find out if the claims are true or not. No initiative has been taken so far, except for Bangladesh Bank's yearly 'CSR Review' in which it summarizes CSR expenditures for a particular period. The expenditure data comes from banks as the Bangladesh Bank made it mandatory for all banks to submit a half yearly report on CSR in the prescribed format within a month's timeframe. Therefore, the problem remains about the validity of the CSR claims.

Another concern is whether banks' CSR expenditure creates capacity or not. As the majority of the expenditures go to the education sector, it should be useful to draw reference from the sector. There is no doubt that CSR expenditure in the education sector creates capacity, but the questions remains on how much capacity is created and whether this capacity is sustainable or not. The scholarships that are given to students at various levels by different banks should be considered. The Dutch Bangla Bank offers the highest amount in this field among all banks; Tk. 2,500 per month to students to study at bachelor level. Is this amount enough? For a bachelor level student at a public university in Dhaka, the cost of living is around Tk. 4,500; although this varies depending on the faculty. For example, students from the Business School need more money per month compared to their colleagues in the Faculty of Arts or Faculty of Social Science. On the other hand, students from the Science Faculty and Life Science Faculty might need more than that of the students in the Business School. The Dutch Bangla Bank scholarship then helps a student with approximately 50 % of their required funding. Once again this is a great help, but not complete.

It raises the question of whether funds are channelled to appropriate fields or not. Sometimes it happens that the client needs money for infrastructure development but the organization offers money to buy computers to enhance IT education. Sometimes clients may not have the room for these computers, so therefore there is no benefit for them in this 'donation'; this particular company would be better off with money to develop infrastructure. Again, when it is of more value to give clothes and study materials to students, it does not make sense to give money for the beautification of schools. So there is a great deal of decision making, but unfortunately stakeholders are not part of this process. Banks alone decide and decisions are usually based on publicity considerations, which often mean CSR initiatives are not serving their purpose. Banks, seeking major media attention, do what the media will pick up on the most rather than what stakeholders need. As a result, stakeholders are on the outside of a program designed for their betterment. They do not know what they can expect from the CSR programs and in what form. Regulatory bodies, like the Bangladesh Bank, do little regarding this issue. The Bangladesh Bank requires all banks under its control to submit CSR expenditure data to them and every year it publishes a document, namely the CSR Review, based on the received data. Reporting banks do everything here; they spend money, they audit it, and they report it. There is nothing about transparency or validation of claims. However, as an official from the Bangladesh Banks days, this is the beginning of long journey.

> This is so far an ice-breaking event. We try to make CSR disclosure a regular and systematic event first. Then we shall look at the broader perspective of the CSR programs. But you should also remember that CSR is voluntary as per the Bangladesh Bank guidelines (executive from the Green Banking and CSR department of Bangladesh Bank, 2014).

Considering that Bangladesh is an emerging economy and CSR is still in its beginning stages, this is a good start. But as the sector is developing rapidly, before it is too late it needs to be ensured that CSR expenditures provide for the greater good of society. Also, there should be a framework for CSR activities and reporting. Surprisingly though, no one is talking about this issue. Regulators are doing the minimum, the media does not discuss the issue, and the society is apparently happy with the short-term orientation of CSR. There is no research on the issue in academia. Therefore, it is easy for the banking sector in Bangladesh to focus on the short term rather than long term, which is contrary to the main objectives of CSR.

5.3.2 Social Audits to Ensure Accountable CSR Practice by Bangladeshi Banking Sector

Social audits can increase stakeholder engagement with banks' CSR initiatives to ensure more transparency and accountability. Stakeholders are diversified and it is always tough to unify them. Inherent characteristics of a social audit can be helpful in this regard, since during the social audit process stakeholders have a chance to get engaged directly. They can ask questions, check documents, listen to CSR experts, and make comments on the information they know about CSR initiatives along with many other activities. It works as a stimulus to join and stakeholders get motivated to join which ultimately help to create transparency. In social audits, social intellectuals and the media can play a big role, as they create connections between various stakeholders.

As already stated there are usually two ways of conducting a social audit: government-led and citizen-led. Because of people's lack of confidence in the politicians of the country, influence of politicians on organizational especially government organizational activities, and high level of corruption, government-led social audit would not produce anything trustworthy to the stakeholders. Therefore, it would be a waste of resources. Rather citizen-led social audit having prominent and well accepted persons in the team would be a good option. One good example of this is Parivartan's social audit in Delhi during the period of 2000–2002, which produced positive outcomes (Kejriwal, 2003). In that audit, they found gross corruption in the system in implementing various policies and projects and after the disclosure of this information to stakeholders, public sector jobholders became more careful and honest; they feared that their dishonesty would be revealed, which eventually reduced the corruption.

However, it is also true that in the culture of Bangladesh, it would be very difficult to gain momentum on something that is not backed by the government (Rahim, 2011). Therefore, a citizen-led social audit, while keeping the central bank in charge would be meaningful; although the central bank is also a government unit, people trust it. Also, as a regulatory authority, the Bangladesh Bank is in the position to conduct social audits on any of the banks. More importantly, as it has easy access to the relevant data from all other banks, it can perform the job smoothly. Therefore the Bangladesh Bank either has the resources to execute this or it has the capacity to develop appropriate resources for the job.

One of the major factors is trust. If there were a trustworthy relationship between corporations and stakeholders, there would not be any need for social audits. Therefore, trust is vital throughout the whole process. Without the guarantee of trust, stakeholders would not become involved, which would ultimately null the main purpose of a social audit. Social audit requires information about the community and if the community is not connected appropriately, it minimizes the possibility of having adequate information. So, the social audit team has to be well connected with the community, otherwise the lack of communication creates a barrier to social auditing gaining success and acceptance. In the context of Bangladesh, because of the high population density, to engage maximum stakeholder and to collect adequate information, there should be a decentralized format of social audit. It would be better to put the opinion leaders from the communities in the audit team. To attract the attention of stakeholders, help from media can be requested, as the media plays a significant role in stakeholders' lives. There has been a revolution in the media sector in the last 10–15 years. Different types of media like satellite television channels and FM Radio channels create a buzz among people and nowadays everyone is connected to at least one form of media. Therefore, it has become ever easier to connect stakeholders with the appropriate use of the media. Media not only connects stakeholders but also provide input for social audits. From media events like news, talk shows, debates, articles in newspapers, and documentary, social audit teams can collect information and through similar events social audit team can inform stakeholders about the outcome of social audit.

Another major concern is the determination of the scope of social audit. It's not always possible to audit every organization or every activity of an organization followed by the need to define the area within which social audit should be performed. The issue deserves more concern especially when social audit process is introduced for the first time in country. One way to define the scope of social audit is the selection of relevant areas and the creation of a priority list. Both area selection and prioritising depend on many factors specific to the community or subject matter being dealt with. Using the banking sector of Bangladesh as an example, one should firstly focus on why banks invest minimum amounts into the environment. Again, social audits can be based on one organization or more. However, before committing resources to social auditing, it would be wise to study the political situation or culture of the land where the process is to be applied. If political will is absent to go with the social audit, it's really difficult for the social audit to be successful. In Bangladesh, government though says in favour, actually are not much interested in social audit type of activities. But considering the fact that Bangladesh is a democratic country and pressure from mass people work here to a large extent as history says, social audit can be a very good tool to establish good governance and transparency. Therefore, it's very important to campaign about the need and usefulness of social audit among the stakeholders so that they become aware and create a public pressure in favour of social auditing.

Social audits can be complimentary to financial audits; in fact it minimizes some of the limitations that financial auditing has. In general, financial auditors' scope of work is limited to cash and other financial transactions; they cannot go beyond that. Also the self interest of financial auditors (Locke, Qin, & Brause, 2007) motivates them not to report irregularities. Again the self interest of auditors leads them to serve the interests of the managers (Antle, 1984); this is a grey area in the auditing process and usually produces negative results for the stakeholders in terms of transparency. Even more grey area is present in the law, for example, the Company Act of Bangladesh does not determine the social responsibility of directors and the accountability of auditors (Rahim & Alam, 2013). Usually the process of auditing is not disclosed, so there is no option of questioning auditors' reports. Therefore, a social audit would be meaningful here, as it would go beyond the scope of traditional audit functions. It can check if a company complies with the codes and standards that are beneficial for society and stakeholders. After a social audit, a financial audit should be completed. Then the independent auditors of the firms can question the company about their social responsibility status. Considering the banking sector and the culture of Bangladesh, the following model is proposed for social auditing in this sector (Fig. 3).

In the proposed model, initially the Social Audit Team (SA Team) meets with the CSR Team from a bank. In the SA team, along with professionals, there should be representatives of stakeholders like intellectuals, prominent bankers, NGO experts, academics, and labour union representatives. As mentioned earlier, the Bangladesh Bank should be in charge of the social audits and the SA team should be headed by a representative of the Bangladesh Bank. On the other hand, the team from the bank should be headed by the CSR manager or person in charge of the CSR section in the bank. In that team there might be a bank director who takes care of the CSR policies of the bank, the internal auditor of the bank, and the head of finance. In the first meeting, the CSR team explains their CSR policies and what they have accomplished so far. They deliver documents to prove their claims. The SA team asks questions, discusses the relevant issues, and talks about how they can perform the next phase. They would require permission to visit the bank and interview bank people if necessary.

As pointed out by Pruett et al. (2005), the visit should be unannounced rather than pre-planned. With a pre-planned visit, the bank would get a chance to prepare, which might bring bias into the audit. The SA audit team then checks the documents, talks with relevant stakeholders, and prepares a draft report on the CSR

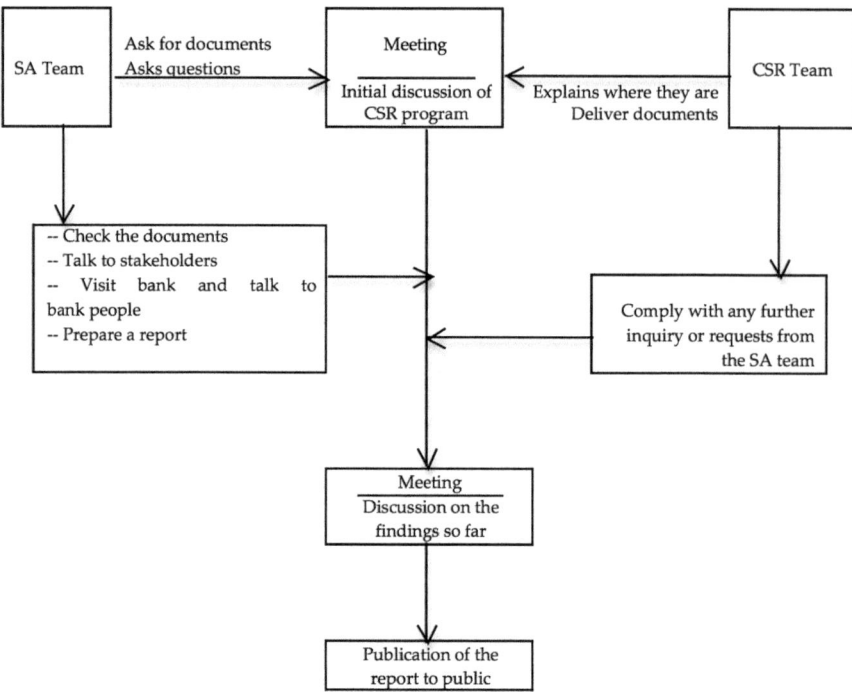

Fig. 3 Proposed model of social auditing in the Bangladeshi banking sector

performance of the bank being audited. In the discussion with the stakeholders, the SA team can invite bank's staff and the meeting can be a closed or open door one. With all the information available to them, the SA team will then meet with the CSR team again. This time they discuss the findings so far and give the CSR team a chance to clarify or defend their position. This is followed by the preparation of the final audit report and the dissemination of that report to the stakeholders. With the final report, the SA team would arrange one or more open-door meeting with the stakeholders in addition to making the report available online. Once again, relevant people from bank including the CSR team would be invited. At the same time, the SA team would invite local administrators like parliament members, relevant ministers and police authorities so that more people can become aware of the bank's CSR programs. It would minimise the distortion in communication and create more transparency.

Therefore, the introduction of social auditing into the Bangladeshi banking sector would be a good initiative. It would give decision makers insight into stakeholders' perceptions on their CSR programs, and stakeholders would get a chance to question the liability of the people in charge of the CSR programs in banks. It would also disclose the fact that banks do not invest in the environment, degradation of which will carry severe consequences in the near future for Bangladesh. Stakeholders would also become aware that banks invest in education in forms that are not always appropriate. It would create pressure on corporations to behave in a responsible way, as they would have the threat of possible resistance from stakeholders if the results of social auditing go wrong. Social auditing could bring people together and unite pressure groups that create change in the system (Mwawashe, 2011). Therefore, corporations would not afford to go wrong with the social auditing and let stakeholders be united against them, as it would be quite expensive for the corporations having threat of nonexistence in the market.

Conclusion
Social auditing has great potential to bring accountability to corporate practices regarding CSR in the banking sector of Bangladesh. This sector contributes hugely toward social causes and it is now an appropriate time to bring these contributions into a framework so that they work efficiently for the betterment of the stakeholders. Major issues are the lack of transparency, minimum or no investment in the environment, and low stakeholder engagement. Social audits can contribute to these challenges by bringing stakeholders in the single platform. It provides the opportunity for stakeholders to compare the CSR commitments of banks and their actual performance on that. It checks environmental and social performances of organizations and compares these with their commitments to create a picture of how close or how far the banking sector is from sustainable banking practices.

Though social audits may appear to be a tool for stakeholders, they are beneficial for banks too. Banks, in the social auditing process, interact with the stakeholders directly and this is an opportunity for the banks performing well with an stakeholders' orientation. The bank can promote its CSR strategies in ways that create a bond between them and their stakeholders, which is mutually beneficial. There are many investors in the market who want their investment to have a social contribution; going through the social audit process, organizations would be able to attract these investors. However, for that reason, emphasis has to be placed on the social audit's design. It should be managed by people who are trustworthy to maximize community participation considering the fact that trust is a requirement for the success of social audits. There have to have a legal basis for social auditing and the outcome of the social auditing process. One of the ways to do that is adding a provision of social audit in the relevant acts governing businesses in Bangladesh. Otherwise, trust won't be created and stakeholders won't be interested to join the audit process.

References

Abbott, W. F., & Monsen, R. J. (1979). On the measurement of corporate social responsibility: Self-reported disclosures as a method of measuring corporate social involvement. *Academy of Management Journal, 22*(3), 501–515.

Afridi, F. (2014). Social Audit isn't enough. *The Indian Express*. Retrieved from http://indianexpress.com/article/opinion/columns/social-audit-isnt-enough/

Ahmed, M. K. (2013). Corporate social responsibility practices of commercial banks in Bangladesh: A case study on Southeast Bank Ltd. *IOSR Journal of Business and Management, 12*(1), 13–18.

Alam, S. M. S., Hoque, S. M. S., & Hosen, M. Z. (2010). Corporate social responsibility of multinational corporations in Bangladesh: A case study on Grameenphone. *Journal of Patuakhali Science and Technology University, 2*(01), 51–61.

Ali, Q. M. (1994). *Social responsibility, consumerism and corporate behaviour: A comparative study of managerial attitudes and marketing action in Australia and Bangladesh*. University of New South Wales.

Almona, C. P. (2005). *A review of the business case for corporate social responsibility in the UK financial service sector*. (Unpublished Dissertation), 1–94.

Antle, R. (1984). Auditor independence. *Journal of Accounting Research, 22*, 1–20.

Azim, M. I., Ahmed, E., & D'Netto, B. (2011). Corporate social disclosure in Bangladesh: A study of the financial sector. *International Review of Business Research Papers, 7*(2).

Azim, M. I., Ahmed, S., & Islam, M. S. (2009). Corporate social reporting practice: Evidence from listed companies in Bangladesh. *Journal of Asia-Pacific Business, 10*(2), 130–145.

Banerjee, S. B. (2008). Corporate social responsibility: The good, the bad and the ugly. *Critical Sociology, 34*(1), 51–79.

Bank, B. (2012). *Review of CSR initiatives of banks-2011*.

Bank, B. (2013). *Review of CSR initiatives of banks-2012*.

Belal, A. T. (1999). Corporate social disclosure in Bangladesh annual reports. *The Bangladesh Accountants, ICAB, 27*(1), 76–81.

Belal, A. R. (2001). A study of corporate social disclosures in Bangladesh. *Managerial Auditing Journal, 16*(5), 274–289. doi:10.1108/02686900110392922.

Belal, A. R., & Cooper, S. (2011). The absence of corporate social responsibility reporting in Bangladesh. *Critical Perspectives on Accounting, 22*(7), 654–667.

Björkman, H., & Wong, E. (2013). *The role of social auditors: A categorization of the unknown*.

Blacconiere, W. G., & Patten, D. M. (1994). Environmental disclosures, regulatory costs, and changes in firm value. *Journal of Accounting and Economics, 18*(3), 357–377.

Boyd, D. E., Spekman, R. E., Kamauff, J. W., & Werhane, P. (2007). Corporate social responsibility in global supply chains: A procedural justice perspective. *Long Range Planning, 40*(3), 341–356.

Buchholz, R. A. (1989). *Fundamental concepts and problems in business ethics*. NJ: Prentice Hall Englewood Cliffs.

Burke, L., & Logsdon, J. M. (1996). How corporate social responsibility pays off. *Long Range Planning, 29*(4), 495–502.

Carroll, A. B. (1979). A three-dimensional conceptual model of corporate performance. *Academy of Management Review, 4*(4), 497–505.

Carroll, A. B. (1999). *Business and society: Ethics and stakeholder management*. Cincinnati: Western Publishing.

CGG. (2009). *Social audit of NREGS (AP) in Andhra Pradesh*. Center for Good Governance. Retrieved from http://www.sasanet.org

Cochran, P. L. (2007). The evolution of corporate social responsibility. *Business Horizons, 50*(6), 449–454.

Cotten, M. N., & Lasprogata, G. A. (2012). Corporate citizenship & creative collaboration: Best practice for cross-sector partnerships. *Journal of Law, Business and Ethics, 18*, 9.

Davis, K., & Bromstrom, R. L. (1975). Implementing the social audit in an organization. *Business & Society, 16*(1), 13–18.

Development, W. B. C. (1999). *Corporate social responsibility: Meeting changing expectations.* World Business Council for Sustainable Development.

Dodd, E. M. (1932). For whom are corporate managers trustees? *Harvard Law Review*, 1145–1163.

Du, S., Bhattacharya, C. B., & Sen, S. (2010). Maximizing business returns to corporate social responsibility (CSR): The role of CSR communication. *International Journal of Management Reviews, 12*(1), 8–19.

European Commission. (2011). A Renewed EU Strategy 2011–14 for corporate social responsibility. Commission to the European Parliament, the Council, the European Economic and Social Committee and the Committee of the Regions. Retrieved from http://ec.europa.eu/enterprise/ policies / sustainable-business/corporate-social- responsibility, Accessed on 1 October 2014

Ferdous, M., & Moniruzzaman, M. (2013). An empirical evidence of corporate social responsibility by banking sector based on Bangladesh. *Asian Business Review, 2*(3), 82–87.

Fontaine, M. (2013). Corporate social responsibility and sustainability: The new bottom line? *International Journal of Business and Social Science, 4*(4), 110–119.

Friedman, M. (1962). *1982, Capitalism and freedom.* Chicago: University of Chicago Press.

Friedman, M. (2007). *The social responsibility of business is to increase its profits.* Springer. Retrieved from http://link.springer.com/chapter/10.1007/978-3-540-70818-6_14

Gray, R., Kouhy, R., & Lavers, S. (1995). Corporate social and environmental reporting: A review of the literature and a longitudinal study of UK disclosure. *Accounting, Auditing & Accountability Journal, 8*(2), 47–77.

Guthrie, J., & Parker, L. D. (1989). Corporate social reporting: A rebuttal of legitimacy theory. *Accounting and Business Research, 19*(76), 343–352.

Hanlon, G. (2008). Rethinking corporate social responsibility and the role of the firm–on the denial of politics. *The Oxford Handbook of Corporate Social Responsibility, 156.*

Henderson, D. (2001). *Misguided virtue: False notions of corporate social responsibility.* Wellington, New Zealand: New Zealand Business Roundtable. Retrieved February 25, 2015, from http://www.iea.org.uk/sites/default/files/publications/files/upldbook126pdf.pdf

Hiller, J. S. (2013). The benefit corporation and corporate social responsibility. *Journal of Business Ethics, 118*(2), 287–301.

Hossain, M., Islam, K., & Andrew, J. (2006). Corporate social and environmental disclosure in developing countries: Evidence from Bangladesh. *Asian Pacific Conference on International Accounting Issues, Hawaii.*

Imam, S. (2000). Corporate social performance reporting in Bangladesh. *Managerial Auditing Journal, 15*(3), 133–142.

Isaksson, L., Kiessling, T., & Harvey, M. (2014). Corporate social responsibility: Why bother? *Organizational Dynamics, 43*(1), 64–72.

Islam, M. T., Islam, F. T., & Ahmed, A. (2013). Performance analysis of selected private commercial banks in Bangladesh. *The Jahangirnagar Journal of Finance & Banking, 1.*

Jamali, D., & Mirshak, R. (2007). Corporate social responsibility (CSR): Theory and practice in a developing country context. *Journal of Business Ethics, 72*(3), 243–262.

Kabir, E. (2003). Corporate social responsibility in Bangladesh. *The Financial Express of Bangladesh.*

Kejriwal, A. (2003). Where did our money go? *India Together.* Retrieved from http://indiatogether.org/moneytrail-rti

Khan, A. F. (1985). *Business and society.* New Delhi: S. Chand & Company Limited.

KPMG, T. (2008). *KPMG International survey of corporate responsibility reporting 2008.* Amsterdam, The Netherlands: KPMG.

Locke, R. M., Qin, F., & Brause, A. (2007). Does monitoring improve labor standards? Lessons from Nike. *Industrial and Labor Relations Review*, 3–31.

McElhaney, K. (2009). A strategic approach to corporate social responsibility. *Leader to Leader, 52*(1), 30–36.

Mwawashe, K. M. (2011). Youth as drivers of accountability: Conducting a youth social audit. *Young Citizens: Youth and Participatory Governance in Africa*, 181.

Natale, S. M., & Ford, J. W. (1994). The social audit and ethics. *Managerial Auditing Journal, 9*(1), 29–33.

Nguyen, C. V., Islam, A. M., & Ali, M. M. (2012). Asymmetric responses of commercial banks to monetary policy: The case of Bangladesh. *Journal of Business and Policy Research, 7*(4), 14–29.

NHSSP. (2011). *A review of social audit guidelines and practices in Nepal*. Nepal Health Sector Support Program.

Owen, D. L., Swift, T. A., Humphrey, C., & Bowerman, M. (2000). The new social audits: Accountability, managerial capture or the agenda of social champions? *European Accounting Review, 9*(1), 81–98.

Patten, D. M. (1992). Exposure, legitimacy, and social disclosure. *Journal of Accounting and Public Policy, 10*(4), 297–308.

Petkoski, D., & Twose, N. (2003). Public policy for corporate social responsibility. *WBI Series on Corporate Responsibility*, 7–25.

Pruett, D., Merk, J., Zeldenrust, I., & de Haan, E. (2005). Looking for a quick fix: How weak social auditing is keeping workers in sweatshops.

Quazi, A. M., & O'Brien, D. (2000). An empirical test of a cross-national model of corporate social responsibility. *Journal of Business Ethics, 25*(1), 33–51.

Rahim, M. M. (2011). Meta-regulation approach of law: A potential legal strategy to develop socially responsible business self-regulation in least developed common law countries. *Common Law World Review, 40*(2), 174–206.

Rahim, M. M., & Alam, S. (2013). Convergence of corporate social responsibility and corporate governance in weak economies: The case of Bangladesh. *Journal of Business Ethics, 121*, 607–620.

Raman, S. R. (2006). Corporate social reporting in India—A view from the top. *Global Business Review, 7*(2), 313–324. doi:10.1177/097215090600700208.

Rana, M. M., Kalam, A., & Halimuzzaman, M. (2012). Corporate social responsibility (csr) of dutch-bangla bank limited: A case study. *Bangladesh Research Publication Journal, 7*(3), 241–247.

Rangan, K., Chase, L. A., Karim, S. (2012). *Why every company needs a CSR strategy and how to build it*. Retrieved from http://citeseerx.ist.psu.edu/viewdoc/summary?doi=10.1.1.302.8709

Steiner, G. A., & Steiner, J. F. (1991). *Business, government, and society: A managerial perspective*. New York: Random House Business Division.

Sushmita, G. (2013). Social audits in India. *International Research Journal of Social Sciences, 2*(11), 41–45.

UNDP. (2011). *A practical guide to social audit as a participatory tool to strengthen democratic Governance, transparency, and accountability*. UNDP Publications.

Utama, S. (2007). *Regulation to enhance accountable corporate social responsibility reporting*. Faculty of Economics University of Indonesia. Retrieved from http://sydney.edu.au/business/__data/assets/pdf_file/0003/56613/Regulation_to_enhance_accountable.pdf

Van Zile, C. (2011). India's mandatory corporate social responsibility proposal: creative capitalism meets creative regulation in the global market. *APLPJ, 13*, 269.

Vian, T. (2008). Transparency in Health Programmes. *U4 Brief, 2008*(9).

Vinten, G. (1990). The social auditor. *International Journal of Value-Based Management, 3*(2), 125–135.

Waddock, S. A., & Graves, S. B. (1997). The corporate social performance. *Strategic Management Journal, 8*(4), 303–319.

Zappi, G. (2007). Corporate responsibility in the Italian banking industry: Creating value through listening to stakeholders. *Corporate Governance, 7*(4), 471–475.

Corporate Social Responsibility Assurance: Theory, Regulations and Practice in China

Yuyu Zhang and Lin Liao

1 Introduction

The importance of corporate social responsibility (CSR) and accountability in China is highlighted by promised economic success, the emphasis on China's integrity in the world market and infamous corporate social scandals. Revealed in the striking Sanlu melamine event,[1] which occurred in 2008, the reliability and transparency of CSR information may be as influential as financial information (if not more) to companies and their stakeholders. While CSR reporting in China is prevalent—in response to stakeholder and regulator expectations—some listed companies intentionally hide negative material information in their CSR disclosures. In a local survey of CSR disclosure quality regarding 2,000 listed companies

---

[1] In September 2008, after months of intentionally hiding information from the public, the infant formula from Sanlu Group, one of China's leading infant formula producers, was reported to contain an industrial chemical ingredient melamine. This caused numerous infant kidney stones and/or kidney failure in China. The melamine incident not only led to the bankruptcy of Sanlu Group, it swept 22 dairy enterprises (including well-known brands) and destroyed consumer confidence in the entire dairy and food safety industry. This was in addition to the high social cost. An estimated 300,000 victims caused considerable pressure on the public medical system. Laid-off workers from the dairy industry required government intervention. The incident reshaped the national regulation system, with the suspension of inspection-free systems and the implementation of the 'Regulation on the Supervision and Management of the Quality and Safety of Dairy', triggering intensive debates on CSR in China.

Y. Zhang (✉)
School of Accountancy, QUT Business School, Queensland University of Technology, 2 George St, 4001 Brisbane, QLD, Australia
e-mail: yuyu.zhang@qut.edu.au

L. Liao
Research Institute of Economics and Management, Southwestern University of Finance and Economics, Guanghuacun Street, Chengdu, Sichuan 610074, China
e-mail: liaolin@swufe.edu.cn

in mainland China, Hong Kong and Taiwan in 2011, 503 experienced negative press releases in relation to social or environmental responsibilities. Few companies made official disclosures. Specifically, in companies producing their CSR reports for the public, 84.6 % did not disclose any negative events (Zhong, Zhang, & Zhai, 2011). Ensuring the credibility of CSR reports has become a significant issue in China. In addition, social reporting standards adopted in developed countries have greatly influenced Chinese companies involved in international trade. Since 2000, many multinational companies domiciled in developed countries have explicitly requested CSR evaluations and certifications from their global suppliers and subcontractors, suggesting the necessity of enhanced CSR reporting quality and independent assurance of CSR reports.

Yvo De Boer, global chairman of KPMG's Climate Change & Sustainability Services, critiqued the situation: "In the twenty-first century, CR reporting is—or should be—an essential business management tool. It is not—or should not be—something produced simply to mollify potential critics and polish the corporate halo". It is "the means by which a business can understand both its exposure to the risks of these (environmental and social) changes and its potential to profit from the new commercial opportunities" and "the process by which a company can gather and analyse the data it needs to create long term value and resilience to environmental and social change". It is "essential to convince investors that your business has a future beyond the next quarter or the next year" (KPMG, 2013, p. 10). This argument implies that CSR assurance[2] is not optional any more. More companies are moving towards deeper integration of CSR reporting in their business strategy and management processes. External stakeholders seek information from auditors, who can provide independent CSR assurance and demonstrate that a company is as serious about CSR disclosure as it is about financial information. Accordingly, the focus of CSR assurance might now be: "why would we not?" and "how do we choose CSR assurance option that meets stakeholders' needs and puts us ahead of our peers?" (KPMG, 2013, p. 12).

This paper is built upon stakeholder theory, and critically evaluates the theories, regulations, practice and literature, supported by first-hand data from China's capital market. It examines the period from 2009 to 2013, adding to the scant literature on China's CSR assurance and providing insightful understanding for regulators, industries and academics.

This chapter is structured as follows: Section 2 summarises stakeholder and related legitimacy and signaling theories frequently adopted in CSR assurance research. Section 3 introduces the current regulatory framework on CSR assurance in China. Section 3.1 presents a demand-and-supply analysis of China's CSR assurance market. From a demand perspective, we address two questions: who

---

[2] In this paper, academia and practice, 'CSR assurance' and 'social audit' are used interchangeably, both referring to the independence assurance of CSR reporting. However, 'CSR assurance' is a more accurate terminology, as the level of assurance provided is usually moderate and cannot be classified as an 'audit', which requires high levels of assurance.

are the stakeholders for CSR assurance in China? Are there any significant economic factors, regulatory changes and social events that contributed to the demand for CSR assurance in China? Longitudinal and cross-sectional analyses illustrate the development and current status of China's social audit market. From a supply perspective, market shares, and the strengths and weaknesses of major social assurance providers, are tabulated and discussed. Section 3.2 concludes the study, with further research opportunities identified.

2 Theories

2.1 Stakeholder Theory

Stakeholder theory emphasises that a firm's continued existence requires the support of various stakeholders, and that the organisation's activities must be adjusted to gain stakeholder approval (Gray, Kouhy, & Lavers, 1995). Freeman, Wicks, and Parmar (2004) raise two core questions addressed by stakeholder theory:

> First, what is the purpose of the firm? This encourages managers to articulate the shared sense of the value they create, and what brings its core stakeholders together... second, what responsibility does management have to stakeholders? This pushes managers to articulate how they want to do business—specifically, what kinds of relationships they want and need to create with their stakeholders to deliver on their purpose (p. 364).

Gray, Owen, and Adams (1996) and Deegan (2000) further classify stakeholder theory into two categories: the ethical or normative, and the managerial position. The ethical or normative perspective of stakeholder theory suggests that all stakeholders have certain minimum rights in an organisation. These must not be violated and should be met regardless of their power. Therefore, a business should be managed in the interests of *all* stakeholders, not just shareholders (Hendry, 2001). In contrast, the managerial perspective of stakeholder theory argues that organisations place more emphasis on the information needs of those stakeholders who dominate organisational survival, as those stakeholders have more influence than others (Friedman & Miles, 2002). Following this concept, whether a particular stakeholder receives information will depend on the power that stakeholder is perceived to have.

Stakeholder theory provides strong explanatory power in CSR reporting and assurance literature. Early research on voluntary assurance suggests that the adoption of voluntary assurance is motivated not only by shareholders, but also by other stakeholders (Chow, 1982). Gray et al. (1995) note that corporate social and environmental disclosure is considered an effective dialogue between an organisation and its stakeholders. Through such disclosure, organisations seek support and approval from stakeholders for continued existence. In other words, corporate disclosure legitimises the company's activities to stakeholders, given their diverse and various expectations; this corroborates with legitimacy theory. Organisations also manage legitimacy by signaling to stakeholders that their behaviour is appropriate and desirable (Suchman, 1995); this is compatible with signaling theory.

For instance, drawing on stakeholder theory, Michelon and Parbonetti (2012) find that board composition—measured as the proportion of influential members—positively effects sustainability, environmental and strategic disclosure. Kolk and Perego (2010) also document that companies located in countries with stakeholder-oriented legal systems and higher pressures regarding corporate sustainability due to public policy, are more likely to issue sustainability reports and get these reports assured.

2.2 Legitimacy Theory

Legitimacy theory focuses on social acceptance to ensure a company's existence and survival. Accordingly, legitimacy is defined by Dowling and PfeVer (1975) as:

> A condition or a status which exists when an entity's value system is congruent with the value system of the larger social system of which the entity is a part. When a disparity, actual or potential, exists between the two value systems, there is a threat to the entity's legitimacy (p. 122).

Thus, an organisation's legitimacy can be considered as a 'social contract' existing between a company and the whole community in which it operates (Mathews, 1993; Patten, 1992). An organisation's existence is threatened whenever society believes there is a violation of that social contract. Where social members are not satisfied that an organisation is operating in an acceptable or legitimate manner, then society will effectively revoke the organisation's 'contract' to continue its operations (Deegan, 2002). Consequently, other than material resources and technology that a company pursues to survive and thrive, it also needs social acceptability and credibility that conforms to the framework of society's norms and values (Scott, 2001).

In the context of CSR assurance, Deegan (2002) defines a CSR assurance as "a process that enables an organization to assess its performance in relation to society's requirements and expectations" (p. 289). CSR assurance can either be a managerial device taking various social pressures away from an organisation, or a strategy undertaken for accountability, which explains the various social impacts on an organisation. Deegan (2002) uses the international sportswear company Nike as an example to demonstrate that CSR assurance implemented with the assistance of the Global Alliance for Workers and Communities is an effective approach in response to community suspicion and concerns about Nike's labour practices. Simnett, Vanstraelen, and Chua (2009) analyse a sample of international firms participating in assurance practices, from sustainability reports between 2002 and 2004. They found that mining, utilities, production and finance industry companies (which are more exposed to environmental and social risks and need to increase user confidence in sustainability reports credibility) are more likely to engage in sustainability report assurance activities. Deegan (2002) and Milne and Patten (2002) identify several issues in relation to legitimacy theory. For instance, do legitimising activities actually change the social perceptions of an organisation? Is any particular party more affected by legitimising activities than others? How do managers understand social concerns and the 'social contract'? Despite such limitations, it is generally accepted that legitimacy theory extends stakeholder theory and provides useful insights to understanding corporate social and environmental activities and CSR assurance.

2.3 Signalling Theory

Signalling theory is also relevant to stakeholder theory in the CSR literature. Signalling theory assumes that one party voluntarily discloses information to another to reduce information asymmetry (Spence, 2002). When two parties have access to different information, the sender must choose whether and how to communicate the information and the receiver must choose how to interpret the signal (Connelly, Certo, Ireland, & Reutzel, 2011). This situation is particularly prevalent in relations between a company's management (insiders) and external stakeholders (outsiders), due to the agency relationship. Dye (1985) and Skinner (1994) argued that managers have incentives to voluntarily disclose not only good news (to enhance corporate image and financial return), but also bad news, as signalling bad news can avoid or reduce reputational costs and litigation, as well as other unfavourable consequences.

Signaling theory has been used in numerous accounting studies to explain organisations' disclosure and assurance practices. Mitchell (2006) and Kanagaretnam, Lobo, and Whalen (2007) showed high quality firms will advise market participants about quality. Similarly, Cong and Freedman (2011) show that firms with good corporate governance systems, who provide good accountability for their activities, will likely provide more extensive disclosures than firms with poor accountability. Directly related to CSR assurance (and based on a dataset of 148 Australian listed companies releasing environmental disclosures, with 74 voluntary CSR assurances, covering 2003–2007), Moroney, Windsor, and Aw (2012) found that the disclosure quality scores for assured companies were significantly higher than unassured companies. This indicated that companies used CSR assurances to signal commitments to social and environmental CSR disclosure. Li and Li (2012) analysed 940 Chinese listed firms participating in CSR report assurance activities from 2009 to 2010. They found that firms who engaged in CSR assurance with an independent third party experienced higher abnormal returns. This suggested that companies intend to informing stock markets about their high quality CSR information, and reacting positively to these signals. Li, Guan, and Li (2013) further demonstrated that companies with negative social or environmental events chose not to attest their CSR reports, avoiding signalling negative effects to the public.

3 CSR Assurance Regulations in China

Generally speaking, there are five CSR assurance standards frequently used in China, including three international standards and two Chinese standards. Those standards are summarised in Table 1.

Table 1 Standards and guidelines on CSR assurance

	Standards	Standard setter	Characteristics in relation to CSR assurance
International Standards	AA1000AS	AccountAbility	• Stakeholder-focused
			• Specifically designed for CSR assurances
			• Core principles: materiality, inclusivity and responsiveness
			• Assurance on principles only without verification of performance information, and assurance on principles and performance information are both allowed
			• Observations and recommendations are included in the assurance statement
	ISAE3000	International Auditing and Assurance Standard Board (IAASB)	• Not stakeholder focused
			• Limited assurance is more reasonable due to the complex nature of CSR reports
			• Emphasising auditor competence and explicitly foreseeing the use of experts
			• Procedures including assurance planning, risk assessment, the assurance evidence collection and evaluation procedures are all followed; analytical procedures are emphasised
			• Endorsed by Federation of European Accountants (FEE).
	GRI	Global Reporting Initiative	Recommendations for reporting entities in their approach to external assurances of sustainability reports
Chinese Standards	CAS3101	China Institute of Certified Public Accountants (CICPA)	An accounting standard relevant to CSR assurance in China. It is similar to the international standard ISAE3000
	CSR-VRAI	China National Textile and Apparel Council (CNTAC)	The industrial verification standard in China

3.1 International Standards and Guidelines on CSR Assurance

Three international standards used in CSR assurance around the world are: *AA1000 Assurance Standard* (AA1000AS), launched by AccountAbility (2008),[3] *International Standard on Assurance Engagements* (ISAE3000), issued by the IAASB (2013) and *Sustainability Reporting Guidelines*, issued by the Global Reporting Initiative (2006). These three standards constitute the basic international framework on CSR assurance and are either individually or collectively referenced by assurance providers (Kolk & Perego, 2010).

Specifically, AA1000AS is stakeholder-oriented with the following characteristics:

- It focuses on what is material to the organisation and its stakeholders, and includes stakeholders in addition to the organisation itself at the core of assurance engagement.
- It is designed for CSR assurance, aiming to enhance reliability and the reporting quality of CSR disclosure.
- Materiality, inclusivity and responsiveness are core principles in the AA1000AS. Based on these three principles, AA1000AS provides two optional types of assurance engagement. The first type only provides assurance about the core principles, without verifying the reliability of the performance data: conclusions regarding the underlying data's reliability are not issued. The second type offers assurance on both the principles and performance information. Therefore, verification of data is essential in the second type, and the assurance provider will make a conclusion for information reliability in the assurance report.
- Observations and/or recommendations are included in the assurance statement as a minimum requirement. This implies that assurance providers' work is expected to go beyond pure assurance engagement, to include professional consultations on CSR activities and disclosure.

In contrast, ISAE3000 (IAASB, 2013) addresses the external assurance services of non-financial reports, but is not specifically designed for CSR assurance engagements. Compared with AA1000AS, ISAE3000 focuses on assurance procedures and technical criteria. For example:

- ISAE3000 does not give a definite requirement regarding the level of assurance chosen for reasonable assurance and limited assurance. However, a sustainability report is a complex subject matter disclosure that combines quantitative

---

[3] AccountAbility is an independent, global, not-for-profit organisation promoting accountability, sustainable business practices and corporate responsibility. It implemented the AA1000 series of standards on sustainability reporting, including AA1000APS (the AA1000 AccountAbility Principles Standard), AA1000AS (the AA1000 Assurance Standard) and AA1000SES (the AA1000 Stakeholder Engagement Standard).

information with qualitative elements, implying that the process is difficult to formalise. Therefore, empirical studies indicate that a CSR assurance engagement based on ISAE3000 might not provide high-level assurance on CSR reports (Manetti & Becatti, 2009). This is particularly comparable with the high level and moderate level of assurances on CSR reports, as specified in AA1000AS.

- Although both AA1000AS and ISAE3000 mention the competence and independence of assurance providers, compared with the openness and flexibility of AA1000AS, ISAE3000 recognises that auditors might lack professional competence in conducting CSR assurance. As such, it explicitly foresees the possibility of using experts who can give evidence for the elements of greater weakness and subjectivity in the reporting process (Manetti & Becatti, 2009).
- Engagements adopting ISAE3000 should follow the basic planning, risk assessment and assurance evidence collection and evaluation procedures proven effective by professional accounting in auditing and assurance services. Comparatively, AA1000AS does not stress risk assessment, nor assurance procedures and evidence.

Due to differences in the characteristics and assurance criteria between AA1000AS and ISAE3000, it is argued that international accounting firms are conservative and technical when using ISAE3000 as assurance criteria. However, other international certification bodies are more likely to adopt AA1000AS, to provide more flexible certification services containing more relevant consultation components. This is clarified in the discussion paper, 'Providing Assurance on Sustainability Reports' published by the Federation of European Accountants (2002). This report lists various certification criteria adopted internationally and employs ISAE3000 as the underlying assurance standards for discussion.

Global Reporting Initiatives (GRI) implemented the first sustainability reporting guidelines (G1) in 2000, revised as G2 in 2002, and G3 in 2006. In 2013, the new G4 became available; this is the most updated CSR reporting framework. According to KPMG (2013), use of GRI guidelines is almost universal. From 2011 to 2013, the rate of N100 companies using GRI rose from 69 to 78 %. The world's G250 largest companies increased this rate from 78 % in 2011 to 82 % in 2013. GRI guidelines provide detailed and 'do-able' principles and guidance on CSR reporting frameworks.

The GRI framework focuses on: (1) how to report; and (2) what to report. Part 1 addresses reporting principles and guidance: reporting principles are disaggregated into principles for defining report content and principles for defining quality. Reporting principles are made applicable by referring to the guidelines and suggested tests in 'Reporting Guidance'. Part 2 clarifies the base content that should appear in a sustainability report. The format of each type of disclosure is outlined. Performance indicators are further defined as: economic, environmental, labour practices and decent work, human rights, society, and product responsibility. 'General Reporting Notes' follows Part 2, providing further clarification on data gathering, reporting form and frequency, and assurance considerations. Fig. 1 presents the basic framework of G3.

Part 1. Defining Reporting Content, Quality and Boundary

Reporting principles		Reporting guidance	
1.1 Principles for defining report content • Materiality • Stakeholder inclusiveness • Sustainability • Context • Completeness		Guidance	Tests
1.2 Reporting principles for defining quality • Balance • Comparability • Accuracy • Timeliness • Clarity • Reliability		Tests	
1.3 Reporting guidance for boundary setting • Control • Significant influence		Guidance	Decision tree

Part 2. Standard Disclosure[a]

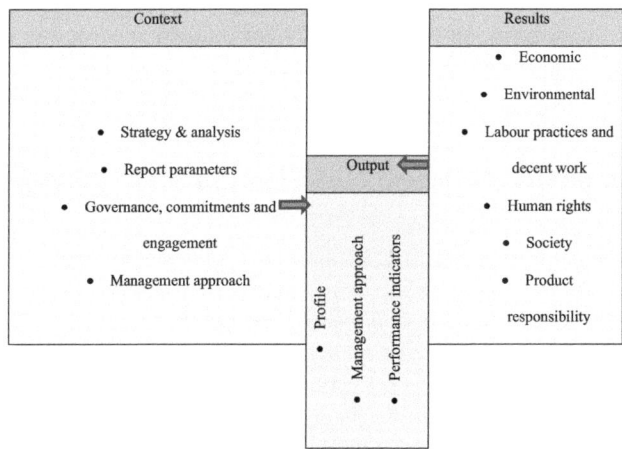

[a]The overview in Part 2 is adapted from Figure 7, G3

General Reporting Notes

- Data gathering
- Report form and frequency
- Assurance

Fig. 1 Overview of GRI Guidelines (G3)

Although GRI guidelines are regarded as the most comprehensive and applicable CSR reporting framework, they are actually guidelines on CSR reporting, rather than an assurance standard. However, assurance providers often make reference to GRI guidelines when conducting CSR assurance for an entity whose CSR report is prepared in accordance with GRI. Specific to assurance, G3 has a separate section titled 'Assurance' (Global Reporting Initiative, 2006, p. 38), which emphasises that:

- GRI recommends the use of external assurance.
- Assurance reports using GRI reporting frameworks are conducted by groups or individuals external to the organisation, who are demonstrably competent in both the subject matter and assurance practices.
- It is implemented in a manner that is systematic, documented, evidence-based, and characterised by defined procedures.
- It assesses whether the report provides a reasonable and balanced presentation of performance, considering the veracity of data in a report, as well as the overall content selection.
- It uses groups or individuals (to conduct the assurance) who are not unduly limited by their relationship with the organisation or its stakeholders, to reach and publish an independent and impartial conclusion on the report.
- It assesses the extent to which the report preparer has applied the GRI Reporting Framework (including the reporting principles) in the course of reaching its conclusions.
- It requires an opinion or a set of conclusions, publicly available in written form, and a statement from the assurance provider on their relationship to the report preparer.

3.2 Chinese Standards and Guidelines on CSR Assurance

Two local standards relevant to CSR assurance practice in China include: *Standards on Assurance Engagement Other than Engagement to Audit or Review Historical Financial Information–Standards on Other Assurance Engagements of Certified Public Accountants of China No. 3101 (CAS3101)*. This document is issued by the China Institute of Certified Public Accountants (CICPA) under Ministry of Finance.[4] *Sustainability Reporting Verification Rules and Instructions (CSR-VRAI)* is issued by CNTAC (2008).

Similar to ISAE3000 in structure and principles, the main topics covered in CAS3101 (Ministry of Finance, 2006) are:

- engagement acceptance
- assurance planning and performing the engagement

---

[4] Similar to ISAE3000, CAS3101 is not specifically designed for assurance services on CSR reports, but focuses more on assurance procedures and technical details.

- use of experts
- evidence
- subsequent events
- assurance documentation
- preparing the assurance report
- other reporting responsibilities.

Several clauses in Chapter Eight of CAS3101, 'Preparing the Assurance Report' (Ministry of Finance, 2006, p. 7), indicate certain issues especially relevant to CSR assurance engagements:

1. Clause 55: *The addressee of the assurance report is the party the assurance report should report to. In normal cases, the addressee of the assurance report should be ALL the expected users of the report.*
   Legally, the parties whom the assurance report normally addresses are considered privy to the assurance agreement; the assurance providers have legal liabilities to them in case of negligence resulting in measurable damages. CSR assurance responds to a large group of stakeholders. In providing CSR assurance services, the assurance provider will consider the major stakeholders' expectations and interests, but will never be able to accommodate the expectations and interests of *all* stakeholders. Although there has not been any legal dispute with CSR assurance in China, the potential litigation risk should be considered by assurance providers. This is especially crucial to accounting firms, which are assumed to have 'deep pockets' with their professional indemnity insurance coverage. Therefore, professional accounting firms should be conservative in the wording and level of assurance provided in their reports.
2. Clause 57: *The assurance report should reference to the criteria used in evaluating or measuring the subject matter, therefore to assist expected users understanding the basis of the conclusion.*
   A CSR assurance can make reference to GRI or other international CSR reporting frameworks as the criteria. Alternatively, it can use local regulations, standards and guidelines as criteria. However, a general concern regarding criteria is that the majority of CSR reporting guidelines are qualitative; local regulations and standards are usually very brief, without applicable guidance.
3. Clause 64: In a limited assurance engagement, the following content should be presented in the overview of work performed:

   - Limitations of the nature, timing and scope in collecting evidence, and, if necessary, specifying the procedures not being performed in the engagement, while being required for an engagement with reasonable levels of assurance.
   - Stating that due to the limitation in assurance procedures and evidence, compared to reasonable levels of assurance engagement, the level of assurance in a limited assurance engagement is lower than the assurance level provided by a reasonable assurance engagement.

Under CAS3101, and due to the complex nature of social and environmental activities and measurement, it would be difficult for assurance providers to collect sufficient appropriate evidence to reach an opinion with a reasonable level of assurance. Together with considerations regarding legal liabilities to a variety of stakeholder groups, the qualitative nature and the lack of applicable guidance in the criteria, accounting firms who use CAS3101 as their assurance standard may enter a limited-level assurance engagement rather than an engagement with a reasonable level of assurance.

CSR-VRAI (CNTAC, 2008), the CSR assurance guidelines for the textile and apparel industry, is comparable to AA1000AS (AccountAbility, 2008) and covers 'Assurance Scope', 'Assurance Principles', 'Assurance Procedures', and 'Assurance Conclusion'. However, its use is restricted to those particular industries, and is not applicable to other industries. In addition, the CSR-VRAI is still voluntary, even though it is comparatively comprehensive and complete.

4 Demand-and-Supply Analyses on CSR Assurance in China

4.1 Research Method and Sample

We adopt a demand-and-supply analytical framework to explore the research questions regarding the current situation of CSR assurance in China. With demand, we attempt to answer two questions: (1) who are the stakeholders and how are they specified in the existing regulations and standards on CSR reporting and assurance in China? To answer this question, we employ a content analysis method to investigate descriptions in international and Chinese regulations and standards on stakeholders (see Table 1); (2) what factors might contribute to the voluntary assurance of CSR reports? A series of descriptive statistics tables are used to facilitate our understanding of this question from various aspects. With supply, we present the current CSR assurance market shares of two streams of assurance service providers: accounting firms and certification bodies. Accounting firms are further categorised as 'Big 4' and 'non-Big 4' accounting firms, and certification bodies are further analysed as international and local certification bodies.

The sample used in this paper consists of 2,064 CSR reports disclosed by listed companies from 2009 to 2013 in China. We chose listed companies as the study sample as the financial and CSR information of listed companies are publicly available in the CSMAR database. In addition, the Shanghai and Shenzhen Stock Exchanges have compulsory CSR reporting requirements to Shanghai Stock Exchange Corporate Governance Index companies, Shenzhen 100 Index companies, companies in the finance sector, and companies with overseas share issues. In contrast, there are no mandatory requirements for the two stock exchanges regarding CSR assurance. A large base of CSR reports and voluntary CSR assurance options existed in the sample, providing a setting to study the demand-and-supply of voluntary CSR assurance. The first year of disclosure in the sample is 2009; this was also the first year in which CSR information was disclosed in the CSMAR database.[5]

4.2 Demand of CSR Assurance in China

4.2.1 Stakeholders in CSR assurance

> Sustainability reporting is the practice of measuring, disclosing, and being accountable to internal and external stakeholders for organizational performance towards the goal of sustainable development (GRI, 2006, p. 3).
>
> Transparency about the sustainability of organizational activities is of interest to a diverse range of stakeholders, including business, labour, non-governmental organizations, investors, accountancy, and others (GRI, 2006, p. 2).
>
> Credibility is a prerequisite for effective sustainability reporting. Credibility can be considerably enhanced through independent external assurance, using accepted professional standards. Reporting organizations and their stakeholders increasingly accept that robust independent external assurance is a key way of increasing the credibility and effectiveness of their reporting, and ultimately their performance (AccountAbility, 2008, p. 6).

Due to the stakeholder-oriented nature of CSR, stakeholders are highlighted in most standards and regulations for CSR reporting and assurance. For example, G3 and AA1000AS both adopt a multi-stakeholder process. Expectations and opinions from a wide range of stakeholders are consulted and agreed upon in developing the guidelines. In particular, G3 includes 'Stakeholder Inclusiveness' as one of the four reporting principles, and requires the reporting organisation to identify its stakeholders and explain how it has responded to reasonable expectations and interests (GRI, p. 10). AA1000AS also emphasises that the target users of assurance reports are stakeholders. It stresses that for a high level of assurance, the assurance provider should also seek more extensive evidence in all areas, including through direct engagement with stakeholders (AccountAbility, 2008, p. 20).

In China, the regulations and guidelines on CSR reporting and assurance often classify stakeholders with human-oriented factors. 'Human-oriented' is a special Chinese terminology equivalent to 'stakeholder-oriented' in relation to human and environmental factors, regardless of any overlap between the two classes. Three lines of regulations comprehend detailed stakeholder descriptions, including:

- **Laws and regulations on environmental protection (e.g., Environmental Information Disclosure Act 2007)**, that provide detailed description of material

---

[5] As CSR information on listed companies in CSMAR is only available since 2009 (to cover the 2008 financial year), we have been unable to include prior year CSR reports and assurance data in this analysis. This is despite an early CSR report, and the first CSR assurance in China being issued in 2006 for China Ocean Shipping Corporation (COSCO).

environmental risks, expected management responses and CSR disclosure, to address the needs and expectations of stakeholders who focus on environmental issues.
- **Shenzhen Stock Exchange Social Responsibility Guideline to Listed Companies** (Shenzhen Stock Exchange, 2006), in which various stakeholder groups (e.g., shareholders and creditors, employees, suppliers, customers and consumers, environment and natural resources and the community) are specifically identified. Expected management strategies and social responsibility disclosures are articulated to each group.
- **The series of CSR reporting and verification guidelines implemented by CNTAC, including CSC9000T (2005), CSR-GATEs (2008) and CSR-VRAI (2008)**, provide comprehensive guidance on how to prepare a CSR report and conduct CSR assurance with a multi-stakeholder process, comparable to GRI guidelines. The 121 CSR performance indicators are divided into five groups, each addressing one stakeholder class (i.e., product safety and consumer protection, employee interests, natural resources and environment, supply chain management and fair market, community and social wellbeing).

In contrast, possibly due to conservatism and legal liability concerns, assurance standards prepared by professional accounting bodies barely address or mention stakeholders. For example, in the explanation paragraph to intended users A16 in ISAE3000 (IAASB 2013), ISAE3000 admits that "in some cases there may be intended users other than those to whom the assurance report is addressed" (p. 29). However, ISAE 3000 includes a disclaimer explanation that:

> The practitioner may not be able to identify all those who will read the assurance report, particularly where a large number of people have access to it. In such cases, particularly where possible users are likely to have a broad range of interests in the underlying subject matter, intended users may be limited to major stakeholders with significant and common interests (IAASB, 2013, p. 29).

Similarly, in the Chinese standard CAS3101, Section 55 states: "the addressee of the assurance report shall be the intended users of the assured information" (Ministry Of Finance (MOF), 2006, p. 8). There is no detailed explanation for intended users. These oversights represent a major weakness of accounting profession-initiated CSR assurance standards.

4.3 Factors Relevant to Voluntary CSR Assurance

4.3.1 The Trend of CSR Assurance

Table 2 presents the trend of CSR reports and assurance in Chinese companies listed in the Shanghai or Shenzhen stock exchanges from 2009 to 2013.

Increasing CSR reporting requirements from the stock exchanges and government agencies reflect greater expectations in the marketplace regarding CSR reporting and assurance. Driven by intensive regulatory intervention from

Table 2 CSR reporting and CSR assurance by year

Year	Listed companies	CSR report (percentage to total annual report)	CSR assurance (percentage to CSR report)
2009	1,817	178 (9.80 %)	4 (2.25 %)
2010	1,944	186 (9.57 %)	4 (2.15 %)
2011	2,326	497 (21.37 %)	19 (3.82 %)
2012	2,570	588 (22.88 %)	23 (3.91 %)
2013	2,668	615 (23.05 %)	33 (5.37 %)
Total	11,325	2,064 (18.23 %)	83 (4.02 %)

Table 3 The number of CSR assurance by company size and by year[a]

Company size	Q1	Q2	Q3	Q4	Total
2009	1	1	0	2	4
2010	0	0	0	4	4
2011	0	0	3	16	19
2012	1	0	4	18	23
2013	1	2	7	23	33
Total	3	3	14	63	83

[a]Companies are divided into four quartiles equally based on total assets. Q1: Smallest firms with total assets less than ¥2,588,577,717; Q2: Medium small firms with total assets between ¥2,588,577,717 and ¥7,013,443,421; Q3: Medium large firms with total assets between ¥7,013,443,421 and ¥22,315,878,989; and Q4: Largest firms with total asset more than ¥22,315,878,989

government agencies and stock exchange regulators in 2008 (see Table 1), Table 2 Panel A shows that since 2009, there has been exceptional growth in CSR reporting in China among listed companies. Reporting has surged from 178 to 615 CSR reports, synchronised with the international trend of increased CSR reporting, as reflected by KPMG (2013, Fig. 7). Until 2013, the ratio of CSR reports, compared with the total of annual reports, reached 23.05 %. This suggests that CSR reporting is becoming a mainstream business practice in listed Chinese companies. Meanwhile, the number of CSR assurances has also dramatically increased during 2009–2013. Specifically, only four CSR reports were assured in 2009 and 2010. This number rises to 19 in 2011, reaching 33 in 2013, indicating the prevalence of CSR reporting and assurance practices among Chinese firms.

In Table 3, we further split all CSR disclosed companies equally into four quartiles, based on total assets. Clearly, the majority of CSR assurances (63) are undertaken in large companies (75.90 %); only three assurances were performed among small and medium small Chinese firms (3.61 %) across this five-year period. In addition, the increased trend of CSR assurance observed in Table 2 is dominated by the largest firms. Specifically, the number of CSR assurances conducted by the largest Chinese firms has grown substantially from two in 2009 to 23 in 2013; however, participation rates by small and medium small firms have remained fairly static (Table 3).

4.4 Geographical and Industrial Characteristics of CSR Assurance

Prior literature has documented that a regional environment can influence companies in making CSR assurance decisions (Li et al., 2013; Perego, 2009). China is known for its unbalanced market development across the country. Generally, some areas, such as Beijing, Shanghai, Guangdong and Zhejiang provinces, have a better developed market together with higher levels of trust among market participants. It is expected that CSR assurance practice will vary among different provinces. Table 4 presents the picture of CSR assurance in China, according to economic zone.[6]

East China, regarded as the most developed area with an advanced economy, legal and social systems, accommodates 92.77 % of CSR reports with an independent assurance. It also features the highest percentage of CSR assurance compared with the number of CSR reports (5.62 %), in which Beijing and Shanghai are ranked as No. 1 and 2 respectively, in terms of the percentage of CSR reports assured. In West, Central and Northeast China, only small percentages of listed companies have CSR reports assured, except for Shanxi, China's 'Capital of Coal', which has 8.11 % of assured CSR reports, and ranks third among all provinces. This picture is further endorsed by Table 4, showing the distribution of CSR assurance by industry. Mining ranks third in the ratio of CSR reports assured, reflecting increased concerns about environmental contamination and work safety issues relevant to this particular area.

The distribution of CSR assurance by industry corresponds to the diversity of regulatory requirement, the size of companies in the industry and social demand for CSR information reliability:

- **The finance and insurance industry** has the highest ratio of CSR assurance, with more than half of CSR reports attested by independent third parties. The China Banking Association and Shanghai Stock Exchange requests that banks and overseas listed companies submit CSR reports. During past decades, financial institutions in China have actively participated in the global financial market. Major Chinese banks and insurance companies—for example, China Industrial and Commercial Bank, Bank of China, China Construction Bank, and China People's Insurance Corporation—have listed in overseas stock exchanges. This internationalisation has pushed China's financial institutions to enhance disclosure quality by providing more comprehensive and reliable CSR information to meet stakeholders' expectations, both locally and internationally.
- **The aviation industry** has the second-highest CSR assurance ratio. Over the years, the global aviation industry has been affected severely by rising fuel prices, the global financial crisis and harsh business environments. Aviation

---

[6] The economic zones are classified based on the 11th Five-Year Plan for National Economic and Social Development Part 5 (2006).

Table 4 CSR assurance in China by economic zone

Economic zone	Province	CSR	CSR assurance	CSR assurance (percentage to CSR report)
East China	**Beijing**	265	36	**13.58 %**
	Hebei	32	0	0.00 %
	Tianjin	45	1	**2.22 %**
	Shandong	103	2	1.94 %
	Jiangsu	110	3	2.73 %
	Shanghai	155	17	**10.97 %**
	Zhejiang	157	4	**2.55 %**
	Fujian	203	0	0.00 %
	Guangdong	285	14	**4.91 %**
	Hainan	14	0	0.00 %
	Sub-total	**1,369**	**77**	**5.62 %**
West China	Chongqing	20	0	0.00 %
	Sichuan	70	1	1.43 %
	Yunnan	51	0	0.00 %
	Guizhou	17	0	0.00 %
	Shaanxi	26	0	0.00 %
	Qinghai	14	0	0.00 %
	Gansu	6	0	0.00 %
	Ningxia	15	0	0.00 %
	Xinjiang	35	0	0.00 %
	Xizang	8	0	0.00 %
	Guangxi	22	0	0.00 %
	Inner-Mongolia	13	0	0.00 %
	Sub-total	**297**	**1**	**0.34 %**
Central China	Henan	82	1	1.22 %
	Shanxi	37	3	**8.11 %**
	Anhui	70	0	0.00 %
	Jiangxi	23	0	0.00 %
	Hubei	55	0	0.00 %
	Hunan	41	0	0.00 %
	Sub-total	**308**	**4**	**1.30 %**
Northeast China	**Jilin**	32	1	**3.13 %**
	Liaoning	46	0	0.00 %
	Heilongjiang	12	0	0.00 %
	Sub-total	**90**	**1**	**1.11 %**
Total		**2,064**	**83**	**4.02 %**

safety and on-board and off-board services to customers have become critical to survival in this industry. With this background, large aviation companies made great efforts in CSR performance, as a public relations and corporate governance tool, to influence the market and customers.

- **The mining industry** ranks third in the list. The Mining Industrial Association in China has invested great effort in CSR performance and disclosure, organising CSR ratings within the industry and holding annual CSR release conferences for the public. Although we are unaware of any compulsory CSR disclosure requirements in the mining industry, a series of laws and regulations regarding work safety have been implemented during the last few years, accompanied by harsh legal liabilities and financial penalties.

In summary, Tables 4 and 5 deliver an important message for regulators and academics in CSR disclosure and assurance. When companies decide to invest in CSR disclosure and assurance, they examine the regulatory requirements, industrial agreements and social pressures from stakeholders and markets, as well as how a CSR investment fits the company's business strategy.

Table 5 CSR assurance by industry[a]

Industry	CSR	CSR assurance	Percentage to CSR reports
Finance and insurance	67	36	53.73 %
Aviation	14	5	35.71 %
Mining	46	9	19.57 %
Petroleum and natural gas exploitation	10	1	10.00 %
Retailing, business and services	73	7	9.59 %
Civil engineering	37	3	8.11 %
Beverage	38	2	5.26 %
Transportation	66	4	6.06 %
Textile and apparel	52	2	3.85 %
Utility	83	3	3.61 %
Manufacturing	267	5	1.87 %
Pharmaceuticals	108	2	1.85 %
Real estate	131	2	1.53 %
Non-ferrous metal metallurgy	85	1	1.18 %
Others	49	1	2.04 %
Industries without CSR assurance	938	0	0.00 %
Total	2,064	83	4.02 %

[a]Only industries with CSR assurance are separately presented in this table. Those industries without CSR assurance are included in the category 'industries without CSR assurance'

4.5 Disclosure Content

Companies with CSR assurance prioritise stakeholders with greater influence over their business success and regulation compliance. Stakeholders with less business and regulation impact (e.g., suppliers and work safety) have minimal disclosure cover. Similarly, CSR system construction and business deficiency are not well covered. This coincides with the scatter of CSR report content coverage. Table 6 presents the areas disclosed in CSR reports and assurance.

As discussed in Sect. 3.1, GRI is the leading reporting framework, but it is not always referenced in Chinese CSR reports. According to KPMG (2013), close to 80 % of N100 companies and G250 companies adopt the GRI reporting framework. In contrast, among 2,064 sampled Chinese CSR reports, only 325 reports (16 %) comply with GRI. However, when GRI is used as an assurance reference, the company is more likely to have its CSR disclosure assured (21.54 %). Note that in Table 6, all 83 CSR reports with independent assurance have addressed shareholder protection, employee interests, customer protection, environment, and public relations in their CSR disclosure.

Table 6 Disclosure content in CSR reports and assurance

Disclosure content	CSR reporting	CSR assurance	Percentage to CSR reports
GRI	325	70	21.54 %
Shareholder protection	2,021	83	4.11 %
Employee interests	2,053	83	4.04 %
Customer protection	1,989	83	4.17 %
Environment	2,036	83	4.08 %
Public relation	1,993	83	4.16 %
Supplier protection	1,477	68	4.60 %
Work safety	1,634	57	3.49 %
Creditor protection	1,217	21	1.73 %
CSR system construction	740	18	2.43 %
Deficiency in business	470	2	0.43 %

4.6 Supply: The CSR Assurance Service Providers and Adopted CSR Assurance Standards in China

Clearly, of 83 CSR assurance reports, 42 (50.6 %) were issued by public accounting firms, and 41 (49.4 %) were from other attestation providers.[7] In 2009, international certification specialists issued three CSR assurance reports: Bureau Verltas (1) and Det Norske Veritas (2), and only one assurance report was conducted by an accounting firm, KPMG. In 2013, this phenomenon changed slightly. Of the 33 CSR assurance reports, 18 (54.55 %) were issued by accounting firms and 15 (45.45 %) by other attestation providers. Table 7 displays the distribution of CSR assurance providers in China.

By comparing Big 4 and non-Big 4 accounting firms, PWC and KPMG seem to lead the CSR assurance market in China, totaling about 50 % of market share. EY and Deloitte have also participated in CSR assurance practice; however, they have seen very little change in their market share over the period 2009–2013. In contrast, non-Big 4 firms have experienced significant growth in market share, more than tripling their share over this period.

Table 7 CSR assurance providers

Public accounting firms	PWC	KPMG	Ernst & Young	Deloitte	Non-Big Four	Total
2009	0	1	0	0	0	**1**
2010	0	1	1	0	0	**2**
2011	1	4	2	1	2	**10**
2012	3	2	2	1	3	**11**
2013	6	2	2	2	6	**18**
Sub-total	**10**	**10**	**7**	**4**	**11**	**42**

Assurance providers other than accounting firms	Bureau Verltas	Det Norske Veritas	Other foreign certification firms	Other Chinese Certification firms	Total
2009	1	2	0	0	**3**
2010	1	1	0	0	**2**
2011	5	1	3	0	**9**
2012	5	1	3	3	**12**
2013	5	2	4	4	**15**
Sub-total	**17**	**7**	**10**	**7**	**41**

---

[7] The literature generally reveals that large international companies tend to use large accounting firms in their CSR assurance. For example, KPMG (2013) found that in 2013, of the 1,099 N100 companies with CSR disclosure and assurance, 67 % used a professional accounting firm as the assurance provider; this was 64 % in 2011.

Table 8 CSR assurance standards used by assurance providers in China[a]

Assurance provider		CSR assurance standards
Accounting firms	International firms	ISAE3000
	Chinese local firms	CAS3101
Certification bodies	Bureau Verltas	AA1000AS, G3, ISAE3000
	Det Norske Veritas	AA1000AS and G3
	Others	AA1000AS and/or G3

[a]Source: Shen et al. (2011) and Li et al. (2013)

Unlike assurance on financial information, which is always conducted by accounting firms, voluntary CSR assurance can also be conducted by certification bodies. By dividing the non-accounting assurance providers into two subgroups—foreign certification providers and Chinese certification firms—only seven reports were assured by local certification firms. This suggests that Chinese local certification firms are still disadvantaged in the CSR assurance market, probably due to a lack of expertise. Two international CSR certification specialists—Bureau Verltas and Det Norske Veritas—dominate the Chinese market. Bureau Verltas leads with 17 assurance reports; Det Norske Veritas follows with seven CSR assurance services.

Table 8 presents a general picture of the assurance standards used by CSR assurance providers in China. International accounting firms usually adopt ISAE3000, and local Chinese firms tend to choose CAS3101. As discussed previously in Sect. 3.1, both ISAE 3000 and CAS3101 are assurance standards specifically used by professional accounting firms with emphasis on assurance techniques and procedures. Meanwhile, certification firms tend to use AA1000AS, as well as G3, as the major assurance criteria. Noting that the focus of standards initiated by the accounting profession (or by institutions specialising in CSR reporting) is different (see Sect. 3.1), the expected coverage and focus of the reports provided by these two streams of providers may also vary. However, the difference between the standards adopted by accounting firms and certification bodies has been obscured during recent years, as evidenced by Bureau Verltas using ISAE3000 in addition to AA1000AS and G3as the CSR assurance criteria.

5 Conclusion

In conclusion, this chapter has provided a comprehensive understanding of the theory, regulations, practice and challenges for CSR assurance in China. By adopting a demand-and-supply analytical framework, the descriptive statistics indicated that government agencies, stock exchanges, accounting standard setters and industrial associations have collectively shaped the current regulatory framework on CSR reporting and assurance in China. Regarding demand, the differences in social and legal environments across China have influenced regional development of CSR assurance. Industries under intensive CSR regulations, and/or social reporting pressures, have CSR reports assured more actively. Regarding supply, the CSR assurance market in China is shared by accounting firms and professional certification bodies. Different assurance standards adopted by these two streams of providers have different foci, potentially leading to different assurance coverage and emphases.

While the prior literature has provided some empirical results in relation to CSR assurance practice in China,[8] scholarly research on CSR assurance is still scant. Areas of future research may include: CSR reporting quality associated with CSR assurance, differences in reporting contents and assurance practices among different CSR assurance standards in China, a comprehensive analysis to different types of assurance providers in China, and the push-and-pull forces of CSR assurance in the Chinese market. We hope our study has built upon the existing literature, providing information for academics and practitioners in better understating CSR reporting and assurance in China, and attracting more discussion and research. We also call for further research in this potentially rewarding area.

References

AccountAbility. (2008). *AA1000 Assurance Standard 2008*. Retrieved from http://www.accountability.org/standards/

China National Textile and Apparel Council (CNTAC). (2008). *China sustainability reporting verification rules and instructions* (CSR-VRAI). Retrieved from http://images.mofcom.gov.cn/csr/accessory/201008/1281060972563.pdf

Chow, C. W. (1982). The demand for external auditing: Size, debt and ownership influences. *The Accounting Review, 57*(2), 272–291.

Cong, Y., & Freedman, M. (2011). Corporate governance and environmental performance and disclosure. *Advances in Accounting, 27*(2), 223–232.

Connelly, B. L., Certo, S. T., Ireland, R. D., & Reutzel, C. R. (2011). Signaling theory: A review and assessment. *Journal of Management, 37*(1), 39–67.

Deegan, C. (2000). *Financial accounting theory*. Australia: McGraw-Hill.

Deegan, C. (2002). Introduction: The legitimising effect of social and environmental disclosures—A theoretical foundation, accounting. *Auditing and Accountability Journal, 15*(3), 282–311.

Dowling, J., & PfeVer, J. (1975). Organizational legitimacy: Social values and organizational behavior. *Pacific Sociological Review, 18*(1), 122–136.

Dye, R. A. (1985). Strategic accounting choice and the effects of alternative financial reporting requirements. *Journal of Accounting Research, 23*(2), 544–574.

Federation of European Accountants (2002). *FEE discussion paper providing assurance on sustainability reports*. Retrieved from http://www.fee.be/images/publications/sustainability/DP_Providing_Assurance_on_Sustainability_Reports1632005451020.pdf

---

[8] Li and Li (2012) demonstrate that a CSR assurance has significant impact on the company's stock return. Li et al. (2013) suggest that a CSR assurance might be used as a signaling tool by the company due to social pressures from media and other major stakeholders.

Freeman, R. E., Wicks, A. C., & Parmar, B. (2004). Stakeholder theory and 'The corporate objective revisited'. *Organization Science, 15*(3), 364–369.
Friedman, A. L., & Miles, S. (2002). Developing stakeholder theory. *Journal of Management Studies, 39*(1), 1–21.
Global Reporting Initiative. (2006). *Sustainability reporting guidelines.* Retrieved from https://www.globalreporting.org/reporting/G3andG3-1/Pages/default.aspx
Gray, R., Kouhy, R., & Lavers, S. (1995). Corporate social and environmental reporting: A review of the literature and a longitudinal study of UK disclosure. *Accounting, Auditing and Accountability Journal, 8*(2), 47–77.
Gray, R., Owen, D. L., & Adams, C. (1996). *Accounting and accountability: Changes and challenges in corporate and social reporting.* London: Prentice-Hall Europe.
Hendry, J. (2001). Missing the target: Normative stakeholder theory and the corporate governance debate. *Business Ethics Quarterly, 11*(1), 159–176.
IAASB. (2013). ISAE 3000 (Revised). *Assurance engagements other than audits or reviews of historical financial information.* Retrieved from http://www.ifac.org/publications-resources
Kanagaretnam, K., Lobo, G. J., & Whalen, D. J. (2007). Does good corporate governance reduce information asymmetry around quarterly earnings announcement. *Journal of Accounting and Public Policy, 26*(4), 497–522.
Kolk, A., & Perego, P. (2010). Determinants of the adoption of sustainability assurance statements: An international investigation. *Business Strategy and the Environment, 19*(3), 182–198.
KPMG (2013), *The KPMG Survey of Corporate Responsibility Reporting 2013.* Retrieved from http://www.kpmg.com/sustainability
Li, Z., Guan, F., & Li, Z. Q. (2013). An empirical research on drivers of corporate social responsibility report attestation: empirical evidence from Chinese listed companies. *Auditing Research, 3*, 102–112.
Li, Z., & Li, Z. Q. (2012). Information content of corporate social responsibility report attestation opinion: Empirical evidence from Chinese listed companies. *Auditing Research, 1*, 78–86.
Manetti, G., & Becatti, L. (2009). Assurance services for sustainability reports: Standards and empirical evidence. *Journal of Business Ethics, 87*, 289–298.
Mathews, M. R. (1993). *Socially responsible accounting.* London: Chapman Hall.
Michelon, G., & Parbonetti, A. (2012). The effect of corporate governance on sustainability disclosure. *Journal of Management and Governance, 16*(3), 477–509.
Milne, M. J., & Patten, D. M. (2002). Securing organizational legitimacy: An experimental decision case examining the impact of environmental disclosures. *Accounting, Auditing and Accountability Journal, 15*(3), 372–405.
Ministry of Finance. (2006). *CAS 3101: Assurance engagements other than audits or reviews of historical financial information.* Retrieved from http://www.mof.gov.cn/zhuantihuigu/kjsjzzfbh/sjzz/
Mitchell, J. (2006). Selective reporting of financial ratios in Australian annual reports: Evidence from a voluntary reporting environment. *Accounting Research Journal, 19*(1), 5–30.
Moroney, R., Windsor, C., & Aw, Y. T. (2012). Evidence of assurance enhancing the quality of voluntary environmental disclosures: An empirical analysis. *Accounting and Finance, 52*(3), 903–939.
Patten, D. M. (1992). Intra-industry environmental disclosures in response to the Alaskan oil spill: A note on legitimacy theory. *Accounting, Organizations and Society, 17*(5), 471–475.
Perego, P. (2009). Causes and consequences of choosing different assurance providers: An international study of sustainability reporting. *International Journal of Management, 26*(3), 412–487.
Scott, W. R. (2001). *Institutions and organizations.* London: Sage.
Shen, H. T., Wang, L. Y., & Wan, T. (2011). Can corporate social report and assurance be effective signals? *Auditing Research, 4*, 87–93.

Shenzhen Stock Exchange. (2006). *Shenzhen stock exchange social responsibility guideline to listed companies.* Retrieved from http://www.szse.cn/main/disclosure/bsgg/200609259299.shtml

Simnett, R., Vanstraelen, A., & Chua, W. F. (2009). Assurance on sustainability reports: An international comparison. *The Accounting Review, 84*(3), 937–967.

Skinner, D. J. (1994). Why firms voluntarily disclose bad news. *Journal of Accounting Research, 32*(1), 38–60.

Spence, M. (2002). Signaling in retrospect and the informational structure of markets. *American Economic Review, 92*(3), 434–459.

Suchman, M. C. (1995). Managing legitimacy: Strategic and institutional approaches. *The Academy of Management Review, 20*(3), 571–610.

Zhong, H. W., Zhang, Y., & . Zhai, L. F. (2011). *Corporate social responsibility reporting in China, A white book.* Retrieved from http://www.cass-csr.org/xiazai/whitebook(2011).pdf

Social Audit: Case Study of Sustainable Enterprise Index-ISE Companies

Dalia Maimon and Cristiana Ramos

1 Introduction

This chapter analyses the incorporation of social responsibility into the audits of 38 companies who, in 2013–2014, were comprised in the Sustainable Enterprise Index-ISE, SRI fund of BM&FBovespa, Brazilian stock market. The ISE family tracks the stock performance of the Brazilian leading companies in terms of economic, environmental and social criteria.

In Brazil, a social audit is voluntary. In the context of an emerging country with a large poor population and a significant environmental impact from farming and mining exports, big firms are increasingly required to contribute to sustainable economic development (Maimon, 2011).

We start with the premise that although the Brazilian business community is identified as sensitive to social concerns compared with other emerging countries, there are few audits that concern the CSR. The absence of a culture of non-financial audits and the high cost may be an explanation. Our hypothesis is that despite the discourse from entrepreneurs, little has actually been developed by companies. The gap between discourse and practices could explain why these economic and sustainable leaders audit very poorly their performance.

The first part of this chapter is a summary of the relevant literature on the evolution of CSR in Brazil. This literature was abundant in field analysis of RS since the democratization of Brazil until today. Much of the literature and the data

---

D. Maimon (✉)
Social Responsibility Laboratory, Institute of Federal University of Rio de Janeiro, Rio de Janeiro, RJ, Brazil

Rua Casuarina 210, CEP 22216-160 Humaitá, RJ, Brazil
e-mail: dalia@ie.ufrj.br

C. Ramos
Center for Studies Social Technology, Rua José Américo de Almeida 375, apt° 203, Recreio dos Bandeirantes, RJ, Brazil

comes from strong business organizations (Brazilian World Business Council-CEBED, Ethos Institute and Group of Institutes, Foundations and Enterprises-GIFE) and is biased.

Regarding CSR tools and audits, the literature is scarce. The majority focuses on the analysis of IBASE Social Balance, apex of social audit in Brazil. Data on the number of respondents to Ethos presented in this paper were never published and were achieved through a demand from our institution. As regards to the Global Report Initiative-GRI, there are some articles on the performance of the electricity sector (Carneiro C. et alli, 2012, Rosa F. S. et alli, 2011). We also got an unpublished historical series of GRI companies.

Regarding the case study, we analyzed 34 questionnaires from 38 companies that are part of the ISE. Through the analysis of questionnaires answered by these companies, the research attempted to explore what are the drivers behind the incorporation of CSR by these leaders in sustainability, whether there is a sectorial trend, or whether they result from a broader movement of economic globalization for the diffusion of social and environmental ethics.

To be able to understand the context, we begin with a summary of corporate responsibility in Brazil and the main tools of management and communication used.

2 From Christian Charity to Social Marketing

It is possible to identify the origins of corporate social responsibility (CSR) in Brazil. This began in 1965 with the creation of the Christian Business Leaders Association- ADCE and the subsequent publication of the "Charter of Principles of the Christian Business Leaders", in São Paulo, the most industrialized province of the country. The participation of entrepreneurs had merely a philanthropic feature with no questions about internal and external business practices. The lack of political freedom and the restrictions imposed by the military dictatorship during this period have restricted the diffusion and the spread of the theme of social responsibility that could be confused with socialist ideas, strongly repressed in this period (Alessio, 2004).

In the eighties, with the democratization of Brazil and the promulgation of the 1988 Constitution, a movement led by the sociologist Herbert de Souza—Betinho-was launched, interacting with entrepreneurs within different philanthropic initiatives. Since 1993, the Campaign Against Hunger, created by Betinho and developed by the NGO, which he presiding over, IBASE, the Brazilian Institute for Social and Economic Analyses became a national point of reference. Particularly in relation to the creation of the Brazilian model of Social Audit Report, which, for the first time, made public the performance of companies in dealing with social issues. The campaign had the broad participation of big state-owned companies such as Petrobras, Banco do Brazil, Furnas and Caixa Economica Federal, followed later on by private companies (www.ibase.br, 2013).

In 1989, the creation of the Group of Institutes, Foundations and Enterprises (GIFE) aimed to promote private social investment by spreading the discourse and actions of American companies with strong neoliberal bias. One year later, the Society for the Incitement to Environmental Management, SIGA, a member of the German network INEM (International Network of Environment Management), which initially used to concede Green Label, distinguished itself by promoting environmental management in the major media on radios and TV (CBN Ecology, CBN Energy among others).

In 1992, the UNCED meeting in Rio led to the diffusion of environmental responsibility. Universities have introduced graduate and under graduate courses, NGOs specialized in environmental issues had being created, the means of communication have opened a space for this new theme, and companies started to introduce new discourses and practices. The meeting also accelerated the international financing, agreements and networks.

In 1995, the Community in the Solidarity created by the anthropologist Ruth Cardoso, then the first lady of Brazil, came to spread the culture of volunteering and prepared a specific legislation for volunteers and the Third Sector. The neoliberal proposal of the president Fernando Henrique administration had reduced the role of the state, privatizing public services and transferring to the Third Sector a range of social actions.

The following year was marked by the adoption of the norm ISO 14000. The Gazeta Mercantil newspaper published 16 booklets titled "Environmental Management—Commitment of the Company", indicating the 'step by step' path to environmental management and highlighting the competitive advantages of companies with ISO 14001 (Maimon, D. (coord) (coord), 1996). In 1999, the magazine Exame launched a booklet called "Guide to Good Corporate Citizenship" that has established itself as a reference for the dissemination of CSR actions by firms.

However, 1997 came to be seen a defining period with respect to CSR. Betinho wrote his article "Public Company and citizens", and the IBASE Social Report got widespread media attention. From the initiative of IBASE, the Brazilian stock market regulator, CVM, joined the movement that sought to encourage the disclosure of the Social Report.

One year later, the Ethos Institute was created in São Paulo, which for a decade governed companies in relation to social responsibility. It is a non-governmental institution, conceived by entrepreneurs and executives coming from the private sector. From a membership of 11 companies at the time of its foundation, the number of members reached 6000, in 2009, representing over 28 % of Brazil's GDP. There is no coincidence that these organizations of mobilization of the business community have emerged in São Paulo—which accounts for 34 % of manufacturing and 12 % of the services companies of the country—urban center that, from the beginning of the century XX, is the dynamic hub of the Brazilian capitalist economy.

Until the creation of Ethos, the axe of CSR was set up in Rio with the Social Report of IBASE and the presence of CBEDS representing the World Business Council for Sustainable Development and having initially a focus on environmental ethics. As a consequence of Ethos' movement, CSR remains aligned with a company's core business, including its marketing objectives. Social and environmental initiatives started to be selected through strategic analysis, not only with traditional shareholders, but also with a network of stakeholders, consumers, workforce, government, media, NGOs, suppliers and others.

In this context, CSR seeks to follow the share value strategies (Porter & Kramer, 2006) and to align the social and environmental agenda of society on the economic interests of companies to generate profits. The determinants of CSR could be to obtain license to operate, the search for legitimacy, participation in the international market, brand enhancement and risk management, among others (Maimon, 2011).

The international movement encouraged BM&FBovespa to create the ISE, an index tracking the economic, financial, corporate governance, environmental and social performance of the leading companies listed on the Brazilian Stock Exchange. It was launched in December 2005 to provide asset managers and investors with a reliable and objective benchmark of the best corporate sustainability practices in the country. In 2004, BM&FBovespa was the first stock market in the world to join the Global Compact, and since 2006 is taking the part of the Brazilian Board as vice president (Barbosa, 2007).

Some initiatives of the financial sector had been seen before; in 2001 when Unibanco launched its first service to identify companies committed to the environment within the country. The service was aimed mainly at allocation for European and North American SRI funds, although not exclusively. In the same year, Banco Real ABN Amro launched two Ethical Funds, the first SRI funds in emerging markets. Since their creation, both funds have outperformed the BM&FBovespa Index (IBOVESPA). In 2004, Banco Itaú launched its Itau Social Best Practice Fund (Fundo Itaú Excelência Social), focusing on Corporate Social Responsibility. BM&FBovespa also launched the Corporate Governance Stock Market Index (Índice de Ações com Governança Corporativa Diferenciada— IGC), tracking companies highly committed to corporate governance (Maimon, 2012).

A large number of other investment funds also adopted this new concept. According to BM&FBovespa there was, in 2010, an amount of almost U$ 1 billion invested in Equity Funds focused on sustainable companies (BM&FBovespa, 2010). In March 2013, US$ 400 million were invested only in ISE companies, a growth of 68 % since its creation in 2005.

From the point of view of state regulation, with the labor President Lula administration, ex unionist, we observed during the 2000s increase of minimum wage and new laws on human rights for labor force with strong impact on social responsibility of the internal public of enterprises.

The Decree No. 6.481/2008 on the prohibition of the worst forms of child labor, states as prohibited 93 activities for persons less than 18 years of age. The Secretary of Labor Inspection of the Ministry of Labor and Jobs makes inspections aimed at combating child labor and the protection of teen workers. In August 2003, the National Commission for the Eradication of Slave Labor (CONATRAE) had the target of monitoring the implementation of the National Plan for the Eradication of

Slave Labor. The Plan contains 76 actions and involves together agencies of the Executive, Legislative, Judiciary, civil society and international organizations (www.MTE.gov.br).

In 2003, the Secretariat of Policies for Women was formed to build the empowerment of women and their inclusion in the society. Three main goals are targeted: Labor and economic autonomy of women; combating violence against women; and programs and actions in the areas of Health, Education, Culture, Political Participation, Equality and Diversity. In 2006, since the Law 11.340/06 (known as the Maria da Penha Law) came into force, violence against women is now treated like a crime.

At the federal level, there is no specific law for moral and sexual harassment, but in some ways this is addressed by Article 483 of the Brazilian Labor Code (CLT). Some provinces and municipalities have specifics laws for sexual and moral harassment.

Additionally, at a macro level, Lula's administration, through a set of policies, takes the lead in fighting poverty, notably with the **Bolsa Família** that helped lift 25 million people out of poverty, and the *Minha Casa Minha Vida*, with the distribution of 1 million of houses for those suffering from poverty, among others.

3 Main Instruments of Social Communication Tools and Certifications

Over the last 20 years we have observed the emergence of a large number of genuinely Brazilian instruments of social responsibility, whose implementation and consolidation have depended on national business leaderships and more recently, with the globalization of the economy, the implementation of international instruments and tools.

Although the first social report was published in 1984 by a state petrochemical company, its impact has been rather slight. The peak in terms of voluntary auditing has been reached, in 1997, thanks to the creation of the social balance of IBASE, as indicated above. The Social Report was prepared by companies and audited by IBASE.

The model proposed by the IBASE Social Report was inspired by the French Bilan Social and relied for its elaboration on partnership with technicians, researchers and representatives of various public and private institutions, and was supported by the stock market Securities Commission.

The Social Balance Model of IBASE has specific characteristics, namely: built on the initiative of a recognized NGO; to demonstrate the transparency and effectiveness of social and environmental activities of companies; separating the actions and mandated benefits from those made voluntarily by companies; essentially quantitative and simple, all companies can publicize its Social Report, regardless of size and sector of activity; and, if it is properly completed, it may allow comparison between different companies, also help to evaluate the same corporation, over consecutive years (Torres & Mansur, 2011).

Besides allowing a systemic vision of a company, this model of social report can also be used as a tool for diagnosis and management, as gathering together important information about the social role of the business, allowing one to follow the evolution and improvement of its indicators: financial information, internal social indicators, external social indicators and environmental indicators.

From 1997 to 2007, while still using the Social Balance model of IBASE, there was an increase in the number of companies preparing and diffusing their Social Report. In the early years of the initiative, the project had significant support. Some important companies presented their annual social reports, such as Inepar, Usiminas, Brasilia Energy Company (CEB) and Light. The National Electric Energy Agency (Aneel) recommended that all electricity organizations adopt the IBASE model. The Municipality of São Paulo created a label based on the IBASE model. Several partnerships have been concluded with the Federation of Industries of State of Rio de Janeiro (Firjan), the Social Service of Industry (SESI), the Foundation Institute of Business and Social Development (Fides), the Association of Analysts and Investment Professionals of Capital Markets (Apimec), the newspaper Gazeta Mercantil and, also, some universities (Torres & Mansur, 2011).

The Fig. 1 below shows that 2003 was the year with the highest number of companies publishing their Social Report. From this year onwards, there was a significant reduction in number of publications.

Later on, in 2000, the Ethos Institute issued and started to implement the Ethos Indicators. These indicators are updated annually and have played an important role in stating a diagnostic of CSR. Indicators allow companies to do a self-assessment of their performance in seven themes: Values and Transparency, Internal public, Environment, Providers, Consumers and Customers, Community, and Government and Society.

The Ethos Institute has also elaborated sectorial questionnaires for Electricity Distribution, Petroleum and Gas, Bakery, Bars and Restaurants, Banks, Mining and Pulp & Paper (www.ethos.org.br) (Fig. 2).

In 2004, ABNT NBR 16001—Social Responsibility—System Management—Requirements had its first edition published, and a second version in 2012. The last version was based on the international ISO 26000 guidelines published in 2010.

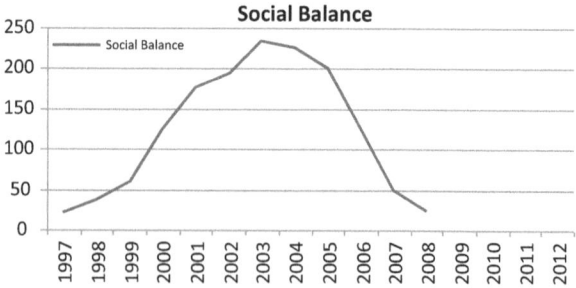

Fig. 1 Number of companies' publishing IBASE's Social Balance model. *Source.* Torres and Mansur (2011)

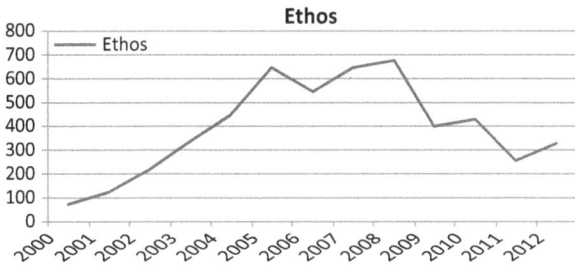

Fig. 2 Number of companies responding the Ethos Indicators. *Source*. Graphique elaborated by the author, Ethos (2013)

The review of ABNT NBR 16001 was undertaken within the Special Study Group on Social Responsibility of ABNT, and was subjected to national consultation. Other countries have also developed national standards for the purpose of accreditation in light of the ISO 26000 directive (www.abnt.org.br).

In recent years, however, the Ethos Institute progressively lost its leadership, and the GRI indicators became more prevalent. In an attempt to regain more space in the business community, it launched a new version of the Ethos Indicators in accordance with the GRI—G4 and the guidelines for social responsibility ISO 26000 and ABNT NBR 16001. Regarding the international instruments, we must highlight the IS0 14001. This standard changed the planning of environmental management, and had a momentum thanks to the greatest diffusion of environmental issues with the Rio's UNCED, in 1992.

In recent times, a topic that has received much attention from companies is the issue of climate change, which has two initiatives that attract attention in Brazil: the GHG Protocol and the Carbon Disclosure Project (CDP). CEBDS launched the partnership of Brazilian GHG Protocol Program in 2008 with the WBCSD, Ministry of Environment and the Center for Sustainability Studies of the Getulio Vargas Foundation (GVCes). The GHG Protocol has enabled the training and monitoring of dozens of companies in the country to make its inventories of GHG emissions, the basis for subsequent action to mitigate emissions. The Carbon Disclosure Project, an initiative of international investors to increase the transparency of GHG emissions and carbon risk, arrived, in 2011, in its sixth edition in Brazil.

The Fig. 3 below, regarding the international tools, shows the fall of ISO14000 and the increasing participation of GRI reports. 2006 saw the peak of ISO certifications with 838 organizations. From 2007 to date, there has been a significant drop in the use of ISO14000 with a sensible replacement by GRI reports. It seems to be an international trend, as Corporate Register estimates that 40 % of all non-financial reports submitted in 2012 were GRI influenced (http://www.corporateregister.com, 2013).

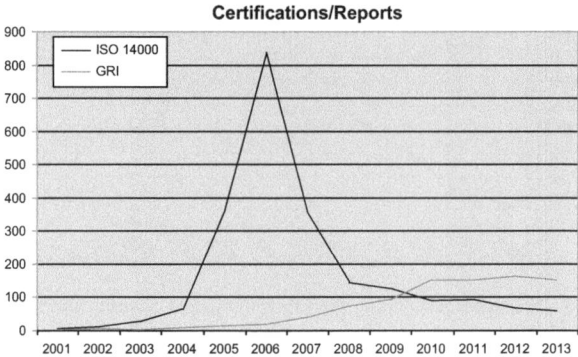

Fig. 3 Number of organizations certificated. *Source*. Graphique elaborated by the author GRI, (2014) and ABNT (2014)

4 Case Study of ISE Companies

The ISE is based on the Triple Bottom Line (TBL) concept introduced by Elkington (1994), which evaluates the economic/financial, social, and environmental dimensions in an integrated manner. The methodology added some corporate governance indicators and criteria to the TBL principles based on the JSE model. Those indicators were gathered in a fourth thematic group, in addition to the TBL, to evaluate companies. The four blocks are preceded by a group of general indicators. These include, for example, publication of a sustainability or social report, endorsement of the Global Compact of the United Nations and damages or risks to health, physical safety or integrity of consumers and third-parties, as well as to public health and safety, posed by the consumption or use of the company's products.

As with other SRIs, the ISE is revised annually to re-evaluate companies according to their sustainability levels.

The ISE project was financed by the International Finance Corporation (IFC), the private sector arm of the World Bank Group, an organization that promotes sustainable private sector investment in developing countries. Developed by the Center for Sustainability Studies of Getúlio Vargas Foundation, a leading business school in Brazil, the ISE can include up to 40 companies that seek excellence in managing sustainability (Corrar, Machado, & Machado, 2008). The Advisory Board of ISE is strongly disputed, and has undergone a lot of changes. In 2013, the board was composed, among others, by BM&FBovespa, IFC- International Finance Corporation, Abrapp-Brazilian Association of Pension Funds, AMBIMA-National Association of Financial and Capital Markets Institutions, MMA-Ministry of the Environment and UNEP-United Nations Environment Programme.

IBASE soon left the board, disagreeing with the presence of the tobacco industry. The Ethos Institute, a NGO which represents ethical business, was suspended from the board in 2009, although Bovespa did not explain the reason behind this move. It happened just after the exclusion of Petrobras (Brazilian Oil Company) from the ISE, in spite of its participation in the Dow Jones Sustainability Index.

By adopting a Positive Screening approach, no sectors are excluded from the index. The most heavily trade-shares in BMF&Bovespa have the opportunity to fill out the questionnaire and participate in the selection process. However, all elements associated with sustainability for each sector will be closely evaluated, including potential risks and adverse impacts associated with nature of products and services of each business.

The Best in Class methodology uses cluster analysis as a statistical tool in the final classification process. Avoiding deviations generated by simply adding the assessment scores of such distinct dimensions as the environment and corporate governance, cluster analysis identifies groups of businesses that exhibit similar performance in each dimension. The final portfolio will be made up of the cluster of businesses that demonstrate the best practices in all five dimensions (Maimon, 2012).

The selection criteria for the ISE theoretical portfolio is done by sending questionnaires to 150 companies pre-selected for their market negotiability. Based on this data, the Board begins the process of choosing the companies with the best classification for a 1-year period, especially considering: Relationships with employees and suppliers; Community Relations; Corporate Governance and Environmental impact and activities.

The criteria for inclusion of companies in the ISE must also meet the following requirements: Be one of the 150 stocks with the highest negotiability ratio; have been traded in at least 50 % of the defined trading days; meet the sustainability criteria endorsed by the Board of ISE. Companies can also be excluded from the ISE for several reasons, such as bankruptcy, significant changing in criteria levels, etc.

The ISE during 2013–2014 consists of 38 companies representing seven sectors (Table 1). The Energy sector, which had been the most representative sector in the 2005–2006 portfolio with 32.1 % of the total, more than doubled its participation, to 65.79 %, in 2013–2014, and now only electricity companies are present. The financial sector declined from 17.9 % to 13.16 % in the same period.

Electricity is highly weighted in the index, despite the fact that the sector is usually considered as having heavy environmental impacts. The Oil, Gas, Biodiesel, Alcohol, Tobacco and Weapons sectors are not represented in this period, although Petrobras, a Brazilian Oil Company, was represented until 2010.

Figure 4, below, compares the performance of the ISE with the Bovespa Index. It shows that both followed the same trends, partly explained by the fact that the ISE companies represent an important part of Bovespa Index. In emerging countries, large companies have become, at the same time, champions in sustainable practices and in trading volumes of shares, having expressive importance both in broad and sustainability indexes.

During the financial crisis, however, the Bovespa Index outperformed the ISE. This could suggest that investors are interested more in speculation for profit, rather

Table 1 ISE: Portfolio distribution by sector

Sector	Portofolio 2005/2006		2013/2014	
	Number of companies	%	Numbers of companies	%
Medical diagnosis	1	3.57	0	–
Meat and derivatives	1	3.57	0	–
Transport construction	1	3.57	0	–
Construction and engineering	0	–	1	2.65
Eletric/Energy	9	32.14	25	65.79
Toll motorway	1	3.57	0	–
Financial intermediaries	5	17.86	5	13.16
Paper and pulp	3	10.71	0	–
Machinery and equipment	1	3.57	1	2.65
Transport material	2	7.14	0	–
Personal care products and cleaning	1	3.57	0	–
Chemicals	1	3.57	0	–
Steel & Metals	1	3.57	1	2.65
Mining	0	–	1	2.65
Dealers	0	–	4	10.50
Total	28	100	38	100

Source: www.bovespa.com.br

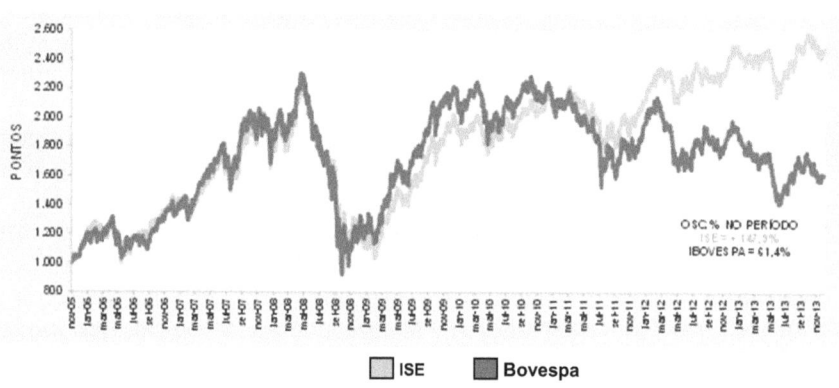

Fig. 4 Evolution of ISE and Bovespa Index. *Source*. http://www.bmfbovespa.com.br

than for minimizing the long-term risks. It can be observed that there is a great variation among yearly returns. Specifically, during the 2008 financial crisis, the ISE decreased sharply, as well as all other Brazilian stock indexes.

After 2011, we can observe a great gap between the two indexes. As a consequence of poor economic performances by Petrobras and EBX, these companies were excluded from the ISE.

Table 2 Social initiatives concerning internal public by sector

	TI	TF	DI	DV	AS	DT	EQ	DL
Electricity	14	15	19	16	22	21	14	10
Financial	4	5	5	5	5	5	3	5
Building	1	0	0	0	0	1	1	0
Mining	1	1	1	1	1	1	1	0
Cosmetics	1	1	1	1	1	0	1	1
Dealers	4	4	4	4	0	4	4	4
ISE%	(25)76%	(26)79%	(30)91%	(27)82%	(29)88%	(32)97%	(24)73%	(20)61%

Source: Table elaborated by the author, ISE Bovespa questionnaires (2013)

The methodology to measure CSR focuses on the analysis of ISE company questionnaires. They are presented in five different dimensions; economic/financial, governance, social, and environmental. During the ISE's selection of companies for 2013–2014, only 33 of the total of 38 published their questionnaire. Our research focuses on the available answers, using governance, social and environment questionnaires.

Regarding social responsibility, we can observe in Table 2 that firms comprising part of the ISE got many actions concerning internal public: 97 % have respect to free union association and right to collective bargaining (DT); 91 % combat all forms of discrimination (DI); 88 % prevent moral and sexual harassment (AS); 82 % Valuing Diversity (DV); 79 % target eradication of slave labor (TF) and 76 % support child labor eradication(TI); 73 % ensure fair treatment and working conditions between employees and contractors (EQ). Almost all companies have internal audits and only three have external audits of these actions. These good performances could reflect the new regulations concerning labor force and the actions of syndicates. The dialog with internal public (DL) is the least representative with 61 %, probably because being voluntary, no regulation address this practice.

It very difficult to conclude about a sectorial trend concerning internal public and also external public, because electricity and financial sectors comprise more than 80 % of the Index. Electricity seems to have relatively less CSR initiatives than the others sectors. Almost all of the companies who completed the survey seek to prevent to moral and sexual harassment (AS) and respect the right to free union association and the right to collective bargaining (DT). The eradication of slave labor (TF), child labor eradication(TI) and fair treatment and working conditions between employees and contractors (EQ) are less well presented, making up only a little over half of the sample.

Table 3 confirms the modest performance regarding the voluntary audit and certifications, 45 % got GRI, and 34 % ISO14000. The SA8000 is represented by only one company.

Table 3 Number of certifications by sector

	ISO14000	GRI	SA8000
Energy	7	9	1
Financial	2	4	0
Building	1	1	0
Mining	1	1	0
Electromotors	1	1	0
Cosmetics	1	1	0
Dealers	0	0	0
ISE	(13) 34%	(17) 45%	(1) 0,02%

Source: Table elaborated by the author ISE Bovespa questionnaires (2013–2014)

Conclusion

Regarding CSR, Brazil started with a strong philanthropic component. In recent years, CSR has remained aligned with a company's core business, including its marketing objectives. Environmental initiatives started to be selected through strategic analysis of relations not only with traditional shareholders, consumers and workforce, but also with a network of stakeholders.

Over the last 20 years we have observed the emergence of a large number of genuinely Brazilian instruments of social responsibility, whose implementation and consolidation depended on national business leaderships, and, more recently, with the globalization of the economy, the implementation of international instruments and tools, especially ISO 14000 and GRI.

In Brazil, Social Audit is still not mandatory and there are no official models or indicators. However, an increasing number of organizations from various sectors are adopting this practice, including the companies analyzed in this research.

It very difficult to conclude about a sectorial trend concerning internal public and also external public of ISE companies, because electricity and financial sectors comprise more than 80 % of the Index. Electricity seems to have relatively less CSR initiatives than the others sectors. Almost all of the companies who completed the survey seek to prevent to moral and sexual harassment (AS) and respect the right to free union association and the right to collective bargaining (DT). The eradication of slave labor (TF), child labor eradication(TI) and fair treatment and working conditions between employees and contractors (EQ) are less well presented, making up only a little over half of the sample.

The analyzed questionnaires show that state regulation seems to be effective when it comes to CSR. With the new regulation of labor force, after the President Lula administration, ISE companies have given more attention to internal public, rather than to external public, although business marketing

(continued)

explores the social and environmental responsibility of external public with very few audit practices.

References

Alessio, R. (2004). *Responsabilidade social das empresas no Brasil: reprodução de posturas ou novos rumos?* Porto Alegre: Edipuc.

Barbosa, P. R. A. (2007). *Índice de sustentabilidade empresarial da bolsa de valores de São Paulo (ISE-BOVESPA): exame da adequação como referência para aperfeiçoamento da gestão sustentável das empresas e para formação de carteiras de investimento orientadas por princípios de sustentabilidade corporativa. 2007.* These of Master Degree (Universidade Federal do Rio de Janeiro, UFRJ, Instituto COPPEAD de Administração).

Carneiro C. et alli. (2012). A divulgação ambiental no Setor de energia elétrica Brasileiro VIII Congresso de Excelencia e Gestão, Rio de Janeiro.

Corrar, L. J., Machado, M. A. V., Machado, M. R. (2008). Desempenho do Índice de Sustentabilidade (ISE) da Bolsa de Valores de São Paulo. Dissertação, Faculdade de Administração, USP, São Paulo.

Elkington, J. (1994). Towards the sustainable corporation: Win-win-win business strategies for sustainable development. *California Management Review, 36*(2), 90.

Maimon, D. (2011). *Determinants de la responsibité Sociale des enterprises brésilienses, Developpement durable, Terrioires et Localisation des Entreprises.* University Bordeaux IV Congress, vol. 1, p. 1.

Maimon, D. (2012). Measuring green growth: Na assessment of Brazilian SRI Funds. *Revista ISEES, 1*, 1.

Maimon, D. (coord) (1996). Gestão Ambiental Compromisso das Empresas. *Revista Gazeta Mercantil*. Rio de Janeiro.

Porter, M., & Kramer, M. (2006). Strategy & Society: The link between competitive advantage and corporate social responsibility. *Harvard Business Review*.

Rosa F. S. et alli (2011). *Global Reporting Initiatives: Survey em empresas de energia elétrica dos Estados Unidos, do Brasil e da Espanha no período de 1999 a 2010*, II Cesar Conference, South America, Ribeirão Preto.

Torres, C., & Mansur, C. (2011). Social—o desafio da transparência. IBASE.

Sites

http://www.abnt.org.br (2014)
http://www.bmfbovespa.com.br (2014).
http://www.cebds.org.br (2014)
http://www.corporateregister.com (2013)
http://www.ethos.org.br (2014)
http://www.ftse.com/Indices/FTSE4Good_Index_Series (2014)
http://www.gife.org.br (2014)
https://www.globalreporting.org (2011)
http://www.ibase.br (2013)
http://www.MTE.gov.br
http://www.sustainability-index.com (2014)
http://www.wbcsd.org (2013)

Corporate Climate Change-Related Auditing and Disclosure Practices: Are Companies Doing Enough?

Shamima Haque

1 Introduction

While many natural factors continue to influence our climate, scientists have determined that human activities, in particular the burning of fossil fuels such as coal, oil and gas which increase greenhouse gas (GHG) concentrations in the atmosphere, are the dominant factor responsible for the changing climate (IPCC, 2007). Scientific evidence also shows that global emissions need to be cut by 80–95 % below 1990 levels by 2050 if we are to avert dangerous climate change and continued disruption to our weather patterns (Carbon Disclosure Project, 2013). One of the key contributors to climate change is the business community whose actions add to the global GHG concentrations. Since the majority of anthropogenic GHG emissions stem from energy use, the manufacturing sector, and the distribution and consumption of goods and services, the role of companies in helping to achieve the required emissions reductions is crucial. Not only are business sectors largely responsible for global climate change, they will also be affected by the potential risks associated with it. There are differential risks that climate change poses on businesses, which in turn affects their profitability and value and threatens their very survival and accountability (Bebbington & González, 2008; Carbon Disclosure Project, 2008; CERES, 2002; Labatt & White, 2007; Rolph & Prior, 2006). Consequently, there are now many international and national initiatives and guidelines provided by government bodies, non-governmental organisations (NGOs) and research organisations, which deal with the threat of climate change, and which raise a range of financial reporting and audit implications for corporations worldwide.

S. Haque (✉)
School of Accountancy, QUT Business School, Queensland University of Technology, 2 George Street, 4001 Brisbane, QLD, Australia
e-mail: shamima.haque@qut.edu.au

Based on a review of media reports, archival documents and a case study on the joint actions of two well-known Australian companies, a bank (ANZ) and Whitehaven coal mine in New South Wales, this chapter explores whether the climate change-related audit and disclosure practices of corporations reflect real change in their corporate accountability practices for climate change. The findings suggest that although there is evidence of companies undertaking social and environmental audit practice and disclosing information in relation to their climate change related performance, there is limited real reform in corporate action. The study suggests that as social auditing is a voluntary activity, it is possibly sometimes used only as a legitimation tool by companies rather than making any real change in their actual practices. Therefore, without appropriate regulation or enforcement of social auditing standards, the accountability and obligations of global companies to mitigate climate change remains negligible. A radical (reform based) approach, such as mandatory monitoring (compliance audit) and disclosure requirements, is necessary to ensure corporate accountability in relation to climate change.

2 Global Concerns and Corporate Responses in Relation to Climate Change: An Overview

In recent years climate change has attracted increasing attention in the international scientific and policy arenas. As science has evolved, growing evidence of anthropogenic influences on climate change has been found. Correspondingly, the Intergovernmental Panel on Climate Change (IPCC[1]) has issued increasingly more authoritative reports about the human impacts on the earth's climate. This has led to the development of a set of policy imperatives in supra-national as well as national settings (Bebbington & González, 2008) which have created a range of reporting and audit implications for corporations worldwide.

The first international agreement on climate change, the United Nations Framework Convention on Climate Change (UNFCC) was established in 1992 at the Rio Earth Summit (The United Nations Framework Convention on Climate Change, 2004). Supported by 166 nations, the convention called for the stabilisation of GHG concentrations in the atmosphere at a level that would prevent dangerous anthropogenic interference with the climate system. Consequently, business organisations were scrutinised for their contribution to climate change by a wide range of stakeholders including the public and governments (Kolk & Pinkse, 2007). Major companies initially opposed international efforts and regulations to control GHG

---

[1] The IPCC is a scientific intergovernmental body set up by the World Meteorological Organization (WMO) and the United Nations Environment Programme (UNEP). It provides scientific, technical and socio-economic information in a policy-relevant but policy-neutral way. It publishes regular Assessment Reports, the findings of which are well publicised and quoted around the world.

emissions (Jeswani et al., 2008, Wehrmeyer, & Mulugetta, 2008; Kolk, 2008; Kolk & Levy, 2001), especially energy-intensive sectors such as coal, oil, steel, aluminium, chemicals, automobiles and paper and pulp. They protested against climate change debate by forming lobby organisations such as the Global Climate Coalition, the American Petroleum Institute and the Coalition for Vehicle Choice (Greenpeace, 1998; Kolk, 2008). Their intention was to undermine the importance of climate science and to prevent the introduction of new government regulation (Greenpeace, 1998).

This climate change debate continued until the mid-90s. By the late 1990s, an increasing number of companies had steadily changed their position from opposition to a more positive approach, and many had started to prepare for regulation (Kolk, 2008; Kolk & Pinkse, 2004, 2007; Pinkse & Kolk, 2007). The Kyoto Protocol, adopted in 1997, contained legal limits on GHG emissions for developed countries. It arguably stimulated this change in corporate strategy, as well as prompting the development of climate change regulation and increasing the pressure from NGOs (Kolk & Pinkse, 2004). Under the Kyoto protocol, the major industrial nations were together required to reduce total emissions of six GHGs[2] to 5.2 % below their combined 1990 emissions by the end of the first commitment period (2008 through 2012). The Kyoto Protocol requires corporations to ensure proper monitoring and verification of its implementation, including stringent and elaborate reporting, review, and compliance procedures. With these increasing reporting and audit requirements many organisations started working with NGOs' on climate change issues as NGOs and business leaders realised that they could not tackle them alone (Pleon Climate Change Stakeholder Report, 2007). This phase has led to the formulation of cross-sector stakeholder partnerships (for example, the Greenhouse Gas Protocol Initiative) and has gained momentum at the beginning of the twenty first century. The climate strategies of major oil (such as BP, Shell) and automotive (such as General Motors, Toyota) companies have changed in response to increasing regulatory and public pressures to adopt a more open position towards climate science and the Kyoto Protocol (Kolk & Levy, 2001: Kolk, 2008).

Adoption of the Kyoto Protocol motivated the development of many new requirements at the international and state level (for example, in Australia, the Mandatory Renewable Energy Target (MRET), 2001). The climate change-related stance taken by corporations then gained momentum with the adoption of an emissions trading scheme by the European Union, which came into force in 2005. To meet the Kyoto commitments, the European Union GHG Emission Trading Scheme (EU ETS) imposes emission limits on utilities and big industrial emitters in the European Union (Jeswani et al., 2008). The first and second phases of the EU ETS run from 2005 to 2007 and from 2008 to 2012 respectively to coincide with the first Kyoto Commitment Period. A further 5-year period (or an alternative commitment period such as 2013–2020) has subsequently been implemented since 2012.

---

[2] The six GHGs are carbon dioxide, methane, nitrous oxide, halocarbons, per-fluorocarbons and sulphur hexafluoride.

In this carbon cap-and-trade system, each member state is required to set an emission cap and manage allocations for all installations in their country, if they are covered by the EU ETS.

With increasing concerns about how to account for emissions trading schemes, the accounting profession has also paid attention to the measurement and reporting framework required to assist different stakeholders such as investors, rating agencies and analysts (KPMG, 2008). The accuracy in monitoring, measuring and reporting companies' actions against climate change has become increasingly important to stakeholders. The international Carbon Disclosure Project (CDP[3]) is a good example of the activities undertaken by stakeholders, specifically institutional investors. Growing awareness of other stakeholder groups including consumers, media, the scientific community, competitors and companies in other industries has also emerged (Pleon Climate Change Stakeholder Report, 2007). The Stern Review (2006) was another important milestone in relation to the climate change debate that identified the economic impact of climate change and urged an immediate global response. This review estimated that if society does not act, the overall costs and risks of climate change will be equivalent to losing at least 5 % of global GDP (Gross Domestic Product) each year (HM. Treasury 2006). In 2012 Rio +20, a follow up conference to the 1992 Rio Earth Summit, appeared to foster a deeper understanding that an effective response to climate change can be an engine for economic growth. One of the top global agendas at the Rio+20 conference was corporate climate change-related disclosures and audit requirements. Consequently, corporate support for climate policies is evident in 'a wide range of positive actions including basic technological change, behavioural change, product and process-based innovations, emissions trading and public education' (Okereke, 2007, p. 484). However, a majority of companies are still at an early stage of taking action against climate change (Haque & Deegan, 2010; Pinkse, 2007).

The above discussion highlights the changing trends of climate change-related global concerns and policies and corporate responses to such concerns, as summarised in Table 1. There is now an increased level of public pressure and consequent policies to ensure better monitoring and reporting requirements by corporations worldwide. Initially companies opposed international efforts and regulations to control GHG emissions by questioning the scientific basis of the issue. However, this opposition has shifted to a gradual acceptance as evident in corporate actions and mechanisms to reduce their contribution to climate change.

---

[3] The Carbon Disclosure Project seeks information on the business risks and opportunities presented by climate change by sending questionnaires to the world's largest companies. This project has the support of a total of 385 institutional investors with a combined US$57 trillion of assets under management (www.cdproject.net).

Table 1 Global concerns and corporate responses in relation to climate change

Major climate change-related global policies	Trend of stakeholder engagement	Corporate attitudes towards climate change
United Nations Framework Convention on Climate Change (1992, came into force in March 1994)	• A wide range of stakeholders including public and government started to pay attention (Kolk & Pinkse, 2007). • Establishment of the Intergovernmental Panel on Climate Change (1988) to provide independent scientific advice on the issue of climate change (First IPCC Assessment report, 1990).	Companies opposed international efforts and regulations towards climate change issue (Jeswani et al., 2008; Kolk, 2008; Kolk & Levy, 2001).
Kyoto Protocol (1997) a. joint implementation b. emissions trading c. clean development mechanisms	• Adoption of Kyoto Protocol stimulated the development of regulation (Kolk & Pinkse, 2007). • Increased pressure from NGOs (PLEON, 2007). • Emergence of cooperation between NGOs and corporations (Pleon Climate Change Stakeholder Report, 2007). • Emerging new requirements at the international and state level. • Second IPCC Assessment Report (1995). • Emergence of cross-sector stakeholder partnerships (for example, Greenhouse Gas Protocol Initiative, 1998). • Third IPCC Assessment Report (2001).	• Companies gradually stopped their opposition against regulation and moved to more proactive climate strategies (Kolk, 2008; Kolk & Pinkse, 2004, 2007).
European Union GHG Emission Trading Scheme (EU ETS) (2005) Rio + 20 (2012)	• Growing stakeholder activism in demanding monitoring, measuring and reporting of climate change information (for example, The Carbon Disclosure Project, 2002).	• Companies now appear more concerned about the risks and opportunities associated with climate change (Jeswani et al., 2008; Okereke, 2007).

(continued)

Table 1 (continued)

Major climate change-related global policies	Trend of stakeholder engagement	Corporate attitudes towards climate change
	• Growing awareness of other stakeholder groups including consumers, media, the scientific community, competitors and companies in other industries. • Stern Review (2006). • Fourth IPCC Assessment Report (2007). • At Rio + 20 conference, corporate sustainability disclosure including climate change audit and disclosure was at the top of the global agenda (UNFCC, 2014).	Participation in voluntary emission reduction programs, assessment, monitoring and disclosure of GHG emission data publicly (such as through CDP) (Haque & Deegan, 2010; Pinkse, 2007).

3 Corporate Climate Change-Related Disclosure and Audit Practices

With increasing global concerns regarding climate change, different stakeholders are expressing their interests and expectations about organisations' climate change-related reporting and audit practices, including those of corporations (Haque & Deegan, 2010; Kolk, 2008; Kolk & Pinkse, 2004, 2007; The Association of Chartered Certified Accountants 2011). These groups include NGOs, consumers, media, scientific communities, shareholders, suppliers and professionals (Kolk & Pinkse, 2007; PLEON, 2007) who seek to hold organisations responsible and accountable for the issue. Focusing on the expectations for climate change-related information, Bebbington and González (2008, p. 707) stated that:

> Investors, policy makers and the public in general, therefore, could be expected to need information from which they can assess the carbon intensity of corporate products and services and estimate the regulatory and competitive risks that a corporation is likely to face. Moreover, there is also a need for information on how the organisation manages GHG emissions (and the risks associated with their approach). This is likely to require non-financial accounting and reporting of and about GHG emissions.

This statement demonstrates the expectations of different stakeholder groups for more information than that currently provided in financial reports. At the same time companies are increasingly expected to monitor and audit their own climate change related performance:

> An organization's entire sustainability program needs to be audited to ascertain that the program is not only meeting all its established goals and targets, but also its voluntary

commitments (e.g. the United Nations Global Compact, Carbon Disclosure Project, Sustainability Strategy, etc) (Ernst and Young, 2011, p. 3).

Bebbington and González (2008, p. 708) suggest that in order to reflect a "true and fair view" of corporate climate change-related performance, it is necessary to have performance monitored and reported accurately. Companies that do not disclose information about their climate change-related activities will be subject to various risks compared to their business counterparts who do disclose. For example, investors who rely on company reports may take action if a company's reporting on its GHG emissions, energy use and energy production statements are shown to be incorrect, insufficient or misleading (Liberty White Paper, 2010). Therefore, it is important to regularly review, monitor, and disclose the climate change-related practices of the company. This is reflected in the focus given to social auditing in recent literature (see for example, Deegan, 2002; Hunter & Urminsky, 2003; Merk & Zeldenrust, 2005), and in corporate practices, as many corporations worldwide have embraced social audits as a part of their social responsibility programs (GRI, 2011; Islam & McPhail, 2011). Previous studies in social and environmental accounting literature highlight that social and environmental reporting via annual reports takes place as a response to legitimacy threats or as a tool for maintaining legitimacy (see for example Patten, 1992; Deegan et al., 2000; Deegan, 2002). These studies suggest that the greater the chance of unfavourable shifts in community expectations, the greater will be the need to attempt to influence the process through corporate social and environmental disclosure. This notion appears equally applicable in the context of social auditing.

Social auditing is the process by which organisations can assess their performance in relation to society's requirements and expectations (Elkington, 1997). It can be undertaken with the aim of establishing whether an organisation is complying with its own or other recognised principles and standards (Gray, 2000). If there are concerns from stakeholders, organisations might be motivated to take such a strategy. Social audits might be undertaken for accountability purposes and to try to explain to stakeholders the various social and environmental impacts an organisation might be creating (Deegan, 2002). Social audits, therefore, can be defined as the process by which an organisation determines the impact of its activities on global climate change and measures and reports relevant information to its wider stakeholder groups. Thus, it can be beneficial for corporations to perform regular or annual audits and disclose information. In this context, they are a tool by which an organisation can plan, manage and measure its GHG accounting and reporting, and monitor both the internal and external consequences of these activities.

The precondition for a social audit is something against which companies can assess their performance (Kolk & van Tulder, 2002). Companies use various standards and guidelines for this purpose, while conducting social audits, and publicly disclose in reporting media such as annual reports, and individual social and environmental reports. There is a steadily expanding body of global forums and initiatives that provide standards for corporations' monitoring and reporting of climate change issues. These include, but are not limited to,

- Climate Disclosure Standards Board (CDSB);
- Carbon Disclosure Project (CDP);
- Institutional Investors Group on Climate Change (IIGCC);
- UN and Coalition for Environmentally Responsible Economies (CERES);
- Investor Network on Climate Change (INCR);
- Global Framework for Climate Risk Disclosure;
- World Economic Forum (WEF);
- World Business Council for Sustainable Development;
- World Resource Institute's Greenhouse Gas Protocol; and
- Global Reporting Initiatives (GRI).

These are all working to provide disclosure and audit guidelines for companies who want to address climate change (Global Reporting Initiative, 2007; KPMG, 2008). For example, to enforce carbon-related reporting in annual reports, seven business and environmental organisations have formed a consortium named the Climate Disclosure Standards Board (CDSB), to create the Generally-Accepted Carbon Accounting Principles (GACAP); this provides a framework, called the Climate Change Reporting Framework (CCRF) for climate reporting in annual reports, similar to the generally accepted frameworks that have been created for corporate financial reporting. The proposed reporting framework focuses on the disclosure of climate issues in company annual reports, such as total emissions, assessment of the physical risks of climate change, assessment of the regulatory risks and opportunities from climate change, and strategic analysis of climate and emissions management (Climate Disclosure Standards Board, 2009). CCRF also specifies a minimum level of auditor involvement. This includes the requirement by the International Standards on Auditing (ISA 720) for the auditor of financial statements to read the information accompanying the statements to identify any material inconsistencies between it and the audited financial statements, and to consider any observed material misstatements of fact in those disclosures (Climate Disclosure Standards Board, 2012). However, CDSB encourages organisations to work with their professional advisors to agree on an appropriate assurance approach to disclosures made under the CCRF by reference to existing assurance standards (Climate Disclosure Standards Board, 2012). These include International Standards on Assurance Engagement (ISAE) 3000 and 3410, the International Organization for Standardization's ISO 14064-3:2006 and AccountAbility's AA1000 assurance standard (Climate Disclosure Standards Board, 2012). CDSB is aware of the demand from preparers and users for climate change-related disclosures to be assured, and is following the development of a standard for assurance of GHG statements by the International Auditing and Assurance Standards Board.

Apart from the initiatives taken by global organisations, government initiatives are also taking place in different national contexts. For example, the Australian government has introduced various mandatory and/or voluntary programs to encourage climate change-related corporate reporting (e.g. the Mandatory

Renewable Energy Target (MRET) and Greenhouse Challenge Plus). Another mandatory carbon-related reporting framework, the National Greenhouse and Energy Reporting Act 2007 (the NGER Act), has been established for Australian corporations to monitor, measure and report GHG emissions, reductions, removals and offsets, and energy consumption and production, from 1 July 2008 (The Parliament of the Commonwealth of Australia: Senate, 2007). Greenhouse and energy auditing is a key compliance monitoring measure under the NGER Act. A failure to comply with the National Greenhouse and Energy Reporting (NGER) Act can result in fines of up to $22,000 for Australian companies (Department of Climate Change and Energy Efficiency, 2011). Companies' chief executives will be held personally responsible for failing to report or failing to keep required records or providing false information, with daily penalties of $11,000 for each day of non-compliance. Failure to address these issues will not only leave organisations open to significant corporate and personal liabilities, but may also jeopardise corporations' competitive advantage, and adversely affect investor and financial institutional confidence (Department of Climate Change and Energy Efficiency, 2011).

Consistent with the increasing number of climate change-related standards and guidelines, many organisations worldwide incorporate these into their own practices, requiring the disclosure of relevant information. A number of research studies have examined the climate change-related disclosure practices adopted by corporations. These studies have identified increased levels of voluntary emission disclosures by companies worldwide (ACCA, 2007; Cowan & Deegan, 2011; Freedman & Jaggi, 2008; Friends of the Earth, 2006; Haque & Deegan, 2010; Kolk, Levy, & Pinkse, 2008; Stanny & Ely, 2008). Although there is a lack of extant research into companies' climate change-related audit practices, recent literature shows evidence of multi-national companies adopting social and environmental audit practices (Islam & McPhail, 2011). Considering the risks posed by climate change, companies appear to be faced with the challenge of assessing their own performance. Many companies are now voluntarily disclosing relevant information to stakeholders through media such as annual reports, CSR reports and corporate websites. For example, ANZ,[4] a leading Australian Bank, states on its website that:

> ANZ is committed to measuring, then reducing, and lastly offsetting the carbon emissions from our operations. We do this by: Measuring our global carbon footprint in a manner that is consistent with NCOS[5]; Reducing our carbon emissions with specific targets for reductions in those areas that represent the most significant impact (i.e. premises energy and air travel); and Offsetting our remaining emissions on an annual basis by purchasing and retiring internationally recognised certified carbon offsets, in alignment with NCOS requirements, within 90 days of measuring our annual global emissions...In 2013 KPMG was engaged to conduct independent assurance over ANZ's environmental data. Assurance

---

[4] ANZ is among the top 4 banks in Australia, the largest banking group in New Zealand and Pacific, and among the top 50 banks in the world (http://www.anz.com.au/about-us/our-company/profile/facts/history/).
[5] National Carbon Offset Standard (NCOS).

was provided in accordance with International Standard on Assurance Engagements ISAE 3000 **Assurance Engagements other than Audits or Reviews of Historical Financial Information,** ISAE 3410 **Assurance Engagements on Greenhouse Gas Statements** and **National Greenhouse and Energy Reporting (Audit) Determination 2009**... SGS Australia was commissioned to conduct an independent assurance of the environmental data on the ANZ website. Assurance was provided using the following protocols from the GRI guidelines, AA1000 Assurance Standard and ISAE3000 (ANZ, 2014a).

This statement is an example of how companies are voluntarily adopting audit practices and disclosing information about their climate change-related performance, despite scepticism about the actual measures taken by corporate managers and the effectiveness of their measures to mitigate climate change. We will examine whether the voluntary audit and reporting practices bring any real change in organisations' accountability for climate change in the next section through a case study.

4 Are Companies Doing Enough? The Case Context

While corporations are making commitments to mitigate climate change, measuring their own performance and disclosing relevant information via reporting media including annual reports, CSR reports, and press releases (Haque & Deegan, 2010; Rankin, Windsor, & Wahyuni, 2011), there appears little change in their actual performance. There are significant concerns from different stakeholders, including media and NGOs, about corporations' irresponsibility with respect to their GHG emissions (Greenpeace, 2013a).

To understand the particular context of companies' contributions to climate change, this study presents a case that examines the joint actions of two well-known Australian companies: a bank, ANZ, and Whitehaven coal mine, in NSW. The interactions of these two organisations attracted the attention of environmental NGOs and activist groups, and their local communities. The case specifically highlights the assessment of climate change impacts of the proposed expansion of Whitehaven.

ANZ Bank has been under scrutiny for its commitment to the environment for the last few years (Wilson, 2007). Since January 2008, ANZ has loaned over $2.3 billion to coal and gas export projects along Australia's eastern seaboard, including $1.1 billion to projects within the Great Barrier Reef World Heritage Area (Greenpeace, 2013b). In 2012 ANZ faced heavy criticism for investing $1.2b in Whitehaven Coal's Maules Creek coal mine in the Leard State Forest in NSW, a place where, according to the incumbent NSW Planning Minister Brad Hazzard, it was 'illogical' to situate an open cut mine (Sydney Morning Herald, 2013). ANZ is the leading lender to Whitehaven Coal's Maules Creek coal mine, which is twice as large as any other new coal mine currently under construction in Australia (The Australian, 2013). Whitehaven's Maules Creek mine is inherently risky and faces growing opposition due to its impacts on health, land use, water, native habitat, and the climate, as the coal extracted from Maules Creek mine will release 30 million tonnes of CO_2 emissions per year. This is 7 million tonnes more than the entire transport sector in NSW (Greenpeace, 2013b). In addition, the Maules Creek coal mine would destroy up to 2,000 ha of the Leard State Forest, home to koalas and other vulnerable species (Climate Citizen, 2013).

The mine was opposed by the Maules Creek Community as well as a number of environmental organisations, including Greenpeace, the Lock the Gate Alliance, the Northern Inlands Council for the Environment, the National Parks Association and the Nature Conservation Council of NSW (The The Australian, 2013; Climate Citizen, 2013). These anti-coal mining activists have been campaigning against ANZ in order to stop its financing of the proposed Whitehaven coal mine. Community concerns regarding the climate change impacts of the Maules Creek coal mine were highlighted by local residents, bank customers, and NGOs, especially considering ANZ's previous commitment to mitigating climate change. Community activists and volunteers around Australia joined Greenpeace in a number of actions to pressure the big bank. For example, a blockade camp was established at the mine site, aiming to delay and eventually stop the project from proceeding (Greenpeace, 2013b). In protest against ANZ's fossil fuel lending policy to Whitehaven, dozens of ANZ customers have also reportedly closed their accounts (Greenpeace, 2013b).

ANZ adopted the Equator Principles on 15 December 2006, voluntarily committing to not lend money to projects that had a negative social or environmental outcome (ANZ, 2014a). The Equator Principles Financial Institutions (EPFI) voluntarily agree not to provide "project related loans and project finance advisory services to projects where the borrower will not, or is unable to comply with, the Equator Principles" (Equator Principles, 2014). The latest version of the Equator Principles, known as EP3, was published on June 4, 2013. It provides, for the first time, risk management tools whereby project finance lenders are able to ensure that climate change is addressed as a key aspect of the identification, assessment, and management of environmental risk in large, complex, and expensive projects (Equator Principles, 2014). In 2007, ANZ announced the launch of the ANZ Climate Change Trust (ACCT), Australia's first wholesale, capital protected climate change investment trust (ANZ media release, 2007). The ACCT is a 6-year fund which will invest in companies that offer products and services that support environmental sustainability and combat the impact of climate change (ANZ media release, 2012). ACCT is also linked to the performance of the Sustainable Asset Management (SAM) Sustainable Climate Fund based in Luxembourg (ANZ media release, 2012). In relation to ANZ's commitment to climate change, the bank's Head of Investor Sales (Institutional), Mr Angus Graham, announced:

> The ACCT will invest in a range of sustainable companies involved in areas such as products for the construction industry that reduce the energy use of buildings, new agricultural systems that help address the effects of drought as well as traditional sources of renewable energy ... The ANZ Climate Change Trust demonstrates that financial and environmental investments are not mutually exclusive (ANZ media release, 2007).

However, it seems those principles do not go far enough in stopping the funding of projects which destroy biodiversity and add substantially to carbon pollution and climate change (Climate Citizen, 2013). ANZ's lending policy to Whitehaven indicates that the bank would be deemed to be failing to comply with its environmental and climate change commitments, including the Equator Principles. Despite these inconsistencies, in September 2012, ANZ was ranked the most sustainable bank globally in the 2012 Dow Jones Sustainability Index (DJSI) for the fifth time in 6 years (ANZ media release, 2012).

ANZ's lending to Whitehaven Coal carries heavy risks, given Whitehaven's declining performance, including a drop in its share price of 66 % since January 2012 and a net loss of $82.2 million in 2013 (Greenpeace, 2013c). On 7 January 2013, a media release, purportedly from the bank, announced the bank was withdrawing its $1.2 billion loan for Whitehaven for the development of the new coal mine at Maules Creek. The announcement also highlighted ANZ's current undertaking of a review of coal and gas investments on productive agricultural lands and areas of high biodiversity (Climate Citizen, 2013). However, the announcement was proved to be a hoax, sent out by an anti-coal activist claiming to represent the ANZ Bank, using bank letterhead (Climate Citizen, 2013). Shortly after the bogus announcement, shares in Whitehaven plummeted on the Australian Stock Exchange, with the company stock losing almost 9 % from $3.52 to $3.21 in a fall that reduced the value of the company by about $314 million. Whitehaven was put into a trading halt, but trade later resumed in the afternoon with the share price closing at $3.50, just 2 cents down on the day (Brisbane Times, 2013).

The ANZ Bank responded with a brief media statement on its website:

ANZ today became aware of a fraudulent media release purporting to be from ANZ falsely stating that funding has been withdrawn from Whitehaven Coal. This media release is a hoax and was not issued by ANZ. There have been no announcements from ANZ regarding Whitehaven Coal. ANZ remains fully supportive of Whitehaven Coal (ANZ media release, 2013).

Consequently, activist groups demanded that ANZ Bank should not fund the project as "the mines do not comply with the Equator Principles for Financial Institutions in relation to cumulative assessment, biodiversity conservation, health, occupational safety, cultural heritage, land conservation and the promotion of renewable energy" (Climate Citizen, 2013). In response to ANZ's investment policy Greenpeace stated on its website that:

ANZ is the biggest investor in polluting coal power in Australia. To solve the climate crisis we need to ensure no new coal power stations are built, as they will lock us into decades more of pollution. We've been calling on ANZ to make a commitment not to finance any new coal power stations in Australia and instead lead the clean energy revolution (Greenpeace, 2013d).

An investigation of ANZ's entire reporting media including annual reports, CSR reports and its own websites revealed that there was no recognition of Whitehaven Coal's Maules Creek mine within ANZ annual reports or on their website despite ANZ being the leading lender to the project. While ANZ's own GHG reductions between 2011 and 2012 amounted to 15, 313 tonnes, coal from the Maules Creek coal mine project will generate annual emissions almost 2000 times greater than those saved by ANZ staff (ANZ, 2012). Despite that, ANZ's 2011, 2012 and 2013 Annual Reports did not discuss any involvement in the Maules Creek coal mine project as well as the possible impacts of its operation on the GHG emissions and climate change (ANZ, 2014a).

Further, an investigation of Whitehaven Coal's annual reports and websites for evidence of ANZ's involvement in the project was also revelatory. Whitehaven's annual report 2009 and 2010 disclosed that ANZ was among its top 20 largest shareholders, but did not recognise whether and how ANZ was involved in the Maules Creek project. This information was not found in Whitehaven's 2011, 2012, and 2013 annual reports either, despite the fact that Whitehaven was financed by ANZ. Whitehaven's annual reports did not mention anything about the likely impact/amount of GHG emissions from the project, nor did they discuss the continued activist/community campaigns or protests against their project, or what corrective actions they might be taking in response to communities' complaints of likely GHG emissions. Thus, Whitehaven did not acknowledge any likely impact of climate change from the Maules Creek project within its reporting media.

ANZ online disclosures suggested that one important set of guidelines the bank rigorously embraced was the Equator Principles. ANZ developed Sensitive Sector policies for Energy, Extractives, Forests and Forestry and other sectors, committing to ensure that social and environmental considerations are incorporated into their lending decisions:

> The Equator Principles is a set of voluntary standards designed to help banks identify and manage the social and environmental risks associated with the direct financing of large infrastructure projects such as dams, mines or pipelines. We have been signatories to the Equator Principles since 2006. The Principles are applied to all project structured finance transactions. Their use provides a clear, structured process to identify, mitigate, manage and monitor social and environmental risks. Use of the Principles across the banking industry means customers are able to provide social and environmental assessments to one standard, acceptable to banking syndicates (ANZ, 2014b).

According to this statement, ANZ should only finance projects within the Principles' scope, developed according to sound social and environmental standards. However, ANZ is a continuing financier of Whitehaven Coal, yet did not acknowledge this within its reporting media. While ANZ promised to implement Equator Principles, it is doubtful to what extent it has really done this. Equator Principles explicitly require banks to assess and disclose each and every aspect of their impacts upon the community and local stakeholders. Although there was a massive community protest on its continued funding to the Whitehaven project, ANZ did not integrate and acknowledge these issues in its reports.

The case of ANZ leads us to conclude that there has not been enough done to make corporations accountable for their impact on climate change. ANZ's audit and disclosure practices appear to be symbolic or ritual strategy for maintaining legitimacy rather than being a means of discharging corporate accountability, or improving the welfare of stakeholder groups. Symbolic legitimation strategy involves organisations achieving acceptance without actually changing the way they perform or their activities. They 'appear consistent with social values and expectations' but no real change has taken place (Ashforth and Gibbs, 1990, p. 180). The commitment made by ANZ about mitigating, managing and monitoring climate change does not seem to actually reflect the underlying processes and motivations. There is an apparent disconnection between the claimed adoption of social standards such as the Equator Principles and the disclosure of information on one hand, and the real change in corporate accountability in relation to mitigating climate change on the other.

Conclusion
This chapter has examined the way the expectations of different stakeholder groups have changed over recent years in relation to corporations' climate change-related audit and disclosure practices, and the consequent increase in climate change-related monitoring and reporting guidelines worldwide. Research has found that with this increasing trend, corporations appear to incorporate global standards and guidelines into their climate change-related audit practices as part of their social auditing, and disclose relevant information. However, through a case study, the chapter has demonstrated that little has been done to create actual corporate accountability in relation to climate change. Corporate discourse on climate change can be termed as a 'symbolic legitimation' strategy rather than creating any change on the ground. Where corporate auditing and disclosure on climate change has evolved over the years, it has not necessarily reflected real action and effectiveness, and therefore has not demonstrated true accountability to society. Hence it is argued that international organisations and government bodies are not doing enough to create change in corporate accountability as they only recommend voluntary disclosure in this area. As social audits are a voluntary activity, possibly sometimes used only as a legitimation tool by companies, one can be skeptical about whether such audits can make a real change in their actual practices. Without appropriate regulation or enforcement of social auditing standards, the accountability and obligations of global companies to mitigate climate change remains negligible. A radical approach, such as mandatory monitoring (compliance audit) and disclosure requirements, is necessary to ensure corporate accountability in relation to climate change. There should be uniform carbon accounting, monitoring and reporting guidelines across the globe. Regulation and mandatory enforcement of social auditing standards is necessary to discipline corporate operations and related disclosures in relation to climate change.

The issue deserves more research attention. More investigation is needed into areas and aspects of carbon emissions measurement, integration, reporting and auditing that may ultimately contribute to a body of evidence

(continued)

which will be compelling, and to encourage, organisations to increase their transparency in this important area.

References

ACCA. (2007). *Climate change: UK Corporate reporting- an analysis of disclosure in UK corporate reports*. Certified Accountants Educational Trust, February, 2007, London.
ANZ. (2012). *Analysis of ANZ's carbon disclosure statements*. http://www.anz.com.au/about-us/corporate-responsibility/environment/targets-performance/carbon/.
ANZ. (2014a). *Corporate responsibility*. www.anz.com/about-us
ANZ. (2014b). *Environment: Equator principles*. http://www.anz.com.au/aus/values/environment/Equator.asp
ANZ media release. (2007). *ANZ launches Climate Change Investment Trust*, 19 November, 2007. http://media.corporate-ir.net/media_files/irol/24/248677/mediareleases/2007/ANZ-MediaRelease-20071119.pdf.
ANZ media release. (2012). ANZ ranked lead bank globally in Dow Jones Sustainability Index September2012, http://www.anz.co.nz/resources/e/b/eb3768804cd4ee389915ffc0c7ad410a/MediaRelease20121409.pdf?MOD=AJPERES&CACHEID=eb3768804cd4ee389915ffc0c7ad410a.
ANZ media release. (2013). *Fraudulent media release regarding Whitehaven Coal*, 7 January, 2013, www.media.anz.com.
Ashforth, B. E., & Gibbs, B. W. (1990). The double-edge of organizational legitimation. *Organization Science, 1*, 177–194.
Bebbington, J., & González, C. L. (2008). Carbon trading: Accounting and reporting issues. *European Accounting Review, 17*(4), 697–717.
Brisbane Times. (2013). *Hoax press release sparks whitehaven Plunge*, January, 2013, http://www.brisbanetimes.com.au/business/mining-and-resources/hoax-press-release-sparks-whitehaven-plunge-20130107-2cc47.html
Carbon Disclosure Project. (2008). *Carbon Disclosure Project 2008: Australia and New Zealand*. Available at: http://www.cdproject.net/reports.asp.
Carbon Disclosure Project. (2013). *CDP Australia and New Zealand Climate Change Report 2013: On behalf of 722 investors representing US$87 trillion in assets*. https://www.cdp.net/CDPResults/CDP-Australia-NZ-Climate-Change-Report-2013.pdf
CERES. (2002). *Value at risk: climate change and the future of governance*. Boston, USA.
Climate Disclosure Standards Board, CDSB. (2009). *Promoting and advancing climate change-related disclosure*. Copenhagen Update COP 15, December 2009, available at http://www.cdsb-global.org/uploads/pdf/cdsb_copenhagen_update.pdf.
Climate Disclosure Standards Board. (2012). *Climate change reporting framework– Edition 1.1, October 2012: Promoting and advancing climate change-related disclosure*. http://www.cdsb.net/sites/cdsbnet/files/cdsbframework_v1-1.pdf
Climate Citizen. (2013) Uncovering the real hoax: ANZ Bank greenwash while financing coal and climate change, January 2013, http://takvera.blogspot.com.au/2013/01/uncovering-real-hoax-anz-bank-greenwash.html
Cowan, S., & Deegan, C. (2011). Corporate disclosure reactions to Australia's first national emission reporting scheme. *Accounting and Finance, 51*(2), 409–436.
Deegan, C., Rankin, M., & Voght, P. (2000). Firms' disclosure reactions to major social incidents: Australian evidence. *Accounting Forum, 24*(1), 101–130.
Deegan, C. (2002). The legitimizing effect of social and environmental disclosures: A theoretical foundation. *Accounting, Auditing & Accountability Journal, 15*(3), 282–311.

Department of Climate Change and Energy Efficiency. (2011). *The National Greenhouse and Energy Reporting (NGER) scheme.* Available at http://www.climatechange.gov.au/government/initiatives/national-greenhouse-energy-reporting/publications/overview.aspx

Elkington, J. (1997). *Cannibals with Forks: The triple bottom line of 21st century business.* Oxford: Capstone.

Ernst and Young. (2011). *Five highly charged risk areas for internal audit.* 2011forumchi.greenbiz.com/.../ErnstAndYoung_Five_Highly_Charged_Risk_Areas_for_Internal_Audit.pdf

Equator Principles III. (2014). *Equator principles III: An introduction and practical guide.* http://www.nortonrosefulbright.com/files/equator-principles-iii-pdf-17mb-111048.pdf

Freedman, M., & Jaggi, B. (2008). Global warming and corporate disclosures: A comparative analysis of companies from the European Union, Japan and Canada, presented in *Critical Perspectives on Accounting Conference*, 2008, New York.

Friends of the Earth. (2006). *Fifth Survey of Climate Change Disclosure in SEC Filings of Automobile, Insurance, Oil & Gas, Petrochemical, and Utilities Companies*, Friends of the Earth, USA. Available at www.foe.org/camps/intl/SECFinalReportandAppendices.pdf.

Global Reporting Initiative & KPMG. (2007). *Reporting the Business Implications of Climate Change in Sustainability Reports*, a survey conducted by the Global Reporting Initiative and KPMG's Global Sustainability Services, www.globalreporting.org/NR/rdonlyres/C451A32E-AO46-493B-9C62 7020325F1E54/0/ClimateChange_GRI_KPMG07.pdf.

Gray, R. (2000). Current developments and trends in social and environmental auditing, reporting and attestation: A review and comment. *International Journal of Auditing, 4*, 247–268.

Greenpeace. (1998). *The oil industry and climate change.* Amsterdam, The Netherlands: Greenpeace International.

Greenpeace. (2013a). *Dealing in doubt: The climate denial machine Vs climate science.* http://www.greenpeace.org/usa/Global/usa/report/Dealing%20in%20Doubt%202013%20-%20Greenpeace%20report%20on%20Climate%20Change%20Denial%20Machine.pdf

Greenpeace. (2013b). *Disconnected ANZ's fossil fuel financing.* http://www.marketforces.org.au/wp-content/uploads/2013/12/Annual-Report-ANZ_FF.pdf

Greenpeace. (2013c). *Greenpeace analysis of BREE's Resources and Energy Major Projects*, October 2013 report.

Greenpeace. (2013d). *Our actions against ANZ.* http://www.greenpeace.org/australia/en/what-we-do/climate/Polluting-our-future/How-Australias-big-four-banks-use-your-money-to-fund-polluting-power/ANZ---The-diritest-bank/Our-actions-against-ANZ/.

GRI. (2011) *GRI report list.* www.globalreporting.org/ReportServices/GRIReportsList

Haque, S., & Deegan, C. (2010). An exploration of corporate climate change-related governance practices and related disclosures: Evidence from Australia. *Australian Accounting Review, 20*(4), 317–333.

Hunter, P., & Urminsky, M. (2003). *Social auditing, freedom of association and the right to collective bargaining.* ILO, Geneva: Multinational Enterprises Programme. Accessed at: http://www.ilo.org/public/english/dialogue/actrav/publ/130/8.pdf.

Intergovernmental Panel on Climate Change (IPCC). (2007). *The physical science basis. Fourth Assessment Report, 17 November 2007.* Spain: Valencia.

Islam, M. A., & McPhail, K. (2011). Regulating for corporate human rights abuses : The emergence of corporate reporting on the ILO's human rights standards within the global garment manufacturing and retail industry. *Critical Perspectives on Accounting, 22*(8), 790–810.

Jeswani, H. K., Wehrmeyer, W., & Mulugetta, Y. (2008). How warm is the corporate response to climate change? Evidence from Pakistan and the UK. *Business, Strategy and the Environment, 18*, 46–60.

Kolk, A. (2008). Developments in corporate responses to climate change in the past decade. In B. Hansjurgens & R. Antes (Eds.), *Climate change, sustainability development and risk: An economic and business view.* Heidelberg/New York: Physica Publishers.

Kolk, A., & Levy, D. (2001). Winds of change: Corporate strategy, climate change and oil multinationals. *European Management Journal, 19*(5), 501–509.

Kolk, A., & Pinkse, J. (2004). Market strategies for climate change. *European Management Journal, 22*(3), 304–314.

Kolk, A., & Pinkse, J. (2007). Multinationals political activities on climate change. *Business and Society, 42*(2), 201–228.
Kolk, A., Levy, D., & Pinkse, J. (2008). Corporate responses in an emerging climate regime: The institutionalization and commensuration of carbon disclosure. *European Accounting Review, 17*(4), 719–745.
Kolk, A., & van Tulder, R. (2002). The effectiveness of self-regulation: Corporate codes of conduct and child labour. *European management Journal, 20*(3), 260–71.
KPMG. (2008). *International Survey of Corporate Responsibility Reporting 2008*, Amsterdam.
Labatt, S., & White, R. (2007). *Carbon finance: The financial implications of climate change*. New Jersey: John Wiley & Sons Inc.
Liberty White Paper Series. (2010). *Climate change: The emerging liability risks for directors and officers, Liberty International Underwriters*. Available at http://www.liuaustralia.com.au/downloads/7170782d-c256-4e00 93cf3fe1585bcf42/Liberty%20White%20Paper%20Series%20_%20Climate%20Change%20Booklet%20(AUS).pdf.
Merk, J., & Zeldenrust, I. (2005). The business social compliance initiative. A critical perspective. Accessed on 28 September, 2013 at: www.cleanclothes.org/ftp/05-050bsci_paper.pdf.
Okereke, C. (2007). An exploration of motivations, drivers and barriers to carbon management: The UK FTSE 100. *European Management Journal, 25*(6), 475–486.
Patten, D. M. (1992). Intra-industry environmental disclosures in response to the Alaskan oil spill: A note on legitimacy theory. *Accounting, Organisations and Society, 17*(5), 471–475.
Pinkse, J. (2007). Corporate intentions to participate in emission trading. *Business Strategy and the Environment, 16*, 12–25.
Pinkse, J., & Kolk, A. (2007). Multinational corporations and emissions trading: Strategic responses to new institutional constraints. *European Management Journal, 25*(6), 441–452.
Pleon Climate Change Stakeholder Report. (2007) *Multi-stakeholder partnerships in climate change*: A leadership agenda towards 2012... and beyond, www.pleon.com, Pleon b.v.
Rankin, M., Windsor, C., & Wahyuni, D. (2011). An investigation of voluntary greenhouse gas emissions reporting in a market governance system: Australian evidence. *Accounting, Auditing and Accountability Journal, 24*(8), 1037–1070.
Rolph, B., & Prior, E., (2006). *Climate change and the ASX100 - An assessment of risks and opportunities*. Published by Citigroup Research, www.sunenergy.com.au/pdf/CitigroupClimateChangeReportFeb2007.pdf.
Stanny, E., & Ely, K. (2008). Corporate environmental disclosures about the effects of climate change. *Corporate Social Responsibility and Environmental Management, 15*(6), 338–348.
Stern, N. (2006). *Stern review report on the economics of climate change*. Available at http://www.hmtreasury.gov.uk/independent_reviews/stern_review_economics_climate_change/stren_review_report.cfm
Sydney Morning Herald. (2013). *Manning, Paddy 2013, mining for controversy*. The Sydney Morning Herald, January 29.
The Association of Chartered Certified Accountants. (2011, May). *Audit under fire: a review of the post-financial crisis inquiries*. http://www.accaglobal.com/content/dam/acca/global/PDF-technical/audit-publications/pol-af-auf.pdf
The Australian. (2013). Whitehaven coal falls to $82m loss. *The Australian*, August 27.
The Parliament of the Commonwealth of Australia: Senate. (2007). *National Greenhouse and Energy Reporting Bill 2007: Revised Explanatory Memorandum*, available on the internet at http://parlinfoweb.aph.gov.au/piweb/Repository/Legis/ems/Linked/18090701.pdf
UNFCC. (2004). *The first ten years* (The United Nations Framework Convention on Climate Change). Bonn, Germany: Climate Change Secretariat.
UNFCC. (2014). *First steps to a safer future: Introducing The United Nations framework convention on climate change*. http://unfccc.int/key_steps/the_convention/items/6036.php
Wilson, T. (2007, July). *No, really—what are the 'Equator Principles'?* Institute of Public Affairs. http://www.ipa.org.au/library/59_2_WILSON_EquatorPrinciples.pdf

Social Audit in the Supply Chains Sector

Samuel O. Idowu

1 Introduction

Effecting the ethos of social auditing in an organisation expects that the organisation in question will take a structured approach in evaluating how well it is doing in its quest to be socially responsible Andersen and Skjoett-Larsen (2009) and Awaysheh and Klassen (2010). The approach expects that the entity will set out a plan of the actions it wants to take and ensure that it follows them through. Carrying out a social audit exercise either in the supply chains sector or in any other sector is similar to carrying out a financial accounting audit of a limited liability company. A financial accounting audit process requires an independent examination of the financial records of assets and liabilities of a company as at a point in time; usually towards the end of its accounting year (or at any point in time by the company's internal auditors) in order to establish whether or not the financial records are accurate and reflect the true state of events as depicted in the company's financial statements which a company wants to put in the public domain for readers and users. A financial accounting audit exercise would also try to identify if there are weaknesses in the internal control systems and record keeping exercise in place within the company which could encourage dishonest and fraudulent employees or others to defraud the company. A social audit is not too dissimilar to this, it is also about establishing the accuracy of the details contained in the social records and activities put in place by the entity being audited, it ensures that the organisation is in control of the social aspects of its operations. The audit exercise could either be carried out by the organisation's internal or external social auditors. The objective

---

S.O. Idowu (✉)
London Guildhall Faculty of Business and Law, London Metropolitan University, 84 Moorgate, London EC2M 6SQ, United Kingdom
e-mail: s.idowu@londonmet.ac.uk

© Springer International Publishing Switzerland 2015
M.M. Rahim, S.O. Idowu (eds.), *Social Audit Regulation*, CSR, Sustainability, Ethics & Governance, DOI 10.1007/978-3-319-15838-9_10

is to ensure that the organisation is in control of this non-financial aspect of its activities to prevent things going wrong.

There are many differences between a financial accounting audit and a social audit. The main difference being that a social audit is generally voluntary in most countries of the world, just like CSR reporting is still voluntary in many parts of the world (except in a few countries which have in place what is often referred to as "Mandatory CSR (MCSR) reporting" namely: Sweden, Denmark, Norway, The Netherlands, France, Australia, Mauritius, Indonesia and India). In these aforementioned countries, reporting on CSR is now mandatory and social auditing of CSR activities and consequently social auditing is also mandatory. It follows then that carrying out of social auditing of CSR activities in most nations of the world is still voluntary. But conducting a financial audit is a mandatory requirement for all listed companies worldwide. There are International Financial Reporting Standards (IFRS) for carrying out this reporting function. Another difference between the two is that many of the issues involved in social auditing might not be tangible but rather qualitative when compared with those issues involved in financial accounting audit which in most cases will be in terms of assets, liabilities and other monetary transactions of the company in question. This difference makes things a bit more challenging for everyone. Despite this, it is a desirable exercise which should help to give credence to CSR reporting and increase users' confidence in the information provided by the CSR report. Undertaking a social audit and coming out of the exercise satisfactorily is tantamount to a motor vehicle being taken by its owner for a MOT (Ministry of Transport) road worthiness test. Passing the test and being certified by the MOT examiner as fit to be used on the road. The MOT test certificate confirms that all being well, the vehicle will pose no danger to other road users if driven carefully and responsibly by a qualified driver. A company which has been honestly audited by a team qualified social auditors and being certified as fit for purpose by this team of social auditors poses no danger to all stakeholders and the environment as at the point of the audit and hopefully shortly after that. Companies are expected to be regularly audited to prevent laxities from setting in.

2 Problems in the Supply Chain Sector

The supply chains sector has been besieged by a series of unpleasant events and scandals since corporate social responsibility has become a 'near mainstreamed' event on the corporate scene around the world. Many irresponsible practices have been unveiled by the media and research studies in the supply chains sector of the global economy; see for example Wood (1996), Amaeshi, Osuji, and Nnodim (2008). Most of these problems have been identified in developing countries around the world, for example in Indonesia, Bangladesh, India, Pakistan, Cambodia, and China, to name but a few. See for example Locke and Romis (2012) who note that over the course of the 1990s, Nike was criticised for sourcing its products in factories and countries where low wages, poor working conditions and human rights problems were rampant. Locke and Romis (2012) argue that these problems were evident in their factories in Indonesia, Cambodia, Pakistan, China and Vietnam. A similar version of these social problems are also evident in the farming and mining sectors in Africa—in the farming of cocoa, coffee, tea, banana and other farm produce sold by popular retailers in Europe and North America and in the mining of many high value natural resources in many countries in Eastern, Western and Southern Africa. The FairTrade Foundation came into being to address some of the social ills which had besieged workers and farmers who operate in the farm produce sector around the world. In terms of problems in the mining sector, the Extractive Industries Transparency Initiative (EITI) was set up to bring about improvements in different areas of the sector including CSR issues.

To ignore some big economic differences the sourcing of supplies by large Western retailers and brand names from emerging economies are making to peoples' lives in these communities and countries, and dwell only on some unacceptable practices by some factory owners in these countries is to present a one sided picture of events, which certainly is not the objective of this paper. One should also not overlook the economic mysteries and deprivations that could be created in the communities and countries where supplies were originally sourced before being diffused to these emerging economies, perhaps for cost reduction/saving reasons but packaged and publicised by the diffuser in the name of corporate social responsibility. Unfortunately, the economic problems created or solved as a result of abandoning or diffusing supplies from one community/country to another are not the main concern of this Chapter, but suffice it to note that some social problems solved by sourcing supplies from one emerging nation in terms of job creation and other economic benefits, would also create many social problems in another nation or nations where those jobs were taken away. The chapter seeks to explore some of the actions taken by different organisations from around the globe to improve the effectiveness of social auditing and reporting in the global supply chains sector in developing nations and examine how the process could add value to the field of corporate social responsibility and genuinely make a big difference in people's lives.

3 Historical Review of Social Auditing in the Supply Chains Sector

In this section, we look at the history of social auditing; how it commenced in general terms, how it has permeated into different sectors including the supply chains sector, how things have moved on since its origins and where we are today.

A search of the literature has revealed that discussions on 'social audit' emanated in the 1950s in the United States of America about the same time the American Economist *Howard R Bowen* posited in his book on "Social

Responsibilities of Businessman" in 1953 which notes that business owes a responsibility to society and business people should pursue strategies which are desirable in meeting societal objectives and values Idowu (2010, 2014). This perhaps suggests that social audit is as old as 'the modern version of CSR' itself since scholars have noted the existence of CSR in societies around the world well before the 1950s Carroll (1999), Lee (2008), Idowu (2008) and Carroll (2008).

The term 'social audit' was said to have been coined and used for the very first time by George Goyder in the 1950s, (by an American, who felt that financial auditing without social auditing provides an incomplete one sided story of events in corporate activities) as noted on the website of the *Social Audit Network* (SAN, Pearce, 2005). According to the SAN which notes that "the need of social audit in the 1950s was to make business more accountable to the community and to ensure that the impacts of business—both beneficial and adverse are understood by society". This suggests that the expectations from social audits around the world today have not changed drastically from that of the 1950s when the term was first coined. Social auditing is still about enabling businesses to be accountable and ensuring that stakeholders understand both the adverse and beneficial impacts of business actions on them and the environment. Goyder's idea, was to enable society to exert some control and influence over corporate activities from the local level where most companies often start to flourish before they expand and some of them transform to becoming multinationals and global corporations. The idea behind the practice of social auditing was originally to make companies more accountable and encourage transparency for their actions in and around the community where they operate in order to enable local stakeholders to understand both the beneficial and not so beneficial impacts of corporate actions on them and also to understand what these companies are doing or intend to do in order to reduce the incidence of the adverse impacts of their operations on their stakeholders and the environment.

That said, the globalisation of the world economy has meant that being accountable for a business impacts (both adverse and beneficial) on society cannot now only be about accountability and transparency to the local community but to the global community. As local accountability was the case during its beginnings, this paper contends that social auditing must now be considered and addressed by corporations taking a more inclusive and global approach in the twenty-first century. This became more so because of some of the reasons noted by Locke and Romis (2012) when they argue that in many of these developing countries from where some notable brands are sourcing supplies of their merchandise, there are many compelling reasons why these Western retailers and household names must employ the use and practice of social auditing in order to take 'the bull by the horn'. Locke and Romis (2012) note the limited capacity of many developing country governments to deal with the salient issues affecting the sector for example these governments' inability to enforce and police even their own labour and health and safety laws. These outsourcers' failure to act in this regard would have resulted in even more serious scandals and embarrassments relating to child labour, sweatshops, human rights abuses etc. for these corporations. Being actively involved in the use and practice of social audit became inevitable. Apart from that, we now live a more civilised world which makes it our concern to fight for the rights and welfare of people who live thousands of miles away from us, even when the social or environmental problems in question have nothing to do with us either directly or indirectly. See for example the activities of the Anti-Apartheid Movement of the 1970s and beyond during black struggle for equality in South Africa and several other examples in the past.

The recent vogue by some companies in the more advanced parts of the world to engage in outsourcing some of their business functions and activities to some developing parts of the globe has heightened the need for social audits to be taken more seriously than ever before in the supply chains and other sectors. Mullins (1996) notes that outsourcing did not formally become identified as a business strategy until 1989 but then, Mullins notes "it was all about cost cutting, landing the big deal and making a big splash with little or no regard for the subsequent inadequacies which have not been carefully thought through before landing the deal". Outsourcing became a business strategy for two reasons, first, organisations were and still not totally self sufficient in meeting their functional needs; hence the need to outsource some of the functions which they had little or no competence in arose Handfield (2006) and secondly for competitive advantage and cost saving reasons. It became apparent later on that issues surrounding outsourcing were not carefully thought through, because as far back as the 1990s, shortly after it became popular there had been various scandals and unacceptable practices which had besieged the sector and resulted in untold reputational damage to many of these Western outsourcers, retailers and brand owners. It became important that the use of social audits as required by CSR would permeate and filter through to the supply chains sector since the need to repair damaged reputations arose and the need to guide against future reputational damage became even more important. It is therefore no surprise to see the rapid growth in the field of social auditing.

4 International Standards in the Field of Social Audit

It also became inevitable that we would see the emergence of international standards and guidelines designed to regulate and ensure effectiveness in the use of social auditing in both the supply chains and other sectors. There are a few international standards which actors in the sector have to take cognisance of; we shall look at three of them in this section of the Chapter.

4.1 Social Accountability 8000 (SA8000)

The SA8000 was issued by Social Accountability International (SAI)—a non-governmental organisation (NGO), multi-stakeholder entity with a mission to advance the integrity and rights of workers irrespective of the industry or sector they work in around the world. The SAI notes on its website that its SA8000 which was launched in 1997 is the world's first auditable social certification standards for decent workplaces across all industrial sectors including those sectors that work in the supply chains. The standard is based on international conventions of the International Labour Organisation (ILO) and the United Nations (UN) and it covers nine important employment and work based issues under the following headings since many of the atrocities committed by irresponsible third world suppliers happen during the course of employment. SAI reviews the standard every five years to ensure its currency and usability.

- Child Labour—forbids any support and use of children in factories and other places of employment.
- Forced and Compulsory Labour—makes it glaringly clear that the practice of forced or compulsory labour should not be used in any industry anywhere in the world. Employees must choose out of their own freewill to want to work for an employer and freely choose to end the association when they want to.
- Health and Safety—this element provides some guide with regard to employers' responsibility for safe and healthy workplaces.
- Freedom of association and right to collective bargaining—it specifies that employees should freely choose to organise themselves into trade unions and collectively bargain with their employers for employment related issues such as pay, working conditions etc.
- Discrimination—this aspect of the guideline bans all forms of employment related discrimination.
- Disciplinary Practices—this part of the guide forbids any form of inhumane treatments of employees.
- Working Hours—this part of the standard specifically notes that the working hours per week should not exceed 48 h.
- Remuneration—this notes that a worker must be paid a 'living wage' which it defines as one which enables the worker to support half the basic needs of an average sized family based on prices in the locality of the worker.
- Management Systems—this requires top management of an entity to explain in writing in workers own language its policy for social accountability and labour conditions and display it in an easily viewable position in the company premises.

Basically, the nine areas covered in the standard are designed to ensure that employers of labour worldwide adopting the standard behave responsibly and comply with international standards on the employment of labour. Addressing all these nine points connotes that the entity has covered areas that could lead to unacceptable employment related practices. The entity will then be required to regularly bring in trained social auditors to inspect and certify whether or not the entity has satisfactorily dealt with all the nine issues without infringing or compromising the position of all its employees, that is simply what the social auditing entails. It is normal for the social auditors to write a report of their findings during the process of carrying out the exercise and suggest where things need to improve if they were to believe that should happen. If generally the auditors are happy with the standards put in place by an organisation in the nine aforementioned areas they might issue a certificate to the entity for a specific period of time valid until the next social auditing of the entity's premises and activities as required by SA8000.

4.2 AA1000: The AA Series of Standards

The AA1000 is a series of standards issued by AccountAbility an organisation set up in in 1995 which prides itself as providing innovative solutions to the most critical CSR and Sustainable Development challenges facing global corporate entities. The AccountAbility standards cover different areas of interest to modern stakeholders. They are designed to help organisations to become more accountable, transparent, responsible and sustainable covering the following issues which are of interest to the global community:

- AA1000(APS)—Accountability Principles Standard
- AA1000(AAS)—Accountability Assurance Standard
- AA1000(SES)—Stakeholder Engagement Standard

Again, like the SA8000 we considered above, the AA1000 series issued by AccountAbility organisation provides a set of guidelines to companies in three different areas of interest to CSR and society in general. The aim is that any corporate entity regardless of their sector that embeds the requirements of the three areas in its operational practices would come out of any social audit with a clean certificate of compliance with the ethos of corporate social responsibility advocate.

4.3 The Ethical Trading Initiative

The Ethical Trading Initiative (ETI) was established in 1998 with the objective "to improve the lives of workers in global supply chains". This objective suggests that the ETI is about helping workers in the supply chains to carry out their operational functions under acceptable working conditions and environments. The ETI notes that to them, ethical trade means all employers of labour in the supply chain must ensure that they improve the working conditions and lives of all those people responsible for making the products they sell. The ETI operates under what it calls 'a base code' similar to the nine SA8000 areas discussed above. The ETI's nine areas are bullet pointed below:

- Employment must be freely chosen
- Freedom of association and the right to collective bargaining must be respected
- Working conditions must be safe and hygienic

- Child labour shall not be used
- Living wages must be paid to workers
- Hours of work must not be excessive
- No form of discrimination should be practiced
- Regular employment must be provided to workers
- No harsh or inhumane treatment is allowed

Source: ETI website

The ETI notes that the above are the minimum requirements not the maximum and adopters of the code are expected to do more in other areas relevant to employment in the supply chains sector not specifically mentioned nine rules as their base code.

In addition to the above mentioned ETI's base code, the organisation notes the following as its strategy for helping all its members in what it describes as its intention to focus activities on five critical areas

- Promoting good workplaces
- Payment of living wages
- Integrating ethics in to core business practices
- Tacking discrimination in the workplace
- Improve audit practice

Source: ETI website

4.4 Voluntary Sector Based Standards

In addition to general global standards available which deal with CSR and Sustainability issues, there are also some voluntary sector based codes and guidelines which companies operating in certain sectors have put together to direct their operations with regards to social, economic and environmental activities. One of such is the code of conduct put together by those companies working in the Electronic industry which they have called the "Electronic Industry Citizenship Coalition (EICC)" code of conduct. These industry based codes are offshoots of many of the international standards and guidelines available to corporate entities internationally, three of them we have mentioned in the preceding sections above. The EICC code of conduct covers five areas, namely:

- Labour
- Health & Safety
- Environment
- Ethics
- Management Systems

These are areas covered by three of the standards explored under international standards in social audits. A cursory look at the five areas covered by the EICC codes demonstrates the specificity of their code to the electronic industry which can only improve their effectiveness in addressing some of the issues that could result in scandals and bad practices for companies operating in the sectors.

5 What Does a Social Auditing Entail?

It would have perhaps been unhelpful to readers who are unfamiliar with what social auditing is all about and what it entails if we left this section out of the chapter. Let us now describe what social auditing is and what it entails. Social auditing is "a process by which an entity accounts for its social performance to its stakeholders and seeks to improve its future social performance". Originally it was wrongly assumed that only governmental organisations, charitable organisations and non-governmental organisations (NGOs) that should carry out social duties. CSR has put an end to that assumption since it has become glaringly clear to everyone that private sector organisations that are socially inactive and irresponsible are unlikely to survive beyond the immediate future. It is generally understood that their operational activities would make them create either intentionally or unintentionally some social and environmental problems which will affect everyone, they therefore need to account for what they are doing or going to do about these problems. This would involve them in undertaking their own social audit of all the events that impact on their stakeholders and the environment.

It is therefore expected that an organisation that is carrying out an audit of it social activities should do the following about its social data:

- Record
- Process
- Summarise
- Report

Going through the four stages mentioned above should allow everyone—stakeholders and the community to be aware of what is going on in the organisation with respect to this area of its activities. Any information derived from this method of accountability to all stakeholders, regardless of whether they are internal or external could be used in planning, control and decision-making. Not only that, it should also provide a sort of yardstick against which organisations either in the same or different sector could be compared. Managers of one organisation could use the information derived from the social report of another company to benchmark and plan their own future activities in the field.

Social auditing requires that an organisation will establish a department which will consist of a team of experts in the different areas to be socially audited. The number of people in the department and whether all that they will specialise in will mainly be social auditing will depend of many factors including the size and complexity of the organisation. For instance a small company venturing into social auditing might not have the resources to employ several people in the social auditing department or make the few people in the department specialise solely on social issues. That a company or business has set up a social audit department should at least be a good start which should hopefully ensure that all those social problems we have been identified above as having been responsible for scandals and reputational damage to some big retailers in many Western countries would have been spotted well in advance and dealt with by management before resulting into what stakeholders and the media would find unpalatable. Another issue that a company embedding social auditing into its activities should not overlook is training and development of its social auditors. As a result of the dynamic nature of our world, things change all the time these social auditors must therefore continuously update their knowledge about the industry and sector they work in. This can only be good for the social auditor, their organisation and stakeholders. The field of social auditing is indeed a dynamic field.

6 Making the Current Social Auditing More Effective

We were unable to see any company's social auditing report on the internet having Google searched a number of companies we are aware source their merchandise from suppliers in third world countries. The absence of these reports on the internet might be for confidentiality reasons. We saw their CSR reports with pictures of workers in many third world countries at work that should perhaps be enough evidence that all is well in regard to social auditing, perhaps not. We were there unable to provide evidence of what companies are reporting in their social audit reports.

We will therefore devote the remainder of this section to actions we believe would improve the effectiveness of social auditing in both the supply chains and other sectors of an economy in order to help identify quickly areas where things are not working well in social auditing for corrective actions to be taken before things get out of hand.

- Sectorial Inspectors—It is proposed that despite all the actions being taken by operators in the supply chains sector with regard to social auditing, it is suggested that each sector or industry should set up a team or body of independent inspectors to police social responsibility activities for each sector. These inspectors should have bases in different continents of the world and be totally independent of companies in the sector or industry.
- Non-Governmental Organisations (NGOs)—Some NGOs have actively worked well over the years to police some of these issues and draw attention to inadequacies and bad practices which effective social auditing would have picked up. The chapter contends that more NGOs are still needed to help in identifying irresponsible practices.

- Legislation—Government regulations might also be a useful tool to ensure responsible practices in the field. Regulations will be necessary in both the advanced country of the diffuser and emerging country supplier.
- Awards for excellent practices—We suggest that some charitable organisation could spring up to award annual prizes to companies with excellent practices in the field of social auditing.

7 The Future of Social Auditing and Concluding Remarks

Let us at this stage of the chapter considers what the future holds for the social audit reporting aspect of an organisation's impact on society. Like many things in general, the future is never certain but there are a few things that this author personally believes will happen in the field of social auditing not just in the supply chains sector but in social auditing generally.

Many countries are going to have to reassess their stance in regard to CSR in general terms take the route of properly regulating social responsibility reporting. Bangladesh and Pakistan like their India neighbour will likely make CSR mandatory and social auditing mandatory, this is because of the May 2013 incident in Bangladesh noted about. Indonesia which has already gone the route of mandatory CSR would strengthen its legislation in the field of social auditing to cover any remaining loopholes.

In the developed world, companies would be required to mandatorily to follow international standards on social reporting similar to that which they follow in International Financial Accounting Reporting. This will be necessary to improve the reliability of the information they publish to the world at large in this area. Many trained specialists are going to set themselves up in practice to provide external social auditing duties to organisations. In terms of International standards on social reporting, a note of warning is necessary here. All the currently available international standards e.g. SA8000, AA1000 etc have need to be codified into one set of standards dealing with different areas of social responsibility reporting. To continue to have several standards dealing with the same or similar areas will cause confusion and connotes a wrong impression that the field needs not be taken seriously. Having said all this, I personal believe that the future looks promising for corporate social responsibility globally. The room to continue to research and innovate in the area is extensive and can only benefit the current and future generations of people wherever they are on planet Earth.

References

Amaeshi, K., Osuji, O., & Nnodim, P. (2008). Corporate social responsibility in Supply Chains of Global Brands: A boundary less responsibility? Clarifications, exceptions and implications. *Journal of Business Ethics, 81*(1), 223–234.

Andersen, M., & Skjoett-Larsen, T. (2009). Corporate social responsibility in global supply chains. *Supply Chain Management: An International Journal, 14*(2), 75–86.

Awaysheh, A., & Klassen, R. D. (2010). The Impact of Supply Chain Structure on the use of supplier socially responsible practices. *International Journal of Operations and Production Management, 30*(12), 1246–1268.

Bowen, H. R. (1953). *Social responsibilities of the businessman*. New York: Harper Row.

Carroll, A. B. (1999). Corporate social responsibility: Evolution of a definitional construct. *Business and Society, 38*, 268–295.

Carroll, A. B. (2008). A history of corporate social responsibility: Concepts and practices. In A, Crane, A. McWilliams, D. Matten, J Moon, D. S. Siegel (Eds.), *The Oxford handbook of Corporate Social Responsibility*. Oxford: OUP.

Electronic Industry Citizenship Coalition (EICC) http://www.eiccoalition.org/

Ethical Trading Initiative. http://www.ethicaltrade.org/about-eti/why-we-exist. Accessed 24 September 2014.

Ethical Trading Initiative. http://www.ethicaltrade.org/about-eti/our-strategy Accessed 24 September 2014.

Handfield, R. (2006). A brief history of outsourcing, from Supply Chains Management Resources. http://scm.ncsu.edu/scm-articles/article/a-brief-history-of-outsourcing. Accessed 3 May 2014.

Idowu, S. O. (2008). Practicing corporate social responsibility in the United Kingdom. In S. O. Idowu, W. Leal Filho (Eds.), *Global practices of corporate social responsibility*. Heidelberg: Springer.

Idowu, S. O. (2010). Corporate social responsibility from the perspective of corporate secretaries. In S. O. Idowu, W. Leal Filho (Eds.), *Professional perspectives of corporate social responsibility*. Heidelberg: Springer.

Idowu, S. O. (2014). CSR: A modern tool for building social capital. In S. O. Idowu, A. S. Kasum, A. Yuksel Mermod, (Eds.), *People, Planet and Profit*. Farnham: Gower.

Lee, M.-D. P. (2008). A review of the theories of corporate social responsibility: Its evolutionary path and the road ahead. *International Journal of Management Reviews, 10*(1), 53–73.

Locke, R. M., & Romis, M. (2012). Improving work conditions in a global supply chain. *MIT Sloan Management Review, 48*.

Mullins, R. (1996). Managing the outsourced enterprise. *Journal of Business Strategy, 17*(4), 28–36.

Stanford Social Innovation Review, Mandatory CSR in India: A bad proposal. http://www.ssireview.org/blog/entry/mandatory_csr_in_india_a_bad_proposal. Accessed 14 May 2014.

Wood, G. (1996). Ethical issues in purchasing. In A. Kitson & R. Campbell (Eds.), *The ethical organisation*. Basingstoke: Macmillan.

Internet Sources

Labour & Worklife, The Labor and Worklife Program at Harvard Law School. http://www.law.harvard.edu/programs/lwp/NLC_childlabor.html. Accessed on 15 April 2014.

Pearce, J (2005), Social Audit Network—Social accounting and audit for the community sector. http://www.socialauditnetwork.org.uk/getting-started/brief-history-social-accounting-and-audit/ Accessed 16 April 2014

https://www.microlinks.org/sites/microlinks/files/resource/files/ML5896_social_audit_handbook.pdf. Accessed 23 September 2014
http://www.socialauditnetwork.org.uk/files/9013/2325/3606/Social_Audit_Toolkit.pdf. Accessed 23 September 2014.

AA1000: An Analysis of Accountability and Corporate Social Responsibility in the Contemporary Context

Priscila Erminia Riscado

1 Introduction

Accountability is a phenomenon that is gaining more space in the discussion of contemporary liberal democracies. Themes that constitute the said phenomenon, such as responsibility and transparency, seem members of analyzes related to the demands and dilemmas presented in proposals for socially responsible actions made by companies in the context by contemporary democratic parts—as the growth of social movements—such as the environmental movement, the civil rights movement, the indigenous movement among others—and the internationalization of companies, e.g.

There is a growing connection between accountability and the adoption of socially responsible corporate actions. It's the aim of this study was to understand the relevance of the theme linked to accountability for companies, since this theme is reflected in the fact that the company is an actor which has undergone profound changes in the current context. The first challenge, proposed below, is to try to understand the meanings of accountability, so that we can analyze the phenomenon from the perspective of the private sphere, which is the object of analysis of this study. To support this analysis, we present even in a brief way—the case of two Brazilian companies that have adopted the AA1000 standard with the objective of improving the relationship between the company and the stakeholders.

P.E. Riscado (✉)
Angra dos Reis Institute, Universidade Federal Fluminense, Rio de Janeiro, Brasil
e-mail: priscila_riscado@yahoo.com.br

2 What Is Accountability?

Accountability is a concept which seems quite far from the consensus, especially with regard to its definition. There are many meanings to the term granted. Because it's a term origin foreign (English), which there's still no one translation that can take account of its meaning in Portuguese. We can understand accountability as a duty of accountability, transparency to present and/or responsibility in management—public or private. In general, the term and most questions related to the topic are usually associated with and are related to public affairs—such as public administration, public institutions, etc. This connection occurs, mostly with beads attached to the State in general. Not the case with this article.

Studies dedicated to the analysis of the subject of literature production about accountability are not rare. Scott Mainwaring and Christopher Welna[1] (2003) point out that much of the literature on accountability was produced within the public administration. The authors cite, for example, the study by Patricia Day and Rudolf Klein, published in 1987.[2] Also recall the relevance of important book on electoral accountability produced by Adam Przeworski, Susan Stokese Bernard Manin (1999).[3] Also record the existence of abundant academic literature on some topics related to political accountability (legislatures, judiciary, regulatory agencies and control of corruption). There are also, according to the authors, among the works that comprise this literature, some studies which specifically focus on electoral accountability. Some books examine a particular slice of accountability—for example, the studie of Woodhouse (apud And Welna Mainwaring, 2003)[4] which examines the accountability of ministers in the Parliament.

Already within the meanings, we can say that accountability—or the act of having a public officer who formally and legally accountable to another actor, in the definition of Mainwaring—is a three-dimensional object. Composed of transparency (answerability, when it comes to a mere disclosure of information); responsiveness (responsiveness, requests for explanation and accountability by engaging acts in an institutional and legal obligation to provide information and answer questions), and ability to sanction or coercion (enforcement, which is the ability, also legal and institutional, to enforce this obligation through penalties and incentives also). These three dimensions must act simultaneously to the accountability happen.

---

[1] This is the book organized by the authors in 2003, to be analyzed in this article. At this point we analyze the introduction of such work, exclusive authored by Mainwaring is titled "Democratic Accountability in Latin America".

[2] The study which Mainwaring refers is the work. "Accountabilities: Five Public Services", by 1987.

[3] The study is the work titled "Democracy, Accountability, and Representation", by 1999.

[4] The work which Mainwaring and Welna (2003) refer is the book *"Ministers and Parliament: Accountability in Theory and Practice"*, by Diana Woodhouse, 1994.

There are many definitions produced about accountability and it's meanings. However, some make more representative in the discussion on the subject. The definition of O'Donnell (1994)[5] for accountability is considered by many experts as one of the most relevant and significant. Like most academic studies on the subject, O'Donnell criticizes accountability from the actions developed within the public sphere. In the current study, the author analyzes the accountability in the context of democratic representation. The idea of representation, he says, always involves an element of delegation: through some procedure, a particular collectivity authorizes someone to speak for them. Eventually, this collectivity is committed for respecting which the representative has decided for acting in this capacity. Consequently, the author says that representation and delegation are not opposite poles. This would be difficult for one to clearly distinguish the types of democracy that are organized around what O'Donnell calls "representative delegation" in which the delegative element is strongly dominant.

In the context of this distinction of representative democratic models that O'Donnell found the idea of accountability: according to the author, somehow the person who represents all citizens is responsible for the way that he acts on behalf of those for whom claims to have the right to speak. In consolidated democracies, one can verify the existence of two types of accountability: vertical and horizontal. Accountability operates vertically when society and the citizens generally charge their attitudes the person who represents all accountable, transparent in operation for which they were chosen—these charges being occurred beyond the times of elections. Have a horizontal accountability is one that occurs within a network of relatively autonomous powers—namely other public or state institutions—which have the capacity to undermine and eventually punish improper ways to perform the tasks that are given to these people who represents all. For O'Donnell, representation and accountability implies the republican dimension of democracy. Because of this, it's necessary to make a careful distinction between the spheres of public interests and private interests of the occupants of the positions for which they were chosen.

The origins of the term accountability and the meanings proposed for this use are the focus of some analysis about the subject. According the study made to Duarte (2010),[6] it's possible to know the origins of the concept of accountability, since the current demand, which has as central objective the accountability and transparency to the original source, linked to the concept of external control. You can also know the elements that make up a process of accountability (transparency, responsiveness and ability to sanction/coercion) and the various types of existing accountability (internal or external, functional or strategic, horizontal or vertical, etc.).

---

[5] The work in which Guillermo O'Donnell presents this definition is the article entitled "delegative Democracy", published in the Journal of Democracy, in 1994.

[6] The work in question is the author's master thesis, entitled "Supervised Decentralization: Accountability and Evaluation of Non-State Public Services and Guidance Results in Public Administration", 2010.

With regard to the origins of accountability aren't possible to verify that this is a really new concept, although the most frequent topic of debate has begun with more force from the late 1980s. As emphasized Duarte, questions relating to external control of the institutions had already been in the works of classical authors of Political Science—as Locke (1983, cited in Duarte, 2010), in the late seventeenth century and Montesquieu (1979, cited in Duarte, 2010) in the eighteenth century, especially when they talk about the separation of powers within the state. The authors also discuss the features of the present debate in the late Middle Ages in England, France and Spain. According to Duarte, it was possible to find at this time mention an idea of cutting bills in these countries.

Mainwaring and Welna (2003) and Duarte (2010) found that accountability has become a major focus of attention and analysis in Latin America since the 80s, in the democratic post-transition period. From that moment, the concern for accountability in the political-state area came out of the field of electoral accountability only and spilled over. In the words of Mainwaring, one of the most important emerging challenges to improve the quality of democracy revolves around how to build more effective accountability mechanisms. Although accountability has become a marked concern in Latin America and other emerging democracies, only recently beginning to an academic debate on no-electoral accountability.

As previously mentioned, various items make up the idea of accountability. External control, present the concept of accountability, is one of these. It's associated with a set of legal mechanisms that aims correction and supervision of the activities of public administration. The context that supports this perception is that management by public agents of the instruments seeking the assistance of the public interest may exceed legal limits and tackle abuses and illegalities. For this reason, become necessary supervision—especially preventive—and control of the acts of public administration. External control is that performed by diversified structure, like the Legislative Power and the Judiciary. As presented by Duarte, the incipient idea of "balance" between the powers that integrate the State as the way to control like a "parent" of the concept of external control. This, in turn, presents an absolute affinity with concepts such as accountability or transparency.

Currently, we can say that when we talk about external control—and consequently accountability and transparency like similar concepts—we see that the meaning of this extrapolates the achievements made within the structure of the power aforementioned. The external control feature is also possible to be observed in civil society organizations in general and in private companies.

3 The Accountability and the Company: The AA1000

In public organizations it is necessary that exist an active participation and social control for occurring high levels of accountability. The demand for accountability—in this case—is largely depend, among other things, on the degree of awareness and citizen participation. Between private organizations, there is other scenario: accountability in private organizations wouldn't be directly related to the implementation of citizenship or some kind of social control, although suffers constant influence of these aspects. In a market perspective, we can say that accountability is mainly aligned with business strategy. Ultimately need therefore be encouraged by the organization by means of the systematic relationship with their stakeholders. More than focus as a political act of power or through law, accountability should happen in this scenario by means of dialogue between the company and its stakeholders. Having as object of study accountability in private corporations means addressing its instrumentality and their contribution with the strategies of corporate social responsibility and forms of relationship between the company and their stakeholders. The accountability only happens in corporations is potentialized by means of dialogue, for the accountability and transparency—concepts suggested for accountability in the public sphere—make sense apart from a legal perspective, helping and guiding the strategies of the business relationship.

The adoption of accountability in the context of private company is directly related to the new demands faced by these companies. According to Duarte,[7] if we make an analysis about accountability from its current demand, we will notice that such demand began in the economic and financial world, as the current growth of the study and the importance of accountability came in large part by the increased practice of financial auditing and supply of stamps and quality certifications (ISOs 9000 and others)—in the end of the 1980s, in which Power (1997)[8] calls the consolidation of a "audit society". From this environment "accountable" established by finance and the private sector, the practice of accountability would spread to other areas, including the state, aided by the coming of what was called "New Public Management". The emergence in the world of finance, a demand for audit and financial accounting and accountability would be responsible for the change in the corporate governance of the corporate world. The private accountability would then have the primary function respond to the legitimate expectations of stakeholders, by means of dialogue, transparency and accountability between companies.

Works such as Michael Power[9] and Ely Corbari (2004)[10] argue that accountability requires, besides a power base with formal rules and mechanisms of self-control, in considering the demands of stakeholders and their self-regulation, a solid foundation of values that guide the business about what should be considered fair and ethical, beyond what are legal actions. Thus, high levels of accountability allow the corporation to comply with the regulation of its conduct, both for what is formally demanded by its stakeholders as the ethical and moral gifts at its base

---

[7] Op. Cit.
[8] The study in question is Power's book titled "The Audit Society: Rituals of Verification", 1997.
[9] Op. Cit.
[10] The Corbari'study is the "Accountability and Social Control: Challenge to the Construction of Citizenship," 2004.

values. From this perspective, the application of accountability in organizations should happen on the business strategies. The perspective of accountability should be understood as a variety of interactive forces, not only as an attribute or isolated mechanism, but one aspect that has a dual pillar: one for power and other moral basis, underpinned by ethical values of the organization.

The debate about accountability also shows how this approach guides the decision-making process in an organization. The decision or choice made by the company in this field is a process by which one of the alternative behaviors is selected and performed—in relation to adopt or not an accountable posture. These decisions, which determine the behavior of a corporation, relate directly to its business strategy. From this point of view, the characteristic of accountability correspond to the company's concern with ethics and responsibility. The accountability as a strategy will permeate the decisions of a company, printing a kind of behavioral pattern to these. Accountability would be able to promote a connection between the various actions of a company, whether directly or indirectly focused on the relationship with their stakeholders.

In general terms, we can say that private organizations have appropriated accountability. Nowadays organizations can count on instruments that help make accountability an integral part of a company's strategy, although the adoption of this posture doesn't originate in this stakeholder demands. The standard of AA1000 Accountability is one example.[11] Published in 1999 in England, introduces the principle of inclusion in this type of instrument, which concerns the participation of stakeholders in the development and achievement of a response responsible and strategic to sustainability in the sphere of action of firms. The engagement of stakeholders, from the use of standard forms in a tool that can be used by businesses to help them promote the inclusion of stakeholders. To support for such inclusion, guidance on how to develop and conduct stakeholder engagement was included in 1999 in AA1000. In 2005, this guidance has become the standard called "AA1000 Stakeholder Engagement Standard", the first international standard that specifically addresses stakeholder engagement published.

The performance of stakeholders from the companies does not present itself nowadays as a novelty. Some of these ways of acting, as in the case of shares of the environmental movement, for example, began some time ago and have trajectory of rising action. However, cannot be said the same about the acceptance by the company that it has become essential to the sustainability and success of a corporation. The AA1000 standard tries, first, help firms understand that stakeholder engagement is a key aspect to its own development and maintenance on the market. Moreover, the standard points out that the engagement between the company and stakeholders can hold two natures: this may be good or poor. The purpose of this

---

[11] Released in 1999, AA 1000 has been created with the objective of helping companies, shareholders, auditors, consultants and certification bodies; it can be used alone or in conjunction with other standards of accountability, such as the Global Reporting Initiative (GRI), norms and standards such as ISO and SA. 8000. To learn more, see http://www.accountability.org/. Date of access: 28/05/2013.

standard is therefore to establish benchmarks for the engagement of good quality. The organization is responsible for creating the standard points that stakeholders are not only members of a community or non-governmental organizations (NGOs). Stakeholders are those individuals, groups of individuals or organizations that affect and/or be affected by the activities of a company, or the products and services offered by them. The organizational performance is associated with the relationship that they develop with their stakeholders. Consequently, also the themes to be discussed by engaging with those.

The standard emphasizes that stakeholder participation should be understood as the process used by a corporation to engage them with the clear purpose to achieve acceptable results. It's now also recognized as a key accountability mechanism, it forces the companies seeking the use of the standard to engage with their stakeholders and to identify, understand and respond to sustainability issues. Moreover, the adoption of the standard is to pass with the corporation to develop concerns to report, explain and be accountable to stakeholders about their decisions, actions and performance.

Related to quality, the standard interpretation is that the engagement of the company must provide some characteristics such as:

- Be based on a commitment to the principles the AA1000APS;
- Clearly define the scope of application;
- Have an agreed decision process;
- Focus on issues relevant to the organization and/or its stakeholders;
- Create opportunities for dialogue;
- Be an integral part of organizational governance;
- Have an appropriate procedure for each of the stakeholders;
- Be timely;
- Be flexible and responsive.

The AA1000 standard intends to act in relations between the company and individuals, along groups of individuals or social organizations affected or that can directly affect the action of a company. The rule tries to address these concerns, making corporations perform better. Moreover, the norm would increase the knowledge of the company and contribute to its operation. With regard to the quality and commitment by the company, the AA1000 could:

- Conduct a more just and sustainable social development, giving stakeholders the right to be heard and the opportunity to be considered in the decision-making process;
- Enable better management of risk and reputation;
- Allow sharing of resources (knowledge, people, money and technology) to solve problems and achieve goals that cannot be achieved by a single organization;
- Allow the understanding of complex operational environments, including where market developments and cultural dynamic;
- Enable learning of stakeholders, resulting in product and process improvements;

- Inform, educate and influence stakeholders to improve their actions. Thus, their decisions will impact on the organization and in society;
- Contribute to the development of stakeholders based on trust and transparency in relationships.

For these benefits to be realized, the engagement must be designed and implemented in a credible way. The AA1000 Stakeholder Engagement (AA1000SES) according establishes, could provide a basis for this.

The starting point for the engagement of a company would often connected to any negative image for the company, which would have generated significant external pressure episode and therefore needs to be addressed urgently by the same. The corporation realizes the needed to engage, to be more transparent and to respond directly to the concerns of stakeholder. Organizations seeking engagement solve a problem: thus, shall seek ways to use this engagement as a preventive measure, rather than to adopt a mechanism opposite to the occurrence of a problem reaction. Companies begin to use it systematically as part of identifying risks in management and find that a better understanding of the results of their stakeholders can make it as easy and responsive operating environment, improving the company's performance. Also according to the description of the standard, companies would be discovering that engagement with stakeholders can be a great source of innovation and forming new partnerships. Large companies have realized that a growing percentage of innovation originates outside the organization, not inside. They found that the stakeholders can be a resource, not just a factor to be controlled.

The last module of AA1000 published in 2011[12] consists of four parts. The first part describes the purpose and scope of the rule, also articulates who are the intended users, while making clear that this standard is geared especially for professionals and engaged owners of the corporation, the standard will be significant use for everyone involved in the company. Clearly this first part, also, that the standard is to be used by companies of all types and sizes, and not only by large corporations.

The other three parts establish the requirements for participation of stakeholders, supported by guidelines to ensure clear and complete understanding of the requirements of the standard. The three parts are: 1) to establish the need for those interested in the work and commitment to ensure that it is fully integrated into the strategy and operations, 2) how to define the purpose, scope and stakeholders of the contract, and 3) what is need for a process of stakeholder engagement quality.

Consider accountability as a strategy within the corporation helps identify and insert stakeholders expectations in how they conduct business. The interaction between accountability and ethical values in the company contributes to the development of corporate social responsibility, it resorts to its guiding function in the

---

[12] The AA1000 standard, as previously said, appeared in 1999, however, after its release, other updated versions of the standard are available-. 2002 and 2003 In general, these reissues, few changes to the core content. Changes take place in increments, in general, with the inclusion of the thematic content of this standard modules.

business strategy, keeping in view the demands of stakeholders. Thus, for the accountability become strategic is necessary from the point of view explained in AA1000, the company seeks your participation around a large network of relationships that fosters social responsibility and sustainability.

The AA1000 standard, like other standards—like the ISO 9001, ISO14000—is a voluntary compliance standard by companies who adopt it. There is no legal obligation to adopt this standard. Companies generally contend that adopt the standards based on the idea that, with the adoption this, be able to become more efficient and transparent. Moreover, be able to intensify relations with their stakeholders.

On the other side, we can observe societal changes. Based on studies such as the Ciro Torres, we find that after the second half of the 60s in the United States of America (USA) and in parts of Western Europe, particularly in France and England, a part of the society initiated an effective recovery for socially responsible behavior within companies. The response to the growth of social movements and the struggle for civil rights was not long in coming: many American and European companies initiated a significant change in dealing with the raw material used, consumers, suppliers and their employees. In the '70s, some companies have realized the strategic importance of publicizing the social actions performed. Therefore, it was still in the 70s that was consolidated in some European countries the need for periodic achievement and annual disclosure statements or reports of so-called social corporate activities. These symbolized the accounts of the company to their stakeholders form.

Since of the 90s this conception of the necessity of a new role played by the company from society deepens. The following definitions shall widely that companies are embedded in complex environment, interacting with institutions and cultural practices, with several social groups "extra company" with the environment and that their actions both influence and are influenced by this context. Therefore, it seemed important that the company contributes to the community in which it's inserted, with the preservation of the environment, with the professional development of its employees and transparent relationships with customers, suppliers and government, among others (Ashley, 2001; Barbosa, 2002). It's under this scenario that gains importance, increasingly, issues such as sustainability and corporate accountability.

While there is over a decade, AA1000 is quite disseminated in Brazil. There has been the adoption of the norm among large companies—such as Light SA and Suzano Papel e Celulose—these companies are the subject discussed in this chapter.

4 The Standard, Stakeholders and Engagement: The Adoption of AA 1000 in Two Brazilian Companies

As we already have seen, AA1000 is a standard that defines principles and processes for accountability, to order, largely to ensure the quality of accounting, auditing and reporting of information of social, environmental and financial character. The existence of the standard can be seen in itself as an indicator denoting that the accountability gained relevance in the contemporary context.

To help us to better understand the relationship between the norm and the company we selected two companies basis in some aspects such as:

- Sized enterprises: all selected companies are considered to be large;
- Nationality: all are Brazilian companies;
- Development projects and/or actions in the area of social responsibility: all companies develop actions and/or projects in the area;
- The presence of companies in the Corporate Sustainability Index (ISE)[13] of BM & FBOVESPA.[14]

Based on these criteria, two companies were selected: Light SA and Suzano Papel e Celulose AS. For analysis of the companies mentioned have been used annual reports and/or sustainability reports. We could capture some information about the treatment between the companies and their stakeholders. The analysis of the annual reports of companies is performed, based on the reports for the years 2007 until 2013.

Light SA[15] is a holding company and is present in 31 cities in the state of Rio de Janeiro, covering a region with more than 10 million people and adding 4 million customers.

In 2009, Light undertook a structured process for promote the stakeholder engagement, but once with the support of experts from the Brazilian Foundation for Sustainable Development—FDBS, and through a methodology developed by them, inspired by the international AA1000.

The adoption of this statement marks a shift in the company: Light modifies the ethics code and redefines its mission and values—important aspects for the development of the company. A typical example is one of the items of the new code of ethics of the company, entitled "Principles of Collective Action." This topic has the core objective redefine the company's relationships with its customers, shareholders, suppliers, society, State and governments.

---

[13] The ISE is intended to reflect the return on a portfolio composed of stocks of companies with a recognized commitment to social responsibility and corporate sustainability, and also to promote good practices in Brazilian companies.

[14] BM & FBOVESPA is a Brazilian capital company formed in 2008 from the integration of the São Paulo Stock Exchange and the Commodities & Futures Exchange operations.

[15] For more information, see www.light.com.br.

On the other hand, the company decides—based on the implementation of the standard AA1000-joining the UN Global Compact.[16] With membership, Light is committed to report annually to the United Nations concrete examples of progress and lessons learned in implementing the ten principles. The Company also undertakes to disseminate the principles of the Compact among its employees, shareholders, customers and suppliers. Light also seeks to contribute to the achievement of the Millennium Development Goals (MDGs), defined from the Millennium Declaration of the United Nations.

To accompany the main practices of Sustainable Development to be performed every year, Light prepares a document entitled "Self-Assessment Report", which details their objectives, target audience, motivation, relevance and viability, as practice management, engagement with the public interest, previous years results and their relationship to the Millennium Development Goals and the Global Compact Principles.

The Sustainability Working Group, reporting to the Executive Board, is responsible for monitoring the reported practice, aiming to identify possible barriers and to advise on any changes in direction, so that the goals are met at the end of each year.

An example of how the adoption of the AA1000 standard was made in the company can be found in the construction of a new methodology for stakeholder engagement. The objective is to improve the relation between the company and their principal stakeholders—such as society, the state, employees, suppliers etc. Let's see an example of this from a passage of the sustainability report of the year 2013 the Light company.

> Since 2009, Light produces its Sustainability Report based on your materiality matrix that guides the construction of all content reported to the electricity sector, the regulatory agencies and society in general. The Materiality Matrix presents relevant topics (materials) that create value for the company and its stakeholders. The Materiality Matrix Light was developed [...] obeying the AA1000 standard. (Light sustainability report, 2013).

The other company is analyzed Suzano Pulp and Paper.[17] The company is part of the Suzano Group, which invests for 85 years in the pulp and paper segment in Brazil. The forest-based company is one of the largest vertically integrated pulp and paper in Latin America, with global operations in about 80 countries. Since 2009—the year the company adopts the standard AA100, this emphasizes the adoption of a management model that considers sustainability in its three dimensions: business, social and environmental.

Again, there was a transformation of formal stance taken by the company in respect of the relationship with stakeholders: from 2008, the company includes in your mission the concern with the needs of the clients so that their activity can promote the satisfaction of shareholders, employees, suppliers and local

---

[16] The Pact is a United Nations initiative and has the objective encourage the business community to adopt internationally accepted values in the areas of Human Rights, Labour Relations, Environment and Anti-Corruption in their business practices. To see more: http://www.onu.org.br/.

[17] To see more, http://www.suzano.com.br.

communities. On the other hand, the company becomes a signatory of the Global Compact, an initiative of the United Nations (UN) for the business community to adopt in their business practices, ten principles in the areas of human rights, labor relations, environment and anticorruption. Still contribute to the Eight Millennium Development Goals, also established by the UN, which compromise organizations, societies and countries to reach by 2015 related to poverty eradication, health, quality of life and respect for the environment objectives. Again, the standard AA 100 is cited as a motivator for development and evaluation of the company's engagement with its stakeholders, causing higher accountability and an increasing in the adoption of a socially responsible attitude.

Suzano Celulose and Paper declares that have changed their way of evaluating their sustainable management from the adoption of the AA1000 standard. The following is a passage from the company's sustainability report, indicates this change.

> Suzano SA included in scope of independent review—from 2009—the following: Data and information included in the Report on the year 2008; The appropriateness and robustness of the underlying systems and processes used to collect, analyze and review financial reporting; Assessment Report in comparison with the following key principles of standard AA1000 Rating (2003): scope, materiality, degree of responsibility (Suzano S.A. sustainability report 2009).

Conclusion

We agree that accountability is directly associated with the development, with the idea of quality of democracy—as is observed by Mainwaring and Welna (2003). Under this view, the authors relate to accountability to the quality of democracy, arguing that one of the most important challenges to improve the quality of democracy revolves around how to build more effective mechanisms for accountability. This leads us to conclude that the convergence between accountability and private company shows up as a formidable challenge for companies. Accountability is an essential tool when we think the reality from the scenario in which we operate—the contemporary liberal democracies. However, we can not observe it from social actions undertaken by companies—at least not in its entirety, when we take the meaning as an indicator of accountability. If we think in interpreting Mainwaring (cited Mainwaring & Welna, 2003), we define accountability as a three-dimensional object. These three dimensions must act simultaneously, so we can assert the existence of accountability. As is seen throughout this chapter, observe transparency, responsiveness and enforcement by companies is revealed as a challenging task in the current context. Therefore, we can observe that the accountability of private enterprise is determined by the agent itself—the company. Accountability relates to these aspects such as transparency and accountability. However, no relationship between accountability and punishment for the company, distancing

(continued)

the adoption of a posture of accountable actions related to the accountability of the company.

Moreover, accountability seems to occur the same problem noted in the attempt to adopt the theory of stakeholders by private companies. From some studies, in particular the analysis by Ebrahim and Weisband (2007), it's possible to see that accountability also represents problematic for company process because sometimes divergent interests will address: the company—especially in figure its shareholders and investors—not to disclose information, be transparent and accountable to all its stakeholders, especially with society and the state. Under this scope, accountability as it appears in the work of O'Donnell (1994), Mainwaring (cited Wełna and Mainwaring, 2003), Duarte (2010), and others translates into a practice difficult to operationalize by the company private. As these authors emphasize relations of accountability in private organizations tend to become complicated due to the fact that they often deal with competing demands. Stakeholder perspectives evolve directly from the organizational behavior literature for subsequent stalemates, started from the practice of accountability.

Still intending to answer different questions, the studies analyzed in this chapter have one thing in common: the attempt to understand the social actions, or socially responsible actions, implemented by private actors—especially businesses—from its effects and its impacts external to the company's own scenario. Moreover, the work sought, directly or indirectly, and understood if there—if there answer to that question is yes—how it gives accountability between these private actors and public context in which they live, and who relate entirely. Although it is not the main purpose or main objective pursued by corporations, is unanimous respect of all the papers reviewed here that the shares of any company produce some kind of impact on the environment in which they are embedded. This relationship does not change when we approach the subject of corporate social responsibility. The results of these actions can have positive or negative effects, even if the initial search is to produce profits for the stakeholders of the companies.

Regarding the second variable analyzed—accountability—it was possible to observe the attempt of large companies that adopt a socially responsible attitude in their practice of seekink, ultimately, a more transparent approach with its stakeholders, in particular society. However, as we observed, accountability developed by large firms differs in form and content indicated accountability in public institutions in general. From the definition of accountability proposed by O'Donnell (1994) it's possible to perceive the public character in this phenomenon. For the author, accountability implies a dimension of republican democracy. Given this definition, how can we think that the big company—yet to develop tools aimed at accountability, such as the preparation and dissemination of sustainability reports, and use these

(continued)

assumptions in its administration and management strategy—can be regarded as a social actor essentially accountable? Although the company is a social actor that interacts with society and with numerous stakeholders—voluntarily and involuntarily—their main goal is to serve their stockholders and their respective demands. Therefore, accountability developed by large companies—and represented by instruments such as sustainability reporting, codes of ethics, among others—seems to hold an essentially private character that makes it a less democratic mechanism than it should be.

We understand that proposed relationships by companies—with regard to accountability, social responsibility, stakeholder engagement etc through some mechanisms, such as AA1000—not effectively represent a transformation in social relations or even a gain for society or for State. However, it's important to note that, even in a formal arena, the establishment and adoption of standards indicate a greater concern for the company in its relations with stakeholders, in particular as regards the transparency of their actions.

Sometimes, corporate social responsibility seems to be an old topic, such amount of space and general news related to it. However, it is still new, if we think that it's a phenomenon that gained continued and growing space in the debate about the companies in the last 20 years, and that growth of the space occupied by social responsibility—both in theory and in practice debate business—no accident occurs concomitantly to the debate about the ways of liberal democracy in the contemporary capitalist context. Moreover, the changes occurring around the world, especially from the 60s and 70s, dating back to responsible for redefining new social demands that will pass directly in the action context of corporations, making these impose new contours to your posture along with the other actors, especially the society.

References

Ashley, P. (2001). *Ética e Responsabilidade Social nos Negócios*. Saraiva: São Paulo.
Barbosa, L. (2002). Globalização e Cultura de negócios". In A. M. Kirschner, E. R. Gomes, & P. Cappellin (Eds.), *Empresa, Empresários e Globalização*. Rio de Janeiro: Relume Dumará/ FAPERJ.
Corbari, E. C. (2004). Accountability e controle social: Desafio à construção da cidadania. Cadernos da Escola de Negócio da UniBrasil. Curitiba, jan/jun 2004. pp. 99–111.
Duarte, A. V. (2010). *Descentralização Vigiada: Accountability e avaliação dos serviços públicos não estatais e da orientação por resultados na administração pública*. 2010. Dissertação (Mestrado em Políticas Públicas, Estratégias e Desenvolvimento)—Instituto de Economia, Universidade Federal do Rio de Janeiro. Rio de Janeiro.
Ebrahim, A., & Weisband, E. (Eds.). (2007). *Global accountabilities: Participation, pluralism, and public ethics*. New York: Cambridge University Press.
Mainwaring, S., & Welna, C. (2003). *Democratic accountability in Latin America*. Oxford: Oxford University Press.
O'Donnell, G. (1994). Delegative democracy. *Journal of Democracy, 5*(1), 55–69.
Power, M. (1997). *The Audit Society: Rituals of verification*. Oxford: Oxford University Press.

Viewed Homepage

Anistia Internacional. Available on: http://anistia.org.br. Homepage viewed on: 02/09/2012.
ABONG. Associação Brasileira de Organizações não Governamentais. Available on: <http://www.abong.org.br/>. Homepage viewed on: 22/08/2012.
Balance Scorecard Institute. *What is the balance scorecard?* Available on: <http://www.balancedscorecard.org/BSCResources/AbouttheBalancedScorecard/tabid/55/Default.aspx>. Data de Homepage viewed on: 15/11/2011
FUNDAÇÃO PROMENINO Available on: <http://www.promenino.org.br/Ferramentas/Conteudo/tabid/77/ConteudoId/9bc30f68-e222-4cfd-9f8a-13063a18a197/Default.aspx#back2>. Homepage viewed on: 28-04-2012
Global Anabaptist Mennonite Encyclopedia Online. Available on: <http://www.gameo.org/encyclopedia/contents/federal_council_of_the_churches_of_christ_in>. Homepage viewed on: 24/01/2011
Global Climate Coalition. Available on: <http://en.wikipedia.org/wiki/Global_Climate_Coalition>. Homepage viewed on: 02/09/2012.
Global Reporting Initiative. Available on: <www.globalreporting.org>. Homepage viewed on: 03/09/2012.
Human Rights Watch. Available on: <www.hrw.org/>. Homepage viewed on: 02/09/2012.
Instituto Ethos. Available on: <http://www1.ethos.org.br/EthosWeb/pt/31/o_instituto_ethos/o_instituto_ethos.aspx>. Homepage viewed on: 14/01/2012.
Light. Available on: http://www.light.com.br/Repositorio/Sustentabilidade/Relatorio%202013/Relatorio%20de%20Sustentabilidade%20LIGHT_2013.pdf
One World Trust. Available on: <http://oneworldtrust.org/>. Homepage viewed on: 27/08/2012.
Pegada Ecológica. Available on: <www.pegadaecologica.org.br/>. Homepage viewed on: 02/09/2012.
Pnuma. Instituto Brasil PNUMA. Available on: <http://www.brasilpnuma.org.br/saibamais/iso14000.html>. Homepage viewed on: 21/08/2012.
Revista Plurale. Available on: <http://www.plurale.com.br/>. Homepage viewed on: 25/09/2010
Suzano Papel E Celulose: http://www.ecodesenvolvimento.org/.../parceiros.../suzano/.../RA2008-Portugues
UN Global Compact. Pacto Global da ONU. Available on: www.unglobalcompact.org.

History and Significance of CSR and Social Audit in Business: Setting a Regulatory Framework

Anjana Hazarika

1 Introduction

Social Audit is one type of audit that verifies information on social responsibility, social performance of the organization. CSR describes the incorporation of social issues such as human and labor rights or community relations into business relations. The aim of CSR is to integrate social and environmental objectives with economic activities. Due to lack of government enforcement on environment, from 1980s onwards, a number of NGOs, Trade Unions and other civil society actors raised doubts on the performance of corporations. Many of the targeted companies set systematic corporate codes of conduct, by establishing environmental management systems. Some companies improved their practices. A number of discrepancies were uncovered by organized stakeholders, especially consumers and civil society actors. These actors have become suspicious about the activities of the corporations. Several studies have found out, false social, and environmental claims on products. As a result, new mechanisms were developed ranging from bilateral relationship between individual NGOs, corporations to multi-stakeholder coalitions.

Social Audit can be used a tool to verify information on CSR initiatives. The focus of verification of information can be based on environmental or social accountability management system, or it can take the form of a performance assessment to a particular set of standards. Sometimes CSR initiatives integrate the process and performance into their audit processes. Social Audit of the company has to verify the validity of the information. The companies have to decide, as to what information are to be provided in their report, and the guidelines or the standards they need to follow, to maintain a regulatory mechanism.

A. Hazarika (✉)
Jindal Global Law School, O. P. Jindal Global University, Sonipat Narela Road, Near Jagdishpur Village, Sonipat, Haryana 131001, India
e-mail: ahazarika@jgu.edu.in

2 History of Social Audit

The history of social audit goes back as far as the 1950s. Initially, the social audit was conceived as a means to make business more accountable to the community. It was also perceived to be a method to communicate, both economic and non-economic impact of business, to the community members. Social audit also refers to a very different kind of evaluation process in which an organization assesses and thereby improves its social performance.

But, "community good" is very hard to ascertain. It is very ambiguous. And this ambiguity led to establishment of a variety of definitions of "social good" and "social responsibility". The mercantile period (sixteenth, seventeenth and eighteenth century) primarily stressed on growth of natural wealth and power, by increasing export. Even in late 1950s and 1960s, an anti-business sentiment surfaced in the United States, but in between, the recession in England (seventeenth century), maintained a certain level of employment. From, 1970s onwards business started recognizing issues of sociability, environment and human resources.

The contemporary history of social accounting is marked by increase in public awareness. Social accounting has evolved into its present stage of development with future prospects for continued evolution. In the mid-1970s, descriptive terms aligned with the social accounting, reflective of changing environmental conditions.

For instance, in the mid-70s, social accounting was interchanged with social responsibility accounting (Anderson, 1977a, 1977b, 1977c) and socio-economic accounting (Belkaoui, 1980; Mobley, 1970).

As a result of some research studies, a concept called phantasmagoric accounting has appeared. It means a constantly shifting, complex succession of things seen or imagined, in support of an opinion that social accounting is like a Kaleidoscope, in that the same pieces turned a little differently from a whole new pattern (Roser, 1979).

By, 1970s, it was established, that social accounting refers to the ordering, measuring and analysis of the social and economic consequences of public and private sector behaviour.

From 1971–1980, it was a pause period, the social and environmental accounting literature was underdeveloped, and the leading North American accounting research journals were inaccessible to social and environmental accounting literature. There were few journals that published articles on social and environmental accounting like Accounting, Organization and Society, Accounting Review, Business and Society, California Management Review.

Accounting, Organizations and Society, was not the first journal of public systematic investigation and exploration of social accounting, but was definitely the first to undertake a systematic encouragement of social accounting (Gray, 2002).

Until, it was 1978, a social audit methodology was developed as an independent worker cooperative training center in England, and it published the first social audit manual—A Management Tool for Co-Operative Working in Free Spreckling in 1981. The manual has been revised a number of times, and in 2000, it was published as social audit toolkit, by social enterprise partnerships. Another significant point was the first social audit reports were published by the Migros Co-operatives, a Swiss multinational organization. The next development was that, the New Economic Foundation (NEF), in convenience with Steathclyde Community Business Ltd (SEB) proposed an alternative method to the Social Audit toolkit.

In 1976, few concepts were identified as social overhead, social income, social constituents, social equity, and net social asset (Ramanathan, 1976). Then in 1970s itself, say around four approaches to the reporting of social activities dominated the literature on social audit. These four approaches are: The Inventory Approach; the Cost Approach; the Program Approach; the Cost-Benefit Approach.

Further, the existing literature looked for relations with social accounting i.e. the accountant's role (Raben & Williams, 1974), accounting for pollution (Marlin, 1973), and information content (Ingram, 1978).

Even attempts were made to classify social accounting into some major areas. It was Dilley (1975) who suggested that there are five overlapping categories of social audit -national social income accounting (macro accounting), social auditory, financial/managerial social accounting for non-profit entities, financial social accounting, and managerial social accounting.

From 1981–1990, there was complexity in the situation of social and environmental accounting, while the first part was directed at increasing sophistication within the social accounting, and then transfer of interest to environmental accounting, towards the second part of the decade (Dierkes & Antal, 1985).

In the 1980s, the public awareness environmentalism increased and this was reflected in writing of some authors, and also broadening of the term social accounting to social and environmental accounting.

A number of attempts were made in 1970s, to develop theoretical module, but no such development got published in 1980s. Well, in 1988, Gray's work entitled 'Towards a theory of Cultural Influence in the Development of Accounting Systems Internationally' was a pioneering work that showed that culture might influence accounting practices. In fact, Gray proposed that theory can link societal and accounting values. He argued that, the value system of accountants is derived from societal values, especially to work-related values and, then the accounting values will influence accounting practices, like reporting and disclosure of information. Willeth (2002), stated that due to varying degrees of external and ecological forces shaping societal values, especially the work related values different accounting systems develop, which reflect and reinforce these values.

As environmental consciousness geared up, the period from 1991 to 2000 saw the advancement of environmental issues within accounting on a broader perspective, including interest from managers as well as accountants.

In 1999, Meline & Adler use of corporate social reporting research methods complemented, by exploring two further areas i.e. what documents to analyze, and how to measure disclosures further confirmed the importance of Social Reporting.

The latest scholarly works are on corporate, environmental and social disclosure practices within the theoretical framework of political economy, legitimacy and stakeholder perspective (Wilmshurst & Frost, 1999). The economic viewpoint is consistent with the research that deals with CSR disclosure in the context of agency theory. Although, political economy theories tries to explain why corporations appear to respond to governmental or public pressure or information.

The stakeholder theory asserts that corporations require stakeholders for their continued existence. The more powerful the stakeholders, the company should adopt more to the situation. Social Disclosure is regarded as part of the dialogue between the company and its stakeholders.

In the post-2000 period, there have been developments in empirical research in the social and environmental accounting. Social Disclosures and corporate reliability came to the forefront.

Kaya and Yayla (2007) re-examine the 35 years of social accounting literature. They highlighted the impact of this literature particularly on social disclosures and corporate supporting. Both the authors reviewed a broad spectrum of literature covering a period of 1970–1990, 1991–1995 and 1995–2006. Primarily, they have reviewed literature in regard to social accounting practices, yet problems around it have also erupted across the world which could be well unearthed by non-researchers. The reporting in Social accounting, Social Audit has become very crucial for the disclosure mechanism of the organization.

Gibbon and Dey (2011) discusses the merits of two approaches to social impact measurement which are currently the subject of debate within third sector i.e. social accounting and audits (SAA) and social return on investment (SROI). Both these approaches have similarities as well as glaring differences as well. The research suggests that SAA in particular includes a more conventional combination of narrative and quantitative disclosures whereas SROI brings more quantitative and reductive outcomes. The authors outline that there is a high possibility of using SROI within an SAA framework, but strong emphasis on quantitative data may enhance the measurement of social accounts. At the end, this study highlighted that the increasing preference to use SROI to SAA might lead to a one dimensional funder and investment driven framework to social impact assessment in the third sector in the near future.

Adams and Evans (2004) analyze two major concerns for securing a strong accountability in social accounting process. These two concerns are: the lack of completeness of reporting, and the lack of credibility reports. The study stresses on the role of social audits in enhancing the completeness and credibility of reporting. This would finally help reducing the audit expectations gap. The authors recommend that this gap emerges because of an over emphasis on the validity of performance data at the cost of addressing completeness and credibility. They argue that this requires an intensive involvement of all the stakeholders in the organization. It also provides an in-depth analysis of the recent guidelines of social auditing like GRI, AA1000 Standard and AA1000S Assurance Standard etc. Finally, the study offers to develop a practical approach to social audit with an emphasis on assurance guidelines which would help narrowing down the gap in audit expectations. In order to secure more accountability Companies were opting for self-regulation. Self-regulation can be one way of drawing out accountability, by the enterprise without competing to any international or private standards.

Graham and Woods (2006) explore that self-regulation by the MNC's of their social and environmental impacts can be solutions to the regulatory capacity problems emerging in developing nations. They argue that growing market pressure globally can provide strong incentives for companies to implement codes and standards for self-regulation. They point out that voluntary schemes can encourage trustworthy and standardized reporting of information. The study emphasizes on the role of the government in bringing effective regulation in the corporate. This can be further strengthened by setting social goals and upholding the freedom of civil society agents to organize and mobilize so that corporates cannot move against the larger interest of the society. Finally, the authors believe that international organizations and existing legal instruments would be of immense help in assisting developing nations for realizing these goals. Adoption of international standards by the social audit mechanism of the companies entails a rigorous monitoring by the corporations. Whether all the arrangements that has adopted the international standards have verified the applicability of the standards, is yet to be determined.

Jamali (2010) emphasizes on the fact that though there has been a significant rise on the proliferation of International Accountability Standards (IAS) in academia yet their adoption and integration in MNC's are not adequately investigated. Depending on the institutional theory and typology of strategic responses to institutional pressures proposed by Oliver, this study uses an interpretive research method to explain a set of MNC practitioner's views on IAS. Further on the basis of this, the author tries to provide some significant insights in regard to probable patterns of strategic responses on these emerging pressures on institutions. The author also outlines the importance and necessity of institutions in regard to IAS initiatives are registered as well as welcomed across, but more responses are on the lines of symbolic conformity.

Unerman, Bebbington and O'Dwyer (2007) provide the introduction of some of the fundamental concepts and theoretical aspects of corporate sustainability and corporate sustainability accounting. This study offers an explanation of the complexity of the current debate on sustainability. The authors in this volume bring forth the sustainability issues and perspectives of the debates in various public, private and NGO's. Finally, the book covers an exhaustive review of past research on sustainability accounting with recommendations for the scope for future research in this field.

3 Significance of Social Audit in Business

Social audit evolved amidst considerable agreements and disagreements. Some view 'social audit' as 'social statement', others consider it as social report; some prefer to call it as socially responsible accounting. To understand the exact nature of social audit, it is required to examine different viewpoints that have been employed to understand the term.

Robert K. Elliot argues that Socially Responsible Accounting is a systematic assessment of and reporting on those parts of a company's activities that have a social impact. (Krishna, 1992). According to him, it is the impact of the corporate decisions on environmental pollution, consumption of non-renewable resources and ecological factors; on the rights of the individuals and groups; on the maintenance of public service; on public safety, on health and education and many other social concerns.

The term 'Social Audit' has been used commonly to denote any form of retrospective review of the impact or contribution of the company in recognized social dimensions. George A. Steiner believed that audit and accounting carry a connotation of quantification and argues that the term social audit should not be used because measurement of performance is very difficult. In some spheres, social auditing has been treated as synonymous with Social Accounting. It is found that Social Audit has been there for 40 years. There has been constant debate regarding the evolution of the concept.

In order to have a very comprehensive social audit, openness should become a part of the system. Organizations are viewed as more open and creative systems, instead of its mechanical and closed bureaucratic systems. Success of any pluralist organizations is dependent upon the ability of the organizations to communicate effectively with other constituents in the society and functions in a coordinated manner. Moreover, it is also visible that society has become very competitive and demanding and aspirations outrun the ability to perform. And if there is a gap between the expectation and performance, then it causes a serious damage to the reputation of the socially responsible enterprise.

Instead of remaining unresponsive to the demands for disclosure of information, corporations can take the lead to develop a system of accountability which is best suited to the needs of business. But there are many obstacles to the development of a workable social reporting system, difficulties in measurement of information, dissemination and implementation of audit results.

Therefore, Social Audit is a tool that needs a systematic and awareness mechanism to reach out to the public for general concerns. The publication of company's contribution in social areas may create awareness among the public. It is always good to inform the public and seek their support, and, make the company's effort more meaningful.

A company usually operates in an environment in which the goals, the norms exist as source of pressure on the company, and also as source of legitimacy. A very important problem of Social Audit is to develop a mechanism through which external goals may be articulated and defined in terms of which can be applied to company's own measurement effort (Beesley & Evans, 1978: 169).

Companies may respond to external concerns, through direct interaction and dialogue. But for this the company needs a medium, and hence, corporate social audit may have to provide a framework for discussion. Another important medium of the relationship between audit and development of goals and norms relate to evolving social norms within the corporate organization. These norms will guide the corporate behaviour and performance of the managers.

The performance of managers in social matters will depend on the personal and corporate norms to which the managers respond voluntarily and not just on the controlling mechanism of the firm.

It is already in the process of establishment that, with continuing measurement and reporting, the potential of audit can be realized. Everyone cannot endorse that the measurement technique and interpretation of reported results will assist the Social Audit process. A selective approach might prove to be more productive, whereby in reporting a continuity of effort is feasible. The more the problem is one of readily identifiable concerns the more the continuous but restricted approach is appropriate.

Social Audit may contribute to new concerns and changing manager's response in a number of ways. Firstly, it is a mechanism through which the concern with the social measurement becomes known in the corporate organization. The argument is that some are actively seeking to measure social impacts that change the manager's environment. This is a necessary pre-requisite of the second purpose of this form of audit, namely developing commitment in the organization both to social measurement and to the social concern that underlie them.

There are similarities between the technical competence required to support responsiveness and other competences that developed in financial planning and control. But the most important concern is, there is a utter necessity for knowledge of new sorts of social phenomena, be it religious, cultural, habits of workers. The most important step would be institutionalization of reporting on these issues. But still the problems remain.

The problems are primarily related to orchestrating commitment and also, sensitizing managerial lot to other people's concern. If Social Audit is just seen primarily as measuring impacts, then its effectiveness must be judged by what it produces? But whether it produces good indicators of corporate performance in relation to social concern is yet to be ascertained.

The principal purpose of social audit is seen as influencing other important processes in the organization, but its effectiveness must be evaluated in terms of its contribution to enhance the insights of others and understanding of others and redirecting their behavior. When the policy issue comes up, then the primary question is what type of audit is conducted and law is used for developing responsiveness to social concerns in the company.

Therefore, a great emphasis should be given to social audit in a company, in the same way as the strategy of other planning is devised. Such a strategy might play an important role in company development. Then the senior management will be concerned about the contribution of social audit, rather than the technical benefits. The company's information strategy will move to the center, instead of it being a measurement of impacts. The information system of the company will reflect on the dual concern of the organizations, i.e. developing social responsibility and also strengthening the policy making.

Social Audit is not just this. It also reflects the state of the arts and willingness of enterprise to disclose information to all its stakeholders. As such, its scope varies from public pressure, government directions, manager's responsiveness, methodology for identifying social concerns, technicality of measures, the development of acceptable format, etc.

Various approaches have evolved to understand the impact of social audit. One way would be cataloguing of what the company is doing in each social programme. It requires identification of a particular area, and then calls for narrative description of what is done or being not done. Here the effort is to explain activities but not to evaluate them in terms of impacts. There is another way to identify and measure the expenditure for social initiative on a qualitative basis. But the problem is, the expenditure will only measure the inputs to social activities and not the outputs. Besides, as the company is spending a lot on social action programme, but whether the company is benefiting from such programme is not clear. One can regard it as a basic step towards evaluation that provides the basis for budgeting. But it is evident that, such an approach does not provide any clues for optimizing investment.

Another method would be to set up goals in different social responsibility areas. It makes evaluation easy, as actual performance can be measured. But again, it is the company's responsibility to compare the social worth of different social projects.

There is another way of approaching the social audit of a company, i.e. the cost benefit estimate. It takes into consideration quantitative and non-quantitative analysis to provide management with systematically developed information. This would help the company to review the effectiveness of the programme. But, ironically this approach does not put forward a comparative perspective, i.e. comparison of performance in one area with performance in another area.

To develop a social audit is not an easy task. Various obstacles may evolve in the development of social audit. The basic questions that come up what activities shall be covered; how performance is measured, so that everyone can accept it, then next would be, assess the expenditure for developing such audit. Why there is lack of public pressure on social disclosures; why there is an absence of legal requirements to develop such audit as the firm is a complex system of social relation, it has to work with diverse interests of social groups. It is true that measurement of social performance is very difficult, but it is the organizational policies, which determine the strategies to realize social audit. That is why; an effective social policy and performance may develop a social audit.

By now, it is clear that social audits, social accountability, corporate citizenship are tools that company employs to identify, and measure the success and ongoing challenges with social responsibility. Social Audit is like other corporate initiatives in terms of budget, assessment, and commitment from the executive. But without adequate measurements of the achievements of socially responsible objectives, the organizations cannot verify their significance, or justify expenditure, and finally address stakeholders' expectations effectively.

It has become established that issues pertaining to social audit, or social accountability (SA), is the way in which company's measure, and report on their social responsibility activities.

Social Audit is voluntary, and there is no law or regulation that requires a specific reporting method or verification of the report. SA is still not subject to internal auditing standards and external assurance practices that accompany financial statements and reports. However, some firms seek independent verification of the SA report. Major accounting firms and other consultants conduct assessments and attestations as to the accuracy and competencies of such reports (Sage Guide Book, 2012: 197).

Verifications are very rare, that is why some feel such reports are mere public relations effort that contains no accurate information. Some critics are of the view that companies may use these reports to just enhance their reputations.

4 History and Significance of CSR

There are lots of debates regarding the origin of the concept of CSR but it will be wise to consider Industrial Revolution or late 1800 as the beginning of the concept. The era before the rise of corporate form of the business, CSR was known as social responsibility (SR). Though CSR began to take shape in the 1950s, but its practice originated from Industrial Revolution. From the mid to late 1800s it was observed that the emerging businesses were primarily concerned with employees and their productivity. Then a very important question sprang up, whether business organizations are making workers more productive or also helping them to fulfil their needs and contributing members of the society.

While tracing the history of CSR, the management historian Daniel A Wren (2005) was very critical of the factory system in 1800s in Great Britain and in America. At that time, the social reform movement of both the countries found the factory system to be the primary source of social disorder, poverty, child labor, exploitation of women. A social welfare movement evolved with a mixture of philanthropy and business. At the same time, industrialist like John Patterson spearheaded an industrial welfare movement. It was found that the industrial welfare movement tried to solve labor problems through welfare schemes and practices. And, such schemes had both social and business implications.

Wren, a management historian noted that some of the early businesses were very generous including patrons of arts, builders of church, and endowers of educational institutions and providers of money for various community projects.

An early practice that was in vogue at that time, was the expenditure incurred by companies for community cause. Morrell Heald, a business scholar cited business programmes at the turn of the century. These programmes suggested some degree of social responsibility taken up by companies, but they were never called social responsibility for instance, George Pullman of the Pullman Car Palace, a community town in Chicago in 1893. Morrell Heald saw YMCA's (Young Men's Christian

Association), has an early social responsibility endeavor. Such an endeavor was not only by the individuals but also by companies as well. In his view executives came in contact with the social workers, and then the new views of social responsibility emerged. Business leaders were exposed to wide ideas of social problems, and as such professionalism arose from the social service communities.

Though socially responsible behavior arose in some cases but it was not a general trend. Towards the end of 1800, a charter was incorporated in those businesses that were socially useful. But in the post-civil war, the situation was very different in the U.S. Large business corporations began to dominate the economy and ironically the business had an influence in the state. And, finally power corrupted the business leaders and they created their own monopolies and defied the rule of trade. Eberstadt argued that business would have never turned back towards responsibility and accountability, for culmination of corporate irresponsibility had led to the collapse of the economic system. Then with the Great Depression an era of unemployment set in along with business failures.

In the post-Depression there was a change in the business–society relations. A Trusteeship phase emerged which upheld the stockholder demand, and created a balance among customers, employees as well as the community. But there was change in perspective; corporations began to be seen as institution, like the government. Even the corporate contributions were legally restricted to the causes that benefited the company. Patrick Murphy (1978), a business scholar created four CSR eras. He stated that the period till 1950s been the philanthropic or charitable era, and then between 1953 and 1967 was an awareness era where the companies were driven towards business community problems. Then, between 1968 and 1973, the issue era came. Companies concentrated on specific issues. And from 1974 to 1978 companies began to seriously address CSR issues. But it was Howard R. Bowen's seminal work "Social Responsibility of Business" that set in the modern phase of social responsibility of business. With his work social responsibility entered into a modern phase. He argues that social responsibility is not a solution to any social problem in business, but it can guide business relations. Since then new ideas of social responsibility of business crept in. Later on, scholars of social responsibility movement, like William C. Frederick (2006) focused on three basic core ideas. These three ideas are—"the idea of corporate manager as public trustees, the idea of balancing competing claims to corporate resources and acceptance of philanthropy as a manifestation of business support of good causes. Corporate giving became a part time exercise and it depended on the whims of the executives who responded to the request by the beneficiary organization.

This was followed by a period of more discussion, debate than action. Business executives were trying to get comfortable with CSR through discussions and debates. Bowen (1953) was ahead of his time by arguing for specific management and organizational change for improving business responsiveness to growing social concern. He 'proposed that there should be changes in the composition of board of directors, a greater representation of social viewpoint in management, more of social audit, social orientation of business managers, development of business codes of conduct, and further research into the social causes". These issues became the base for further discussions, debates and overall reflections.

The formalization period of CSR started in 1960s. During this period, Keith Davis (1960) set forth his argument for CSR, by referring to social responsibility as, businessmen's decisions and actions partially taken beyond the firm's direct economic or technical interest. He asserted that some socially responsible business decisions can also bring about long run economic profit.

Clarence C, Walton (1967), a prominent business and society thinker understood business responsibility from a managerial context. He has put forward models of social responsibilities based on intimate relationship. Such relationship should be first understood by the top managers as the corporations and the related groups pursue their respective goals. He emphasized that corporation's social responsibilities also include a degree of voluntarism, and this voluntarism, draws out the linkage between their voluntary organizations and the corporations, such voluntarism may involve economic costs. But he was not sure of the measurement of the economic returns out of such linkage. CSR did not stop with voluntarism rather idea of conventional wisdom came up. Johnson's (1971) "conventional wisdom" is aligned with a multiplicity of interests. It does not strive only for larger profits for its stockholders, rather takes into account employees, suppliers, dealers, local communities, and the nation at large. As he refers to "multiplicity of interests", his attempt to incorporate others became a precursor to the stakeholder interest.

A new phase started in the CSR movement, when Committee for Economic Development (CED) asserted on the changing nature of business-society relations. In 1971, CED published its report on CSR, known as "Social Responsibilities of Business Corporations", which primarily stressed that business enterprises functions by public consent and its purpose is to serve the needs of the society. The social contract that existed between business and society was changing in substantial manner. The CED articulated a three concentric notion of social responsibility, the inner circle, intermediate circle, and the outer circle. The inner circle responsibility was more into the efficient execution of the economies, products, jobs and economic growth. The intermediate circle encompasses responsibility to exercise the economic functions with a sensitive awareness of changing social values and priorities, for instance, environmental conservation, hiring and relations with employees, and more expectations of customers for information, fair treatment, and protection from injury. The outer circle argues for newly emerging and still amorphous responsibilities that business should assume to become more involved in improving the social environment.

The CED's views were very significant, as it was composed of people from diverse walks of life. Besides, it also reflected a practitioners' view, with the changing social contract relations between business and society. It is also believed that, CED's articulations were a response to the time, especially in the late 1960s and 1970s. During this period, many social movements took place focusing on the environment, employee's safety, consumer concern, and moreover, the employees were in a state of transition from special interest to formal government regulations. Another important development during this phase was on the application of CSR.

But, the CSR movement like any other movement too had a line of dissent. As early as 1950s, CSR drew criticisms from different angles. Theodore Levite (1958) took the lead. He warned that businesses by pursuing corporate objectives, such as CSR, rather than the well-defined traditional aim of profit, might face some serious consequences. Suppose, managers of the firms have the discretion of making value judgment on which social issues to pursue which not to pursue, and then what will be the implication of such an objective? Then such objective will reduce the single motive of maximizing the manager's utility.

Milton Friedman (1962) too had a limited view of corporate social responsibility. In his view, social responsibility of business was to increase profit, stay within the rules of the game, there was no space for social responsibilities of business, and for those are state responsibilities. Friedman criticized CSR from two angles. In one way, individuals can have responsibilities, not business as a whole, and the other is, any action by a business that does not aim for profit, will lead to a burden on business. Friedman was vehemently against such motives of business. In his opinion, CSR is subversive, for there is only one social responsibility of business, i.e. to utilize the resources and engage in activities that are designed to increase its profits, as long as it follows the rules of the game.

Amidst these critical debates a broader perspective on CSR grew, George Steiner (1971) was part of this perspective. He argued that corporate social responsibility goes beyond mere economics, it represents a concern for the needs and goals of society. He believed that in order for the business system to grow it has to function in a free society. And corporate social responsibility is characteristic of a movement that represents a broad concern for business role in supporting and improving the social order.

A new phase with new concepts, ideas entered the domain of CSR. S. Prakash Sethi (1975) became a part of this new phase. He argued for dimensions of corporate social performance. His version of corporate social performance implies 'social obligations', 'social responsibility' and 'social responsiveness'. In Seth's scheme, social obligation in corporate behavior is a response to market forces or legal constraints. Here he implied both economic and legal responsibilities. But to him, social responsibility is beyond social obligation. The corporate behavior has to be in congruence with the prevailing social norms, values, and expectations of performance. At that time, there was a drive to direct attention away from corporate social responsibility. Preston and Post (1975) took the lead in this direction and referred to Dow Votow's (1973) commentary on social responsibility. In Votow's opinion, social responsibility is positive and good, but it doesn't convey the same meaning to everyone. To some, social responsibility means legal responsibility, to others it is just a socially responsible behavior. While others, equate it with a charitable contribution. Some embrace it as a synonym to legitimacy. Others see it as a fiduciary duty imposing higher standards of behavior on businessmen than on common people at large. Preston and Post opted for public responsibility rather than social responsibility, to stress the importance of the public policy process, rather than individual opinions and conscience.

A new trend evolved since then, instead of restricting to the views of business theorists, research studies were conducted to seek the opinions of employees and their perceptions of corporate social responsibility. It was a very general study. Carroll developed an all-encompassing perspective towards CSR. In his view, corporate social performance was embedded in CSR. His basic argument is that, managers of firms in order to involve in corporate social performance (CSP), must know the basic definition of CSR, an understanding of the issues for which social responsibility existed, and a specific philosophy or strategy of responsiveness.

To Carroll, social responsibility of business encompasses the economic, legal, ethical, and discretionary expectations that society has if organization's at a given point in time regarded it as very essential (Carrol, 1979). The economic responsibility that Carroll specifies is something which business does for itself and legal, ethical, and discretionary is something, which business does for other. There was a change in CSR movement since 1980s and as result a research oriented drive evolved in the CSR movement. Voluntarism and obligations of the corporations became broader, as it extended beyond the traditional duty to shareholder, such as customers, employees, suppliers and the neighboring communities. And, then the CSR movement leaned towards a process oriented theme.

Jones (1980) understood CSR as a process. He pointed out that it is difficult to assess the outcome of socially responsible behavior. To assess the impact of CSR, Frank Tuzzolino and Barry Armandi (1981) tried to develop a new way to assess the input of CSR. They proposed a need- hierarchy approach towards CSR, patterned after Maslow's need-hierarchy theory. The core of this approach is, organizations also need to fulfil certain criteria. They argued that organizations have certain needs like physiological, safety, affiliative, esteem, and self-actualization needs that parallel those of humans as depicted by Maslow. The CSR movement was not restricted to needs driven approach.

Then performance dominant phase started in businesses. In the 1980s there was a growing acceptance of the notion of corporate social performance. Even, scholars have tried to relate social responsibility and business ethics. Besides, scholars like Edwin M. Epstein (1987) tried to combine social responsibility with business ethics into what he called as corporate social policy process. According to him, the rule of the corporate social policy process is the institutionalization within business organization of the following three elements, business ethics, corporate social responsibility and corporate social responsiveness.

It is evident that a new consciousness was growing. Freeman's stakeholder significance was one such consciousness. Freeman (1984) recognized the contributions of other segments of the business enterprises, besides the shareholder. He pointed out; in addition to shareholders there are creditors, employers, customers, suppliers, and the communities at large, who are equally responsible for the growth of the enterprises. Thus, the stakeholder approach set forth the growth of strategic perspective in CSR.

Globalization too brought about new changes in the economy. Many international companies appeared in the economy; As a result, new concepts such as corporate reputation, community partnerships, corporate social policy, became an integral part of large companies. In terms of management policy, some changes were visible such as strategic giving, international donations, voluntarism among the employees, cause-related marketing. Even the beneficiaries of CSR initiatives changed. It includes "education, culture and the arts, health and human services, civic and community, international donors, community partners and NGO partners" (Muirhead, 1999).

New organizations were formed to mediate between the companies and their CSR initiatives. In 1992, a non-profit organization called Business for Social Responsibility (BSR) was formed to represent the initiatives and professionals having responsibility for CSR in their companies. With BSR, CSR became very broad and new themes such as environment, governance, human rights, accountability, workplace practices, and marketplace were understood to be a part of CSR. As such, many companies gained reputations for adopting ethical CSR practices.

Thus from 2000, CSR became a global phenomenon with the coming of international institutions like the Organization for Economic Cooperation and Development (OECD). It was observed that voluntary initiatives in CSR have become a major trend in international business. There have been divergences in commitment and management practice, even in areas of application such as environment, human rights, corruption and labor standards.

It is also revealed by OECD that, considerable management expertise has been achieved in legal and ethical companies. This is due partially to the institutionalized support that is emerging in terms of day to-day company practices, management standards, professional societies, and specialized consulting and auditing services. The OECD did not reach definite conclusions, but it has assessed the benefits for companies on the costs of CSR initiatives, for companies and for society and found out to be immense.

In the 1970s, CSR was element of discussion in the business circle, but it broadened its involvement from community to socially responsible products, processes and employee relations. Jeremy Moon, a known business scholar drew up an analysis of CSR, how it evolved in UK, and also in the European Union. He argued for new elements for CSR i.e. socially responsible products, processes and responsible employee relations.

Moon (2005) found out that, there has been a growth in CSR staff in companies, embedding CSR in corporate systems via standards and codes, increased social reporting and growing partnership between companies and NGO's. Another big leap was the institutionalization of CSR by corporate managements in the UK. Such an institutionalization of CSR by corporate management in the UK strengthened its position with the United States and other developed countries in the world. There was a growth in the senior level management, board level responsibilities, and reporting, and thus increased external stakeholder relations.

A very secular thinking also grew in the field of CSR, and 'Steven D. Lydenberg sees CSR as major secular development, with extensive re-evaluation of the role of corporates in society (Teach, 2005). According to Lydenberg, the re-evaluation process is more evident in Europe, where the stakeholder responsibility is more readily assumed. But not all are so optimistic and positive on CSR. David Vogel

(2005), feels that CSR will not be successful, until mainstream companies do not accept it as a very positive phenomena, especially those aspects of CSR that are critical to company's future performance.

5 CSR and Social Audit: Assessing the Standards for Regulatory Framework

Social audit and CSR has to be all comprehensive, so that the organizations can stand tall as an open and transparent system. Organizations are not just economic systems known for its financial performance, but it includes an entire gamut of performances like environmental, social, and cultural etc. Though it is believed to start with the community inside the firm, but it has to cater to the needs of the outside community. Companies need to devise a strategy to respond to both internal and external concern, and the social audit and CSR can assist them to develop a system of accountability. The success of any organization lies within its capacity to communicate effectively with other constituents of the society. If there is a gap between expectations and performance, then it might cause a heavy damage to the organization. At this juncture, CSR initiatives can assist the organizations in developing a network which can connect them to the stakeholders.

Social audit creates an awareness mechanism to reach out to the public because business organizations are not just economic organizations only, they are also social organizations. To be a social organization, it is essential to engage with its stakeholders say workers, investors, stakeholders, customers and the greater community as a whole.

CSR audit or social audit can be assessed, measured and reported based on the guidelines set by the international bodies and reporting initiatives. OECD, UN Global Compact, International Labor Organization (ILO) and Global Reporting Initiatives (GRI) has setup guidelines that are regulated as the standards recognized both at international and national level. Though they provide a voluntarism in setting up the standards, but these organizations can provide the basis for developing a regulatory framework. So that companies can seriously adopt CSR and social audit, for the benefit of the company and its stakeholders.

6 Trans-governmental Bodies-Reviewing the Standards/ Guidelines

6.1 UN Global Compact

The Compact remains relevant in today's world, because it is about building certain integrity into the process, so that a sufficient degree of legitimacy is built in, and its benefits are shared by large chunks of world population.

The utmost challenge the Compact faces involves building a foundation of equity and prosperity for all. Well, before advancing on the challenges ahead, one needs to know about certain characteristics of the compact that place it in a unique position and advances on such causes. Like, the Compact permits the involvement of multiple actors, not necessarily from business, in venture both at the macro and the micro level. The biggest challenge that the Compact faces at the macro level will be either to act or counteract the wave globalization has played in the world market.

The Compact could he a pragmatic response to number of problems that nation's faces, but it cannot be a substitute for political will and state-led action. The Compact aims would be a realistic response to the failure of governments and governance. Now, the governance model prescribed by the compact is not implicit in the design of the initiative.

Further, the Compact needs to scale up its activities in order to exert remarkable influence on policy making and business modules. Though, a good number of organizations including 1,000 companies have joined the Compact, but this is negligible compared to 60,000 Municipal Corporations existing today. Even, the small and medium enterprises (SMEs) are either not aware of the Compact, or have not become involved with its activities. And, these businesses are source of employment in today's time, and efforts have to be made to reach out to them. Other significant players could be the NGOs, Civil Society Organizations, and private organizations.

The Compact needs to be highly visible in the activities of many others multinationals. It should also be able to manage a diverse network of actors. Actors in the network bring with them views and ideologies that can be opposite to other participants. Hence, the tensions between actors need to be adjusted at a creative and production level.

Last, but not the least the Compact rests heavily on the internal capacity of the United Nation (UN) to orchestrate strong set of networks. This is more challenging, as the compact has an extremely ambitious goal. The Compact seeks to establish a model of corporate citizenship that aims to influence the activities of business both within the organizations and outside, so that it can respond to social needs.

The nine principles of Global Compact are primarily in the areas of human rights, labor and the environment, and it derives universal consensus from: the Universal Declaration of Human Rights (UDHR), the International Labor

Organization (ILO), Declaration on Fundamental Principles and Rights at Work, and the Rio Declaration on Environment and Development.

6.2 Importance of Human Rights in Business

Human rights issues are important not only for the individual, but also the organizations they are part of. As part of its commitment to the global impact, the business community has a responsibility to uphold human rights in the workplace and within its sphere of influence.

Business should strive to guarantee that its operations are consistent with the legal principles applicable in the country where it operates.

6.3 How to Promote and Raise the Standards of Law?

Businesses those are operating outside their country of origin have opportunity to promote and raise standards, especially enforcement of human rights issues.

6.4 How to Address Consumer Needs?

Consumers are aware of where their goods came from and also the conditions under which they are made. The company can adopt very active approach to human rights that can reduce the negative impacts of bad publicity from consumer organizations and interest groups.

6.5 How to Manage Supply Chain?

Companies need to be fully aware of potential human rights issues, if they are into global sourcing and manufacturing.

6.6 How to Accelerate Worker Productivity, Integrity and Retention?

Companies need to treat workers with dignity and then their work would be more productive. The social and environmental record of companies will also attract the new recruits.

6.7 How to Build and Sustain Community Relationship?

The Companies that are global players has a wide audience, and if these companies address human rights issues can bring positive accolades within local communities, as well as in the broader global community, in which the companies operate.

6.8 How to Incorporate Human Rights Into Company Policy?

It requires developing a company policy to support human rights, and initiate a health and safety management options. Then, the company has to provide training for the staff on human rights issues, and discuss human rights impacts with affected groups. It also requires improving working conditions in consultations with the workers and their representatives. The company requires enduring non-discrimination in personal practices by avoiding direct or indirect forced labour or child labour. Besides, it is the company's responsibility to provide basic health, education and housing for workers and the families. It has to avoid forceful displacement of individuals, groups or communities, and ensure rehabilitation and resettlement of the project affected communities.

The firms in order to build a strong relationship, has to comply with the governance mechanism of the country. To endure the business relation, the firms have to support the efficient governance system. Besides, the efficient government supported by strong legal order, be it an international body, like the UN Global. Moreover, the firms do not need to affirm to two efficient governance systems at the same time.

If we go by the current statistics, Global Compact is one of the most popular corporate accountability mechanisms among the business houses, both participants from the developed and developing countries.

But due to the absence of screening of new participants and the enforcement mechanism, there is a high risk for companies, that they might use this global compact membership as a means to improve their images and not for real improvements in social and environmental issues.

It is argued that UN Global Compact is primarily suitable for larger firms. Smaller firms do not have the resources to meet the UN Global Compact requirements, and the firms that sell their goods in countries with efficient government will benefit least from UN Global Compact membership.

Initially, the UN Global Compact attracted large firms to become members of it. But in 2008, due to the failure of compliance, it delisted many firms, especially when these firms failed to submit the required communication on progress. Here, a member must give information, how to improve its performance in four areas including human rights, labour rights, environment, and corruption.

If we see the high number of delisting relating to social reporting of companies, it can be perceived that the value added to the companies is not worth. Yet, the social reporting is on the rise, fuelled by demands from-institutional investors. There is a rise in the standard of reporting requirements, but more and more firms are discouraged with reporting.

To become a UN Global Compact member, it is a simple process; aspiring member should express a commitment to it and its ten principles. Besides, the company must emphasize a willingness to engage in partnerships to advance broad UN goals and publicize its participation in the Compact, and to agree to submit an annual compliance report.

As the Compact is a voluntary institution, it is not regulatory. That is why; it relies on public accountability, transparency and disclosure to complement regulations.

Few points out that, it is easy to become a member, but the Compact has no regulatory mechanism to hold the corporations accountable. As a result there are massive violations of rights by some members. Even, the performance standards are not clear. But, there are others who do not support this view; rather they argue that the ease and flexibility is the key strength of the Compact. The Compact can stimulate debates and discussions on new trends of Social Responsibility, Social Accountability and Social Audit practices.

The Compact has set a single global standard that looks beyond, any national, sectoral and regional standards. Though it is a voluntary initiative but never gets in the way of other future initiatives, rather it stimulates discussions and debates. Companies that operate in their home countries, a UN Global Compact membership may be perceived as less valuable.

The Compact is not a standard to measure corporation's compliance against some predefined indicators. Companies are allowed great flexibility to report on their progress. They must disclose openly, but there are no standards to follow. There is a problem here; most firms undertake their social responsibility initiatives to fit the Compact, instead of focusing on the initiatives that could best benefit the company and the society. It is essential to have a thorough evaluation of a company's necessity and demands concerning social responsibility. This can lead to the company critically evaluating the kinds of initiatives to be part of, and the reporting it wants to engage in. It may also happen that, the company may decline to be a part of certain initiatives and reporting system.

Small and medium-sized firms do not have the economic resources for proper documentation. In fact, many large buyers and retailers demanded that SMEs suppliers should adequately control the social and environmental performance of their own suppliers in less developed countries.

It so happens that SMEs often lacks political clout, and even few buyers, and incentives for the supplier to improve the social and environmental performance are limited. Large firms also complain about not having enough opportunities to change supplier practices. But they are at an advantage, because they have more resources to follow up with supplier than the SMEs.

In recent years, large MNCs are consolidating their supply chain so that they work with larger and few suppliers. Even some experts feel the SMEs are being excluded from global supply chains. But, if we see from a managerial perspective, SMEs are becoming more integrated into global supply chains. If we look from another perspective, international competitiveness of SMEs, has increased. Supposedly, large suppliers avoid western SMEs, and then many western economies would be affected, because the SMEs contribute significantly to the growth of the economy. Exclusion of the SMEs from the global supply is not a positive signal; rather it should be given some serious thinking.

The firms that are domestically oriented are unlikely to benefit from the UN Global Compact membership, because they primarily produce or sell in their well-regulated home markets.

Then there are firms that are operating in countries with well-developed and efficient regulations. So these firms may avoid joining the Compact. It is also believed that reporting does not add value to the business. But, this is not wholly true, rather if the firms have a strong reporting mechanism, it may strengthen the ties between the company and its stakeholders.

In 2011, the Compact started a new approach to reporting i.e. the Global Compact Differentiation Program. This program tries to categorize business participants based on their level of reporting and disclosure and the progress made in integrating the Global Compact and contributing to broader UN goals. The differentiation program is somewhat less demanding for the SMEs. It can be argued that the large buyers can work with small suppliers and assist them in meeting requirements.

But, the domestically orienting firms are not going to benefit from the reporting requirements of the UN Global Compact because they are less demanding. Hence the Compact should concentrate on those types of companies that use socially responsible initiatives as a substitute for missing the regulatory standard, instead of recruiting all types of companies. It should also initiate a network that can integrate the SMEs, the biggest players.

7 International Labor Organization

International Labor Organization (ILO) forged in 1977 and revised in 2006, because it is the only 'tripartite', UN agency, for bringing together representatives of governments, employers and workers to shape policies and programmes. ILO has mandated since its formation, a periodic assessment of its policies and programmes. Its member countries are primarily concerned about the social behavior of multinational enterprises. ILO is essentially based on the Declaration of Fundamental Rights, and Right at Work. These two, forms the core of ILO's labor standards. Unlike, other trans-governmental organizations, ILO do not have any implementation mechanisms. Its standards serve as a guiding force for the corporate organization's social accounting, social responsibility initiatives.

The purpose of the Declaration is to encourage positive contribution form the multinational enterprises, to make economic and social progress, minimize and resolve difficulties arising from their operations. The Declaration (adopted in 1998) has set out principles in the area of general policies, employment, training, and conditions of work, industrial relations, governments, employers and the worker organizations are recommended to observe the principles on a voluntary basis.

In the area of general policies, it is required to obey national laws and respect international standards; in terms of employment, it is essential to provide employment opportunities, treatment and security of employment; in terms of training, it is required to develop policies for vocational training, formation of skills, in terms of conditions of work and life, it is required to develop adequate conditions of work, wage benefits, and health and safety and in terms of industrial relations, it is essential to ensure, freedom of association and collective bargaining.

ILO conventions apply only to its member states which ratify them, but the Declaration is relevant to all member states. Hence, the Declaration represents a political commitment by governments to respect, promote and realize the fundamental principles.

The Declaration, an idea that evolved from the Employer's Group as an ILO proactive action to promote positive evolution towards universal fulfillment of internationally recognized values in the world of work. The Declaration was significant, because it was preferred as a positive approach to the social cause. It was felt that, by focusing on the fundamental principles, every member should promote the responsibility for creating and maintaining minimum national standards, which would remain—at the government level—rather than on individual organizations.

ILO Declaration does not have implementation mechanism. It has diffused the declaration, by drawing on its tripartite structure. Lack of implementation mechanism, in the ILO Declaration has led to one of biggest challenge, i.e. whether it has the standing to entice multinational companies to behave in a socially accountable and responsible way.

Hessel (2008a, 2008b) pointed out that there are new patterns of norm-setting and international guidelines, which not only lead to higher expectations regarding the responsibility of companies, it can activate a form of indirect regulation. That is why, he argues that international guidelines, besides being legally unenforceable, can still serve as a norm-setting standard that define an acceptable behavioral norm, below which companies are advised not to fall.

The standards that are set by the ILO Declaration are of practical use, when the multinational companies are willing to let them influence their social accounting, reporting and responsibility policies. It may also happen that, the companies may feel that the Declaration does not offer any value on how to operationalize social reporting at the individual firm level. This would eventually impede the dissemination of the Declaration.

Then, if the declarations are of no practical use to the companies, it is less likely to be used by the multinational companies. Thus, there is a less likelihood that it could become a norm-setting standard for the social accounting of business. Then, the capacities of the ILO Declaration to ensure responsible business would be severely undermined.

ILO Declaration to operate effectively, it has to be known sufficiently in the business community. Then, the individual companies will be able to consult them, while developing their own social policies (social reporting or social accounting or social responsibility). It is also very important that the labor governance is a huge issue, and if it expands and reaches a mass of addresses, then the law-standard firms come under pressure to raise their standards and for the firms who has still not adopted a labor friendly governance system may face a high competition in the globalized market.

For the ILO Declarations to operate effectively, the Declarations have to be widely accepted among the stakeholders of the multinational companies. It so happens that social reporting may be due to direct pressure from socially active actors, like the NGOs, civil society organizations and even the state. There is no one single approach to social accounting, or social report that the multinational companies refer to.

The multinational companies often refer to issue specific social responsibility instrument. If the social responsibility instruments embody the ILO Declarations, then there is high likelihood that it reaches the large mass of companies necessary for the ILO Declaration to serve as a form of indirect regulation.

But, if the ILO Declaration is not taken seriously, by industry and instruments, then its recommendations are unlikely to reach companies to develop into a norm-setting regime. Ultimately, the capacities of the ILO Declaration to guarantee socially responsible business conduct would be severely undermined.

To relate improved labour conditions to trade relations is not new. With the ILO Declaration of Philadelphia 1944, it was finally recognized by the international community that work is a part of life, and it is crucial for the development and dignity of human beings. Economic development implies empowerment with freedom, safety and dignity.

In a globalized economy, it requires the world community to respond to the challenges by developing legal instruments on trade, finance, environment, human rights and labour. The ILO promotes international labour standards by contributing to the legal framework. And this will lead to economic growth and development with the creation of decent work. The ILOs tripartite structure is backed by government, employers and workers. International labour standards lay down the basic minimum social standards agreed upon by all players in the international economy.

The international labour standards draw a framework on social standards that help governments and employees to avoid lowering labour standards. Because lowering the labour standards will lead to low wage, low skill workforce and also prevent in developing a stable high skilled employment. This will be difficult for the trading partners to strengthen their economy.

Compliance with the international labour standards, often results in improvement in productivity and economic performance. Higher wage and better working time standards, can translate into better and satisfied workers. Employment protection, social protection (employment scheme) can encourage workers to take risk and to innovate, and facilitate market flexibility. Active labour market flexibility can make liberalization and privatization sustainable.

Freedom of association and collective bargaining if practiced to a significant extent, may lead to better labour management consultation and cooperation which will reduce labour conflicts and enhance social stability. Foreign investors are in search for the beneficial effects of labour standards. It is also found that the foreign investors rank workforce quality and political and social stability above low labour costs.

The fast growing economies cannot anticipate economic crisis. The disastrous effects of the crisis were aggravated in many of the economies and active labour market policies and social dialogues were very much in demand. It can be argued that social dialogue, freedom of association, and social protection systems would provide a better safeguards against economic downfall.

The ILO standards could develop a strategy to reduce poverty. Legislation and functioning legal institutions ensures, enforcement of contracts, respect for procedures, and protection from crime. A market which is governed by a transparent and a fair set of rules is more efficient and the labour market is not different from the market. It is an established fact that fair labour practices that set out international labour standards applied through a natural legal system can provide an efficient and stable labour market.

In developing countries workforces are more active in the informal sector. Moreover, such countries often lack the capacity to provide effective social justice. International labour standards can act as effective tools. Because the standards apply to all workers, both formal and informal. It is found that the extension of the freedom of association, social protection, occupational safety and health, and other measures required by international labour standards also proved to be effective strategies in reducing poverty in both the sectors.

These labour standards urge for the setting up of institutions and mechanisms which can enforce labour rights. If these standards are combined with well-defined rights, and then the legal institutions can formalize the economy and create a system of trust and order which is the strength of economic growth and development.

International labour standards were not developed within a single day. It is an outcome of continuous process of discussion, deliberation among governments, employers and workers in consultations with experts from around the world. These discursive processes epitomize the consensus at the global level and how a particular labour problem could be handled worldwide and reflect the knowledge and experience from all over the world.

Globalization has created immense opportunities for millions of workers and employers across the world. But it has displaced workers and enterprises and caused financial instability in certain regions. Even the inequalities between the richest nations and poorest nations have grown exponentially. Continuous development may prove to be neither sustainable nor desirable. The ILO has developed a comprehensive decent work agenda that aims at promoting social dialogue, social protection and employment creation, and also respect the international labour standards.

8 Organisation for Economic Co-operation and Development

The Organisation for Economic Co-operation and Development (OECD) Guidelines offer principles and standards in areas of responsible business conduct-employment, industrial relations, human rights, environment, information disclosure, combating bribery, consumer interests, science and technology, competition and taxation. Multinational enterprises headquartered in adhering countries are recommended to observe the Guidelines also when operating outside the OECD area.

The OECD Guidelines encourages the positive contributions that MNCs can make to economic, environmental and social development, and this may minimize the difficulties to which their various operations may give rise to. The adhering governments are committed to promoting the Guidelines among MNCs operating in or from their territories, but the companies are only recommended to follow the guidelines. Hence, the successful implementation of the OECD Guidelines has to do with the member states. Companies in turn are engaged via governments. Each adhering government is required to establish a National Contact Point (NCP), to establish link with the business community. National Contact Points have been organized by the adhering states in different countries. Some are located in a single ministry, or multi-ministry structure, social organizational or civil society organizations or NGOs.

The NCPs has two primary tasks. First, NCPs should promote the OECD Guidelines in their adhering countries, so that the national business community, employee organizations, NGOs are aware of and can understand the OECD guidelines. Second, NCPs handle complaints about companies breaching the OECD Guidelines.

If an issue is brought to the notice of the NCP by an NGO, or trade union, the NCP then uses a mediation and concentration procedure. It makes an assessment of the allegations, and then examines it, in the light of the provisions contained in the guidelines. If the NCP finds out that the issue does not merit further examinations, it must substantiate its decisions. But, if it finds out that the case merits further consideration, the NCP will assist the affected parties to resolve the issue.

To operate effectively, it is imperative that the OECD guidelines and their recommendations are precise. If the recommendations are precise, then the civil society actors, NGOs, trade unions can pinpoint violators in corporate behaviors and report them to NCP for investigation.

Thus, if the OECD Guidelines are not formulated precisely to enable the social actors to substantiate complaints, then the capacities of the Guidelines to ensure responsible business conduct would be undermined.

8.1 Proactive Implementation of the Guidelines

The OECD Guidelines to operate effectively, the adhering governments should implement the guidelines in a proactive manner.

There are two factors, which are of particular importance here. First, the financial and human resources made available to the NCP, and, second, the institutional status of the NCP. It is very much understandable that, NCPs need large amount of finance and human resources to handle complaints of companies allegedly breaching the guidelines. Moreover, NCPs are organized in varying forms by adhering governments.

The social actors should be prepared to file complaints before the NCP. The social actors-civil society organizations or NGOs should be capable and also willing to launch a complaint before the NCP. These social actors should be strong financially, to establish an alleged corporate misconduct. The NGOs and trade unions must also have confidence and consider the NCP as a honest forum.

But, there could be some disadvantages here. The multinational companies may face economic disadvantage upon publication of a statement made by the NCP. Suppose the economic costs of a public report by the NCP are less than the costs of complying with the OECD's social and environmental standards, then the companies are less likely to comply.

The OECD Guidelines are prepared by more than 40 OECD countries and the adhering governments. It is also the government backed corporate accountability, reporting instrument. It includes a complaint mechanism for adhering alleged violations of the guidelines. But, the outreach of the OECD Guidelines is limited as they are only applicable to companies operating from one of the 46 OECD and adhering companies. Some clauses have weak language and some expectations are less ambitiously formulated than the Global Compact. While the specific institutional mechanism provides an opportunity for civil society organizations to lodge complaints about alleged violations of the Guidelines. The effectiveness of this instrument in providing positive outcomes is very limited. Even, the natural Contact Points has been diverse in handling the complaints.

Companies may contribute or be directly linked to adverse impacts through their business relationships. The adverse impact of the company can be in the following form: If the company is causing any adverse impact, that it is expected that the company should stop, prevent, mitigate and remedy the adverse impact it has caused.

The OECD Guidelines and their complaint process are unique. Its complaint procedure provides an opportunity for civil society organizations and trade unions to address corporate misconduct and seek resolution of conflicts for affected parties. Though, the OECD Guidelines are not legally binding on companies, but the OECD and its adhering governments are legally bound to implement them and also establish a National Contact Point to listen and examine complaints.

The purpose of the complaint is to resolve alleged breaches of the Guidelines, and facilitate dialogue between the parties. The government backed complaint procedure is a unique aspect of OECD. But, it has been found out that the civil society organizations and trade unions have mixed experiences with the National Contact Point (NCPs) around the world. Even, the remediation process is long and, then positive outcome is not guaranteed.

9 Reviewing the Standards of Private Organizations

GRI draws heavily of its parent organization, CERES established in 1989. CERES is a coalition of investors, public pensioners, trustees, non-governmental, environmental organization, public interest organizations, and labour unions. Its mission is to improve environmental performance and accountability among U.S. firms by promoting socially responsible investment and shareholder activism and seeking voluntary commitments from industries and the specific codes of conduct (the CERES principle).

CERES draw its ideology from two streams of ideals that is the consumer, investor and shareholder activism in the U.S. dating back to the 1970s and the corporate social responsibility (CSR) movement during the 1990s. Its founding mission reflects the rising interest both among the policy makers, some social activist, in a more collaborative manner with industries, and also escalating the reliance on forces of the market as regulatory instruments.

The nineteenth century anti-slavery, anti-gambling, and other social campaigns, linked money power with moral principle in the US. The 1960s was a period of social turmoil, against gender inequality, campaign for decisive civil rights, and protest against Vietnam War. Even in 1969, the United Church of Christ resolution to use moral principle to guide its investment was a decisive step. As a consequence, several socially responsible forums emerged. These developments spurred demands for reliable information about corporate activities. These lead to the growth of organizations that stressed on performance, in terms of transparency, environment, community relations, minority, and women employment.

The 1980s, the social reformist movement took a responsible step with the formation of Social Investment Forum (SIF), in Boston, with Joan Bavaria as President. Since then, social activism in business gained a huge momentum. Information on Non-Financial performance of firms also became the central theme for pursuing the market based social activism. And, subsequently, CERES was born as a result of the activism in SIF. The SIF directed one of its motives towards partnership with environmental activist organizations, with explicit code of environmental conduct, asking companies for formal endorsements, and regular reports on performance.

The 1992 UN Conference on Environment and Development in Rio lead to a change in the balance of power between global corporations, governments, and the society. The corporate sector presented itself as not only part of the environmental problems, but also as an integral part of the solution, and an equal partner of government and NGO's. Thus, the social activism through markets and the socially responsible movement reinforced each other, for company's information on environmental and social performance, and their unwillingness to disclose it.

The mid 1990s there was a flow to produce standardized reports on company's performance. In 1997, the CERES started the global reporting initiative, to create a global framework for the voluntary reporting of the economic, environmental, social reporting of corporate, and then other organizations.

GRI was originally convened by the collision for environmentally responsible economics (CERES) in partnership with the United Nation Environment programme (UNEP). GRI has evolved into a multiple stakeholder initiative incorporating participations from corporations, NGO's, Trade Unions, accountancy bodies, and other interest groups. Its unequivocal aim is to enhance the quality of sustainability reporting. The early supporters of the guidelines envisioned that GRI could become an agent of change within the reporting organizations. This will provide the ground to create a self-assessment and reflection process that could engage deep learning within organizations.

GRI did not just certify any audit performance. Its aim was to create a universal reporting frame work, in which a discourse about sustainability performance could be carried out, and which could be used by others to form judgment about the socially formulated standards.

The guidelines offer a comprehensive set of reporting principles and structured report contents that stresses on economic environmental and social performance dimensions. The environmental performance draws upon the extent of environmental practices of the organizations; social performance indicators addresses key areas of labour practices and decent work, human rights and also issues related to community, bribery and corruption and product responsibility. The central importance of stakeholder dialogue in informing the reporting process is made quite clear.

The primary goal of reporting is to engage with the stakeholders. Because the report alone cannot provide the value that influence the decision and behaviour of both the reporting organisations and its stakeholders. There are few strategic principles namely five that guided the process of GRI: inclusiveness, multi-stakeholder participation, consultation and empirical testing, use of internet, and transparency.

Being guided by these strategic principles, an atmosphere of neutrality will prevail within the organization and elicit the best ideas, assume that the product serves its goal, to overcome systemic barriers. Its aim is to create legitimacy, and an evolving social network. Though, one may point out at the variance of use of these principles, but it can be considered as one of the success of GRI to bring together diverse constituency to work together.

The organizational structure of the GRI was very efficient in the beginning. It consisted of a secretariat, a seventeenth—member steering committee, a member of decentralized working group coordinated by the steering committee and communicated mostly electronically.

In the later part of the 90s, the steering committee of GRI consisted of NGO's, think tanks, environmental organizations, investors, corporates, the representatives from United Nations Environmental Programme (UNEP) as well.

Initially, steering committee of the GRI was from Europe and U.S, but later on expanded with representatives from Japan, India and Columbia but it was the internet that made GRI to become a global player. GRI embarked on making maximum use of the internet. It allowed the large, widely dispensed participants to form virtual committees and consultation forums.

All the meetings were made public, all the documents and discussions were made widely available through GRI websites. In the meantime, the practice of writing and re-writing of drafts by different GRI working groups became a sophisticated system. GRI initially concentrated on environmental reporting, but later on it expanded its scope to the sustainability indicators, the social performance indicators, economic performance indicators, and environmental indicators. The social performance indicators focus on how an organization contributed to the well-being of its employees, customers, labours, human rights, governance and product responsibilities and safety practices. The economic performance indicators focus on the organization and its host community's financial performance, by focusing on its economic impacts on customers, employees, suppliers. The environmental indicators draw upon environmental performances, both at the present, and for the future generations. It governs such topics as: resource conservations, waste management, controlled and restoration of environmental risk, waste disposal, recycles, greenhouse gases, renewable energy and wildlife conservations.

Organizations prescribing the GRI guidelines as standard for socially responsible reporting, may get prominent NGO's and charitable institutions on board, participate in the multi-stakeholder development process, finance and also making the promise that their social and environmental issues would find a place in regular business reporting activities. These organizations regard GRI as a voluntary, friendly and alternative to governmental regulations. In the light of growing chaotic problem in supply and demand of information about the sustainability performance, GRI standards stand promising enough to deliver efficiency.

The dialogues with individual stakeholder sharpen the GRI's vision to the stakeholder worldview. Even for the companies GRI reporting will lead to better informed decisions on all aspects of sustainability. Customers and consumers will be able to influence markets through their purchasing and to make better informed personal choices of goods and services. The financial sector will have the required information for performing bench mark, and also to calculate future financial risk related to non-financial performance of companies. People will be upbeat on the issue of labour conditions, globally, and also to hold company accountable.

Initially, GRI was lacking in some of the resources, which were essential to establish it was a global standard platform i.e. connection within influential actors, legal competency, financial resources and knowledge. Though the GRI founders were in short supply of all these but they managed to get directly or indirectly access to them. They were well connected to the world of influence by a network of alliance. Even to win the support of the socially progressive charitable foundations, GRI explicitly stated that it would lead to social change. Prestigious international institutions such as UNEP and major corporations have long standing ties with CERES.

GRI adopted an inclusive participatory process, with the aim of tapping the intellectual resources, experiences, and technical expertise. Companies participated in the process to gain social legitimacy. They preferred quantitative over qualitative indicators and continue questioning cost and benefit of sustainability reporting.

Small enterprises find the guidelines highly demanding. But the users of GRI reports, such as shareholders activists, institutional investors and consumer activist's organizations find them insufficient. Some pointed out that GRI reports do not give an adequate picture of the progress towards sustainability. In an another related framing trade off, it was found that the members of many GRI working groups were kept on the technical field, and avoided explicit discussions on sustainability and accountability. Eventually these working groups never realized the importance of in-depth learning.

The instrumental functions of GRI apply to the quality and reliability of information and its usability for the immediate needs of the users. But, what is missing in such criticism is that, it is the recognition of the larger vision of the founders, through which GRI would establish a discourse, practices, norms, and a new language. This was lost on many participants, and was not carried on to the next generation of GRI participants, or users of the Guidelines, or even the users of the GRI Reports. This does not imply that, GRI will not play that role in the future, nor it is indicated that no social learning will take place due to GRI.

It is indeed acceptable that GRI led to the emergence of new language and new concepts. But with the emergence of higher order learning among the GRI participants, a shared vision for GRI may have emerged. This would in turn lead to stronger sense of shared enterprise, strengthening of the emerging institutions. Another important issue was that of financing GRI. The GRI supporters regard the guidelines as a public good, which is used by all societal actors. It is something which is freely available to any organization, and can be used to measure and report its contribution to sustainable development. But to coordinate the multi stakeholder, engagement in general and sector specific guidelines is a costly affair.

In another trade-off, companies to facilitate rapid development and to use the Guidelines by them left no scope for quality control to the reports or the process used to produce them. GRI has been facing number of challenges which are grounded in the strategies adopted by its founders. These strategies were probably the most appropriate under the circumstances, and also generated a tremendous success, but also left behind some unresolved tensions. Regardless of this, GRI has made an immense contribution in the ongoing discourse on accountability, reporting, corporate responsibility, and on the appropriate role of business, government, civil society in sustainability transition.

GRI establishes synergies between various codes, management standards, performance standards. It assists organizations in displaying their commitment, for instance, to the Global Compact and ISO Standards through its GRI Reports. Then why sustainability reporting is essential? Because it establishes new measures of corporate success, and as such requires a complete change of mindsets of the corporates. In recent times, the focus of shareholder has shifted from short term profitability to longer term sustainability of a company.

Sustainability constitutes a set of new rules in the game of business. For a company to survive, and exist in a changing environment, it requires new leadership skills. This will help the company to develop the ability to reflect sustainability concerns in everyday business.

GRI claims to enable transparency which has become a norm in stakeholder relations, investment decisions, and other relations in the market. In order to establish such kind of transparency, GRI aspires to provide, a globally shared and understood concepts, consistent language, and agreed medium for communicating clearly. The GRI Guidelines have come a long way since its inception. The Guidelines display both growth and change; for instance through intensive stakeholder interaction, the 2002 GRI Guidelines have evolved into, G3, a draft version of the third generation of Sustainability Reporting Guidelines, available from 2006 onwards.

Performance indicators are grouped together by GRI in three categories covering the economic, environmental, and societal. Economic indicators are concerned with an organization's impact on the economic resources of its stakeholders, and economic systems both at local, national and global levels. Economic responsibility is not just profitability of the company. It needs to consider how the company is contributing to sustainable growth with the local and international context.

Environmental indicators are concerned with the impact on eco-systems, land, air and water. It also includes the environmental impacts of products and services, energy, greenhouse gas and other emissions, effluents, impacts on biodiversity, recycling, pollution, waste reduction, environmental expenditure, and firms and penalties for non-compliance. Social Indicators are labour practices such as, health and safety of the employees; human rights such as child labour; other social issues like bribery, corruption and community relations and product responsibility.

In the new G4 Guidelines of Global Reporting Initiative, the indicators of the value chain assessments are too complex for multinationals and too burdensome for smaller organizations. But this argument totally negates the view that reporting is a process, and it entails a learning trajectory. It is hard to believe that the total numbers of indicators of the GRI Guidelines are seen as a burden. But there were recommendations on how to report on material issues.

There were also complaints regarding the verdict on the use of commission in the past, especially with supply chain reporting. But after the release of GRI G3 Guidelines, there have been numerous changes. Lack of data has been replaced by availability of more data's, and the cost of reliable information has gone down, due to rigour of information technology. Already, business is flourishing due to the availability of the enormous amount of reliable data's from the suppliers, but it does not mean that, there is no problem. Today, more willingness is there to reveal data's, if supply chain strategies are more collaborative, instead of just a mandatory code of conduct.

The GRI G4 Guidelines is a reaction to non-delivery of reporters since the release of G3 in 2006, primarily towards the failures to go beyond legal boundaries. The value chain assessments are complex, especially for companies with diverse products and business models. It is in fact very difficult to accept that an organization that has developed its business models would not be planning its future impacts, both positive and negative of their business. This is especially true today, because when we think of a circular economy, there are shifts from ownership to use, scarcity of resources and supply of capable workplace.

Sometimes organizations complain of sustainability reporting as too costly. These are companies that spent a huge amount on their annual reports which is not even read by industry specialists except by some important shareholders.

There is a claim from some circles that, GRI Guidelines uses too many indicators, but the reporting process in the GRI Guidelines normally avoids the use of more indicators. Now there are more information available to start building impact based reporting by developing micro–macro link. Even the reporter that thinks about value cycle impacts can anticipate a better picture of whether they are a part of the problem or part of the solution or part of both. That is why reporting needs an impact based meaning that closes the gap in sustainability context in order to be meaningful.

The top management of the companies must realize the position of the organization towards sustainability, long term targets, to motivate the organization and build reputation with customers and other stakeholders.

10 AccountAbility 1000 (AA 1000)

To achieve accountability in leadership, organizations require new frameworks to help them identify, understand and respond to strategic opportunities and risks. The Accountability Principles and AA1000 series of standards provide a basis of principled leadership. The Accountability Principles are the foundation for the other two standards in the series i.e. the AA 1000 Stakeholder Engagement Standard and the AA 1000 Assurance Standard. The flexibility of the principles implies that organizations can use the values they articulate and then adapt them to the needs of their context. And, this requires, understanding and leadership because application varies across time, context and organization.

The AA 1000 Series of standards are voluntary and it is created through a highly legitimate multi-stakeholder process, and it allows all voices and comments to be heard. The AccountAbility Principles are adopted by organizations to develop an accountable and strategic response to sustainability.

The credibility of the AccountAbility Principles lies in its comprehensive nature and the flexibility of these applications. These principles asserts that an organization should actively engage with its stakeholders, fully identify and understand the sustainability issues that will have an impact on its performance i.e. economic, environmental and social. Instead of making it just prescriptive, it allows the organizations to focus on what is material to its own vision, and then provides a framework to identify and act on real opportunities as well as managing non-financial risk and compliance. The AA1000 Standards are based on principle of inclusivity, materiality, responsiveness and disclosure.

10.1 Principle of Inclusivity

The commitment to inclusivity has been basis of the AA 1000 series and standards and it has retained primacy in the 2008 standard. Inclusivity implies allowing the stakeholders to participate throughout the decision making process. The principle of inclusivity is supported by the AA 1000 Stakeholder Engagement Standard. This provides a flexible framework for planning, implementing and evaluating stakeholder engagement process. Giving utmost importance to, listening to and responding to the concerns of others will always remain fundamental to stakeholder engagement. While practicing inclusivity, organizations identify and understand stakeholders, their views and expectations.

10.2 Principle of Materiality

The relevance and significance of an issue to an organization is all about the principle of materiality. It is an issue that will influence the decisions, actions and performance of an organization and its stakeholders.

The principle of materiality has been revised and updated to embrace and understand new developments. Materiality strongly aligns to business performance through analysis of both the relevance and significance of issues. Then determining what issues are material is crucial to the success of an organization's ability to deal with sustainability challenges.

10.3 Principle of Responsiveness

Principle of responsiveness concerns an organization's response to stakeholder demands which eventually affects its sustainability performance. This can be realized through rigorous engagements and communication with stakeholders. The principle of responsiveness emphasizes on the understanding behind the processes used to develop responses themselves. As communication is a part of responsiveness, and it automatically links to the use of reporting frameworks and guidelines.

10.4 Credible Disclosure

The importance of assurances in sustainability reporting is increasing. In a world with a diverse set of stakeholders, there is a high demand for higher levels of assurance about the products and practices of companies, not only their financial performance. Sustainability reporting and assurance has gained prominence as today's financial reporting is failing to provide data on the drivers that create business value. Sustainability reporting is open to abuse and misuse.

AA1000 Assurance Standard (AA1000 AS) was revised to provide a comprehensive method for organization's to account for its management, performance and reporting. Besides, it needs to evaluate the adherence of an organization to the Accountability Principles and the reliability of complementary performance information. It provides a rigorous framework for sustainability assurance and also enough flexibility to adapt to the needs of individual organization. AA1000AS provides enough findings and conclusions on the current status of an organization's sustainability performance as well as suggestions to encourage continuous improvement.

AA1000AS was concerned with nature and extent of adherence of the organization to the AA1000 Accountability Principle, and the quality of the information publicly disclosed on sustainability performance. Reporting organization's that adhere to AA1000AS (2008), assume responsibility for the impacts of their actions, products, decisions and policies. AA1000AS also provides a platform to align the non-financial aspects of sustainability with financial reporting and assure through its understanding of materiality. These principles are set to be of practical use when the organization's are willing to let them influence their social account.

If the organization cannot align the non-financial aspects of sustainability with financial reporting, the practicality of materiality principle will be undermined. Again, if the organization fails to understand the integration of sustainability intangible with financial reporting, then the usefulness of reporting and assurance model will lose its validity.

Sustainability reporting is often misused. That is why the importance of assurance in sustainability reporting is gaining momentum. Stakeholders are also demanding a higher level of assurance in products and practices and the predictive value. Without active disclosure practices, on the management, performance and reporting, the organization's may fail to retain the confidence of its stakeholders.

Conclusion

CSR is not just responsibility to business only, but to all its stakeholders. Social audit assists an organization and draws out mechanisms to present a clear picture of the social performance of the organizations. This is possible by either exercising a self-regulatory method, or otherwise strategize the social accountability or social audit, on the basis of the standards and guidelines devised by international bodies, both government and private.

From the above analysis, it is clear that these guidelines too have its strengths and challenges. That is why the organizations have to review the standards rigorously, and adopt those which can be very effective in measuring the social performance. Moreover, stakeholders are not just happy with the financial performance of the companies, while pursuing the economic goal. Organizations are negligent of its social and environmental role. So, even if the organizations want to strengthen their core goal, they have to be concerned about their other roles. Business may create awareness among the public about their economic responsibility.

Both CSR and social audit grew out of business's inclination towards social cause. The performance social audit and CSR can be acclaimed to be at the peak, when measurements are used based on certain standards. These measurements to be more effective have to adopt a regulatory perspective. For companies to adopt a regulatory perspective, the best possible way would be to refer to the guidelines, standards of trans-governmental and private bodies.

To improve the conditions inside or outside the firm is not new. At the international level this concern is visible under a heading called Social clause. But, the movement towards the inclusion of a social cause has been very slow. Some developing countries have expressed their concern of industrialized nation's relations to workplace conditions. This is probably high, because of the export success of the industrialized nations, and of course, the growing for protectionism, which is a real reason for unemployment. That is why, whatever negotiations take place at the international level, has to be handled carefully, with benefits and costs for both developed and developing countries.

(continued)

Trans-governmental organizations, like ILO finds it difficult to support research and analyses of its labour standards. Another very significant point is that, since its establishment, the ILO has not been able to pass any of its conventions or recommendations on Labour standard into any form of international law. However, many countries have included labour standards set by ILO in their national laws. Many developing countries and their companies have accepted the ILO Labour standards in their domestic laws. It is argued that, ILO has maintained a slow pace in ensuring the application of labour standards, creating new instruments, than in conducting research on the impact and value of its labour standards. But, yes, it has gone ahead by identifying a minimum set of standards that can be included in a social clause, especially in 1998 Declaration on Fundamental Principles and Rights at work. These standards are within the domain of freedom of association, the right to collective bargaining, the minimum age for the employment of young persons and freedom from discrimination of employment and from forced labour. But, given the weak administrative machinery in the developing countries, application of the ILO standards is nearly impossible. Another important point is that, ILO raises the cost of labour and hardly considers the economic benefits to the institution that has to pay the increased charges.

ILO, together with the OECD has moved towards adopting set of labour standards. Several hundred companies have signed the UN Global compact. But, most of the companies ignored the standards set by the compact. The UN Compact has set up a number of working groups to act as a mediator to solve these issues. But this couldn't take off because, the working groups lacks the financial resources and in-depth analytical work.

A number of moves have been made by the transnational and private bodies, to make the companies adhere to a set of standards for non-compliance. Legislation, in the implementation of the standards is different, for one country because, it would penalize those companies headquartered there to the benefit of companies with headquarters elsewhere. A code is not a permanent global guideline, because the societies are changing too fast. That is why there should be a rigorous and continuing process of dialogue between government and enterprises. Such a process has already started, and it was OECD's governance forums which took the lead. The UN Global compact, GRI and ILO's tripartite conferences and meetings, also tried to bring together representatives of employers, trade unions and governments. Enterprise like, Chambers of Commerce have represented the ILO for generations, and large corporations have shown less interest in this endeavor. The other stakeholder groups, such as trade unions, consumers, non-unionized workers, retailers, suppliers, distributors, and shareholders, had not been involved by these forums.

(continued)

Keeping in mind the advantages and challenges of the standards set by both the trans-governmental and private organizations, it requires revisions, reforms, and adaptations. These standards to be used as instruments of corporate social audit by both private and public organizations, requires a thorough analysis, in depth research on its applicability. The UN Global Compact to build up a certain degree of legitimacy in its human rights standard has to draw out certain procedures, so that organizations who comply to these standards can understand the value of it. Reaping the benefit by applying this standard, should be documented, and disclosed to aspiring members of the compact.

The standard should act as a principle that can generate deliberations, and enact democratic practices in the workplace. Enforcing the standard by the global players to gain accolades from the local community is not enough; rather small and medium enterprises should be welcomed to apply the standards. This will create a wide network of connection between the big and the small players. This network could prove beneficial to different types of organizations, and in the future, it can standup against divisive nature of large organizations.

The financial crunch of the SMEs should be taken seriously by the Compact, and the powerful countries. Adoption of human rights should become a part of company policy, so that the company understands value of practicing human rights.

Besides the trans-governmental bodies, the private reporting bodies too have gained momentum with their standards being adopted by MNCs or any large corporations.

GRI, a multi-stakeholder initiative has come a long way, in incorporating participation from corporations, NGOs, Trade Unions, other interest groups. It aims to deliver a quality sustainability reporting and, has been enhanced by its adoption of performance indicators. The performance indicators are supported by a set of guidelines. The reporting principles stresses on economic, environmental and social performances of organizations. The environmental and social practices of organizations are very important, because it encompasses, issues pertaining to the environment, labour practices, community and product responsibility. GRI through its performance indicators have tried to direct organizations social reporting practices. The Social Reporting can become reliable, when it is able to engage with the stakeholders. But this is also not sufficient. There has to be a constant monitoring at different stages of policy making and implementation. The monitoring mechanism should be introduced by the organization, to check and balance the contradictions persisting within and outside the organization. Hence, GRI can very well strengthen the social accounting system of the organization.

(continued)

Companies that use the guidelines for their development purpose may also utilize it for reporting purpose very cautiously. There is no quality control to the reports. If the companies are using the guidelines to direct their social audit or accounting mechanism, must ensure that there is authenticity in the reporting mechanism. GRI should also be taken into confidence, by the companies while setting up a social audit and responsibility system based on its standards.

The credibility and assurance of the standards are very important, for the companies have to take the confidence of its stakeholders. AccountAbility 1000 is a private body established on the strength of its standards that assist organizations to identify and respond to opportunities, and provide a basis of accountable leadership. The credibility of the AccountAbility principles lies in its all-encompassing nature. These principles assert that organizations should adopt a multi-stakeholder engagement that can assess and understand sustainability issues and its impact. AA1000 commitment to inclusivity has to be ascertained on the basis of AA1000 series and standards. Inclusivity applies to stakeholder participation, and drawing out a framework for planning, implementing and evaluating stakeholder engagement process. While implementing the principles, the organization has to observe a clear policy of involving both the primary and secondary stakeholders, so that all voices, opinions, and views are heard and discussed. The principle of materiality of AA1000, is primarily based on relevance and significance of an issue. It is an issue that plays a decisive role in the performance of an organization and its stakeholders. Materiality principle determines the relevance and significance of the issue, on the performance of the organization. A rigorous method has to be exercised to determine the validity of the issues for the performance and growth of the organization.

The organization response to the stakeholder demands is very crucial for the reporting process. The principle of responsiveness of AA1000 argues for rigorous engagements and communication with stakeholders. The organizations social accounting, or social audit process, based on the principle of responsiveness, has to go deep into the response pattern of the stakeholders, and try to examine the cause of the responses, and its implications.

Without a clear disclosure pattern within a firm, it is very difficult to keep abreast about the policies and strategies to the stakeholders. The diversity of the stakeholders has increased, and as such there is a high demand for higher levels of assurance from the company about the products and practices, not just their financial performance.

There is a growth in the understanding and rationality of the stakeholders too. They are not just happy with the financial performance of the company. That is why sustainability reporting is on the rise. Financial reporting has failed to provide adequate data's on drivers that can create business value. It

(continued)

has to undertake continuous process of analyzing issues, which are not purely financially rewarding, but socially effective.

The standards set by the trans-governmental and private bodies have gone through process of consultations, networking and deliberations to ascertain whereby organizations both public and private can set their Corporate Responsibility Reporting, Social Accountability, or Social Auditing, based on their guidelines and standards.

References

Adams, C. A., & Evans, R. (2004). Accountability, completeness, credibility and the audit expectations gap. *JCC, 14*(Summer), 97–115.
Anderson, R. H. (1977a). Social responsibility accounting: Time to get started. *CPA Magazine*, pp. 28–31.
Anderson, R. H. (1977b). Social responsibility accounting: Evaluation, objectives, concepts and principles. *CPA Magazine*, pp. 32–35.
Anderson, R. H. (1977c). Social responsibility preparedness: Measurement and how. *Cost and Management, 53*(5), 12–16.
Beesley, M., & Evans, T. (1978). *Corporate social responsibility—A reassessment*. London: Croom Helm.
Belkaoui, A. (1980). The impact of socio-economic accounting statements on the investment decisions: An empirical study. *Accounting, Organizations and Society, 5*, 263–283.
Bowen, H. R. (1953). *The social responsibilities of the businessman*. New York: Harper.
Carrol, A. B. (1979). A three-dimensional conceptual model of corporate social performance. *Academy of Management Review, 4*, 497–505.
Committee for Economic Development (CED). (1971). *Social responsibilities of business corporations*. New York: CED.
Davis, K. (1960). Can business afford to ignore social responsibilities? *California Management Review, 2*(Spring), 70–6.
Dierkes, M., & Antal, A. B. (1985). The usefulness and use of social reporting information. *Accounting, Organization and Society, 10*(1), 29–34.
Dilley, S. C. (1975). Practical approaches to social accounting. *CPA Journal*, February, 17–21.
Epstein, E. M. (1987). The corporate social policy process: beyond business ethics, corporate social responsibility and corporate social responsiveness? *California Management Review 2 Spring 87, 29*(3), 99.
Frederick, W. C. (2006). *Corporation be good, the story of corporate social responsibility, Indianapolis*. pp. 40–59.
Freeman, R. E. (1984). *Strategic management: A stakeholder approach*. Boston: Pitman.
Friedman, M. (1962). *Capitalism and freedom*. Chicago: University of Chicago Press.
Gibbon, J., & Dey, C. (2011). Developments in social impact measurement in the third sector: Scaling up or dumbing down? *Social and Environmental Accountability Journal, 31*(1), 63–72.
Global Reporting Initiative. *The external assurance of sustainability reporting, Research and Development Series*. The Netherlands: GRI.
Global Reporting Initiative, Sustainability Reporting Guidelines. (2006). 5F735235CA44/O/G Guideline ENU.
Graham, D., & Woods, N. (2006). Making corporate self-regulation effective in developing countries. *World Development, 34*(5), 868–883.

Gray, R. (2002). The social accounting project and accounting organizations and society: Privileging engagement imaginings, new accountings and pragmatism over critique, Accounting. *Organizations and Society, 27*(7), 687–708.
Hessel, A. (2008a). *The CSR policy instrument handbook*. Berlin: Hertie School of Governance.
Hessel, A. (2008b). The evolution of a global labour governance regime. *Governance: An International Journal of Policy, Administration and Institutions, 21*(2), 231–251.
Ingram, R. W. (1978). An investigation of the information content of (certain) social responsibility disclosures. *Journal of Accounting Research, 16*, 270–285.
Jamali, D. (2010). MNCs and international accountability standards through an institutional lens: Evidence of symbolic conformity or decoupling. *Journal of Business Ethics, 95*, 617–640.
Johnson, H. L. (1971). *Business in contemporary society: Framework and issues*. Belmont, California: Wadesworth Publishing Co., Inc.
Jones, T. M. (1980). Corporate social responsibility revisited, redefined. *California Management Review*, Spring, 59–67.
Kaya, U, & Yayla, H. E. (2007). Remembering thirty five years of social accounting: A review of the literature and the practice, *Munich Personal RePEc Archive* (MPRA) Paper No. 3454, http://mpra.ub.uni-muenchen.de/3454/.
Krishna, C. G. (1992). *Corporate social responsibility in India. A study of management attitude*. Delhi: Mittal Publication.
Levite, T. (1958). The dangers of social responsibility. *Harvard Business Review, 36*(5), 41–50.
Marlin, J. J. (1973, February). Accounting for pollution. *Journal of Accountancy*, 41–46.
Milne, M. J., & Adler, R. W. (1999). Exploring the reliability of social and environmental disclosures content analysis. *Accounting, Auditing & Accountability Journal, 12*(2), 237–56.
Mobley, S. C. (1970). The challenges of socio-economic accounting. *The Accounting Review, 45*, 762–68.
Moon, J. (2005). CSR in the UK. An explicit model of business- society relations. In A. Habisch, J. Jonker, M. Wegner, & R. Schmidpeter (Eds.), *CSR across Europe* (pp. 51–65). Germany: Springer.
Muirhead, S. A. (1999). *Corporate contributions: The view from 50 years*. New York: The Conference Board.
Murphy, P. E. (1978). An evolution: Corporate social responsiveness. *University of Michigan Business Review, 30*(6), 19–25.
Preston, L., & Post, J. E. (1975). *Private management and public policy: The principle of public responsibility*. Englewood Cliffs, NJ: Prentice-Hall.
Raben, T. M., & Williams, C. W. (1974). Social accounting: The accountant's role. *Federal Accountant*, December, 3–8.
Ramanathan, K. V. (1976). Toward a theory of corporate social accounting. *The Accounting Review, LI*(30), 516–28.
Roser, S. R. (1979). Relevance of social accounting information for decision makers: An empirical study. (Doctoral Dissertation, USA: University of Nebraska).
Sage Brief Guide to Corporate Social Responsibility. (2012). London: Sage.
Sethi, S. P. (1975). Dimensions of corporate social performance: An analytic framework. *California Management Review, 17*(Spring), 58–64.
Steiner, G. (1971). *Business and society*. New York: Random House.
Teach, E. (2005). Two views of virtue. *CFO, December*, 31–34.
Tuzzolino, F., & Armandi, B. R. (1981). A need hierarchy framework for assessing corporate social responsibility. *Academy and Management Review, 6*, 21–8.
Unerman, J., Bebbington, J., & O'Duyer, B. (Eds.). (2007). *Sustainability accounting and accountability*. London and New York: Routledge.

Vogel, D. (2005). *The market for virtue: The potential and limits of corporate social responsibility.* Washington: The Bookings Institution.

Votow, D. (1973). Genius becomes rare. In D. Volow & S. P. Sethi (Eds.), *The corporate dilemma.* Prentice Hall: Englewood Cliffs, NJ.

Walton, C. C. (1967). *Corporate social responsibility.* Belmont, California: Wordsworth Publishing Co. Inc.

Wilmshurst, T., & Frost, G. (1999). Corporate environmental reporting: A test of legitimacy theory. *Accounting, Auditing and Accountability Journal, 13*(1), 10–26.

Willeth, R. (2002). *An empirical assessment of Gray's accounting value constructs.* Working Paper No. 2002–03.

Wren, D. A. (2005). *The history of management thought* (5th ed.). Hoboken, NJ: John Wiley and Sons Inc.

Defining a Methodology for Social Audit Based on the Social Responsibility Level of Corporations

Adriana Tiron-Tudor, Ioana-Maria Dragu, George Silviu Cordos, and Tudor Oprisor

1 Introduction

This chapter describes what is involved in social audit. Social audit stems from the intersection of sustainability and corporate social responsibility practices. The call for social auditing to be carried out by corporate entity is popular worldwide, from socially responsible investors and analysts, to society, environmentalists, or other internal/external stakeholders; everyone is interested in the value adding capability of non-financial information. Social audit has, therefore, become a proper 'tool' used for assessing the organisations' impact on the environment and society. This new type of audit encourages corporations to disclose sustainability/CSR information and the currently available standards in this area provide specific guidelines when reporting on the non-financial aspects of an entity. The most prominent guidelines in the field of social audit are represented by SA8000, AA1000, and ISAE3000. Out of these three standards, we argue that the SA8000 to be the most representative. Therefore, we have extracted a disclosure checklist from this standard and have included it in our social audit framework. Subsequently, we have tested the framework on a sample of 825 companies, selected from a population of 1,400 global corporations.

Social audit has had a positive evolution over time, both from a historical perspective, and practical/applicability side. We consider that this trend will continue, as corporations start to focus more on non-financial information disclosure. However, we argue that the degree of sustainability/CSR information should be measured differently from one industry to another, since some business sectors' impact on society and environment would be more pronounced than others. We employ a descriptive analysis that determines the disclosure index of SA8000's

A. Tiron-Tudor • I.-M. Dragu (✉) • G.S. Cordos • T. Oprisor
Faculty of Economics and Business Administration, Babes-Bolyai University, Lacramioarelor 4, Cluj-Napoca, Romania
e-mail: ioanadragu@yahoo.com

most relevant information criteria. Apart from the industry factor, our case study includes the country perspective, through the NCRI (National Social Responsibility Index), the HDI (Human Development Index), and the Good Governance Index (GGI). The objective of our investigation is to promote a possible standardized framework for social auditing. The use of a large sample by the study provides a strong argument that the framework can be applied, as it has been tested on more than 1,000 companies and demonstrates the consistency and relevance of our study.

We aim to contribute to the development of social audit literature and practice, encouraging future research studies on this topic, thus providing an understanding for the by stakeholders for social audit information disclosure.

2 Theoretical Background

2.1 Corporate Sustainability

The Brundtland Commission Report 1987 notes for the very first time in "Our Common Future" (Kuhlman & Farrington, 2010; Tovey, 2009) the notion of sustainability. It refines the view that economic development has to adjust its progression to the limited resources of our planet. In the Brundtland report, sustainable development was defined as "meeting the needs of the present generation without compromising the ability of future generations to meet their own needs" (World Commission on Environment and Development, 1987, p.43, cited by Baker, 2006, p. 20).

The three pillars approach has been the most common attempt used to break down the concept of sustainable development. Robinson and Tinker (1998) consider that the three components (economic, environmental and social) have a direct effect on each other and can never be used in isolation.

Tovey (2009) argues that this issue may depend on what we choose to sustain and ultimately underlines the importance of the environment. We argue that the three pillars have the same value, therefore, should be equally treated by corporations.

Sustainability adoption has been highly debated in the socio-environmental literature (Nidumolu et al., 2009; Eccles et al., 2010a; Kolk & Perego, 2010; Vidal and Kozak, 2009). Nidumolu et al. (2009) outline the main elements involved in the process of corporate sustainability adoption namely: compliance becoming opportunity; sustainability of value-chains; products and services developed on a sustainable principle being a new approach for business models. Academics have presented a series of case studies for illustrating best practices of corporations such as Wal-Mart, Clorox, HP and others, in terms of sustainability. The aim of this chapter is to demonstrate that sustainability determines innovation and progress for the business environment. The ACCA has continued to strengthen the idea of connecting innovation to sustainability by underlying the importance of innovation for corporate reporting and information disclosure (ACCA, 2012). Other studies for instance (Lozano, 2009) indicate the contrary, demonstrating that the evidence for innovation within sustainability pillars is not relevant. Nicol (2000) also attempts to develop a framework for sustainability that could support the adoption of sustainable development practices for water projects both from political and practical perspectives. Kolk & Perego (2010) studied sustainability assurance on 250 corporations from the Fortune Global 500. These scholars investigated the drivers for social and environmental assurance. Their findings demonstrate that sustainability assurance is determined by the level of sustainability practice and stakeholder engagement. The impact of stakeholders upon sustainability assurance statements can be derived from the fact that investors are interested in non-financial information for measuring risk and future outcomes (Kolk, 2003). Mandelbaum & Friedman (2009) state the limitations of the sustainability adoption process, in the form of stakeholder communication barriers or in the attempt to practice sustainability for marketing purposes, rather than for ethical reasons. The theory was confirmed by Selvi et al. (2010), who present the benefits of implementing corporate social responsibility. Starting from a sample of the most profitable companies from a particular country and the enterprises that are socially responsible, the authors use the Spearman coefficient to determine the relationship between company reputation and corporate social responsibility. Their findings show that the relationship between the variables is a positive one.

Eccles et al. 2010a, 2010b underline the importance of stakeholder engagement for the reputation, trust, market share and performance of any company. Fonseca et al. (2011) set the main coordinates for sustainability reporting in Canada. Further on, discussions on trends towards non-financial information show that sustainability reporting registers positive evolutions (Kolk, 2003), because more and more companies begin to apply it. According to data from the Fortune Global 250 between 1998 and 2001 there is a considerable increase regarding sustainability reporting (Kolk, 2003). Research on corporate responsibility and sustainability reporting (Michael, 2009) in the area of real estate reveals that corporate reports have changed over time, making improvements on sustainability and CSR disclosure. Michael, 2009 studied eight corporations in the UK and Australia, by creating a matrix with the purpose of evaluation and comparison of the corporate responsibility and sustainability reports. The methodology used involves a characterization of each criterion from the matrix as being or not being disclosed within the analysed reports. These criteria were chosen by these authors in accordance with the GRI Sustainability Reporting Guidelines and GRI Standard Disclosure, and refer to: headquarters location, strategy and analysis, organizational profile, report parameters, governance, economic, environmental, social and governance performance metrics. The findings of the study confirm the initial objective—that CSR and sustainability reporting has improved over time; the researchers note, at the end of the paper, certain limitations in the form of input data, subjectivity, human error or misinterpretation exist which most be borne in mind.

Other academics for example Gray, 2006 have tried to prove that there is a correlation between reporting on sustainability, the environment and social dimension and shareholder value creation. A research on corporate online reporting (Htaybat, 2011) involving Jordanian companies assumes the computation of an un-weighted index for disclosing financial and non-financial information. Others (Fahmi & Omar, 2005) have concentrated on reporting practices regarding the information for minority shareholders, while relating to listed companies.

According to the social and environmental performance, countries can be classified in terms of the following as not by Eccles et al., 2012:

- Sustainable countries;
- Unsustainable countries;
- Sustainable companies' countries;
- Sustainable investors' countries.

Although Japan and the U.S. present low levels of disclosure for sustainability information (Eccles et al., 2010a), they are known to have sustainable investors (Eccles et al., 2012). China seems to have the same tendency of maintaining low disclosure levels for sustainability reporting (Eccles et al., 2010a) as an unsustainable country (Eccles et al., 2012). Brazil, South Africa, and Sweden remain as the top sustainable countries, and South Africa witnessed high progress regarding sustainability integration (Eccles et al., 2012, 2010b; Vidal and Kozak, 2009; KPMG, 2008). Germany and the U.K. proved to be committed to sustainability (Eccles et al., 2012) as sustainable countries.

According to the GRI (2013), since 2001, the degree of sustainability information published in reports issued by companies has increased considerably (95 % of worldwide top companies issue sustainability reports). In addition, since 2012, the number of companies that include external assurance statements on sustainability information registers a positive trend. See Fig. 1:

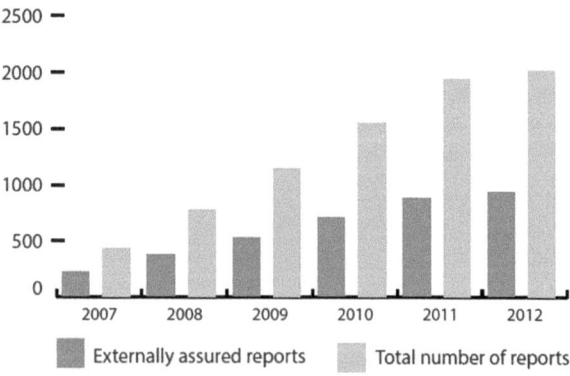

Fig. 1 GRI Reports vs. Assured Reports. Source: (GRI, 2013)

2.2 Corporate Social Responsibility

Corporate social responsibility incorporates the following elements (Leonard & McAdam, 2003: 27):

- human rights;
- workplace and employees (occupational health and safety);
- unfair business practices;
- organizational governance;
- environmental aspects.

Scholars and academics have always been interested in corporate social responsibility, defined as the responsible behaviour employed by companies, even when profit interests should be put at first (Marsden, 2000). Stakeholder engagement and the needs of the future generations represent primary CSR elements (Dahlsrud, 2006). Therefore, corporate social responsibility originates from the organisations' impact upon society and environment (Hopkins, 2003). Pinney (2001) states that a corporate socially responsible behaviour is generated by less environmental and societal damage, and inducing instead a positive impact.

Benabou and Tirole (2010) discuss three dimensions of CSR related benefits, either direct or indirect. The first dimension involves the connection between corporate social responsibility practices and profit generation. The second dimension refers to the role of stakeholders and the visions of sustainability, in contrast to profit interests. The final CSR dimension states that social-environmental benefits should be obtained, even at the cost of profit increases.

The process of CSR implementation assumes an actual integration of both societal and environmental factors into activities and operations conducted by companies. Further more, this implementation should take place according to stakeholders' needs and expectations (Lea, 2002). The integration perspective is strengthened by Lanoizelée (2011) who mentions that "corporate responsibility integrates companies' environmental and social concerns into all their activities" (p. 79).

Lanoizelée (2011) describes the CSR implementation from the perspective of competitive advantage. Engaging in three different accounting theories, the neo-institutional theory, the agency theory, and the legitimacy theory, this academic examines a set of listed companies from the "CAC 40" French stock market index. CSR disclosure analysis was performed by determining the frequency of terms and words used in the text of the reports. The results show that no sustainable competitive advantage can be recognized from the companies' reports, while corporate social responsibility practices seem to be in contradiction with the corporations' interests. Therefore, we can state that the gap between CSR theory and practice, or between CSR discourse and its implementation, is still present.

Szegedi (2010) outlines the development of CSR, in an attempt to characterize the notion of corporate social responsibility from literature debates, or myths versus reality. The scholar discusses previous studies as well as a series of interviews and surveys, and uses descriptive analysis to find the true essence of CSR. Thus, it is demonstrated that corporate social responsibility represents a must for any company that is engaged in the pursuit of profit maximization, as the latter is influenced by CSR.

Gyves and O'Higgins (2008) studied the benefits generated by CSR, especially the mutual ones, for companies as well as stakeholders and society. Both the neoclassical theory and the social legitimacy theory were tested, and the employed methodology involves the interview technique. Their findings prove that shareholders' needs and expectations can be accomplished through a voluntary adoption of corporate social responsibility practices. While Gyves and O'Higgins (2008) argue for the voluntary side of CSR, Björn (2012) pleads against mandatory CSR disclosure, the main argument being that it will not lead to higher transparency in corporate reporting. In addition, Björn (2012) uses the integrative social contracts theory and a qualitative research perspective to demonstrate that the "comply or explain system" is not effective for the corporate business environment. In addition, another study on norms and compliance (Merce and Simon, 2009) presents CSR standards in their evolution. Analyzing the literature review in the field of CSR, Merce and Simon (2009) concludes that, in general, corporations are aware of the standards stipulations, although some aspects still remain unclear. This academic also considers that international bodies from the CSR field tend to issue new emergent and not certifiable standards.

The legitimacy theory was also used for understanding the CSR concept (Claasen & Roloff, 2012). Claasen & Roloff (2012) interviewed various stakeholders to find how CSR can contribute to the organizational legitimacy. They found that CSR practices can address and engage with some of the stakeholders' expectations from a certain perspective, but at the same time, they can do damage to others.

Falkenberg and Brunsæl (2011) studied the connection between CSR and performance. On the ground of instrumental theories, these academics inspected, using a sample of two companies, four different criteria, namely: strategic disadvantage, strategic necessity, temporary strategic advantage, and strategic advantage. The results were inconclusive, so this study has not provided enough evidence of any relationship between CSR and performance. Therefore, we can add that corporations will not be motivated to adopt CSR practices, in the absence of a specific link to their performance. Another study that attempts to connect CSR to performance was conducted by Oba (2011) who uses the market value as an indicator. The researcher performed an analysis upon various annual reports and scored the CSR policies mentioned in the documents. Different results were obtained, however no positive relationship between the two main elements (CSR and market value) was found.

Martinuzzi (2011) investigates competitive advantage in the chemical, construction and textile industries. The scholar initiates both content analysis and the interview method, his research being based on the competitiveness theories. The findings show that companies are indeed willing to adopt a CSR behaviour in all the studied sectors. Balboni & Balboni (2008) developed an empirical study on CSR level and state the impact on CSR practice. Tarí (2011) also discusses the state influence upon CSR, but from the perspective of literature review.

2.3 Corporate Sustainability Versus Corporate Social Responsibility: Similarities and Differences

According to management literature, both corporate social responsibility and corporate sustainability refer to social and environmental aspects (Montiel, 2008). Academics have found that it is difficult to separate sustainability from CSR (Montiel, 2008; Garavan et al., 2010). Both sustainability and corporate social responsibility are microeconomic elements and together with corporate citizenship, bring a contribution to sustainable development (macroeconomic level). In fact, corporate sustainability and corporate social responsibility have different perspectives, sustainability being more oriented towards sustainable development, while CSR is oriented towards corporate social performance—or business impact (Loew et al., 2004).

Although, at first sight, it seems that CSR Reports and Sustainability Reports are one and the same issue and can be common substitutes (GRI, 2013), this is not the case, because, in fact, sustainability is included in CSR (Aceituno et al., 2012). Corporate social responsibility represents the impact of business practices on the social environment and the economy (and how the activity of an organization affects its stakeholders), while sustainability is the result of the three-pillar principle (social, environmental, and economic dimensions). CSR defines the companies' social, economic and environmental impacts.

2.4 Discussion on the Value of Non-Financial Information

Nowadays, the needs and expectations of investors have shifted. If, in the past, they were mainly interested in financial information, they are now considering the value of the non-financial information. Clements & Brown (2012) argue that shareholder value is determined by environmental, social and governance (ESG) integration within long term strategic financial matters. In addition, investors are looking for the whole organisations' performance, which includes not only financial aspects, but also social and environmental issues.

The relevance of the non-financial information is outlined by Radley (2012), who argues that both investors and analysts use it in the decision-making processes and for investment schemes. The same report introduces the notions of socially responsible investor (SRI) and socially responsible investment analyst (SRI analyst). The socially responsible investor considers the investment performance from the financial factors along with the non-financial impacts. The socially responsible investment analyst investigates both financial and extra-financial matters. According to the investors and analysts opinions, the main categories of non-financial information that are relevant for the decision-making processes involve five elements, namely: governance, natural resources, social and community capital, human capital and intellectual capital.

Fig. 2 ESG and economic performance. *Source*: adapted from Signes et al., (2013)

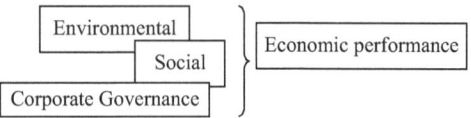

Corporate sustainability was also discussed by Signes et al. (2013) by analysing the environmental, social, and governance (ESG) influence on economic performance (Fig. 2). Scoring the environmental, social and governance ratings, the scholars found indirect correlations between economic performance and ESG factors—the low performers register high ESG ratings, while high performers have low ESG scores.

Another viewpoint on CSR and corporate sustainability was given by Veltri & Nardo (2013). They mentioned a "single ad hoc integrated document" that originates from separate environmental, social and sustainability reports. The report communicates information both internally and externally, so one can deduce the approach of separating between internal and external stakeholders. The research analysed frameworks from intellectual capital and CSR literature. The used frameworks were the GRI's G3 and Meritum reports conducted in terms of the intangible global report; this includes human, structural, and relational capital, as well as environmental and social information (Fig. 3).

2.5 Overview on Social Audit

To these authors, social audit is conducted on the non-financial aspects of an entity (it gathers corporate social responsibility and sustainability related information). As we have previously seen, the non-financial information is valuable to the socially responsible investors; they thus need social audit reports to give credence to information or for adding credibility to reports issued by companies.

Social audit should be conducted on the CSR and sustainability information presented in the reports issued by corporations. As previously mentioned, the connection between corporate social responsibility and corporate sustainability is that CSR defines the impact of business practices on the three pillars (Fig. 4). Therefore, social audit has to include both CSR and sustainability information.

Social audit, also referred to as social accounting, is defined as the reporting process in which companies disclose their impact on society and the environment (Percy Smith & Hawtin, 2007).

Scholars and academics have investigated the evolution of social accounting as a mixture of sustainability, accountability, stakeholder model, and other mechanisms meant to control the impact of organisations' activities on people and the planet (Gray, 2001; Zadek, Evans, & Pruzan, 2013; Unerman, Bebbingto, & O'Dwyer, 2010).

After several "lessons", such as the financial crisis or catastrophic events caused by environmental damage, organisations have understood that they have to align to

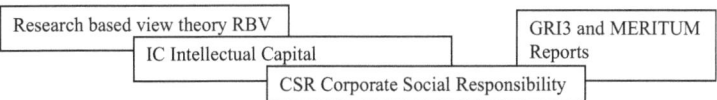

Fig. 3 Framework for an intangible global report. *Source*: adapted from Veltri & Nardo (2013)

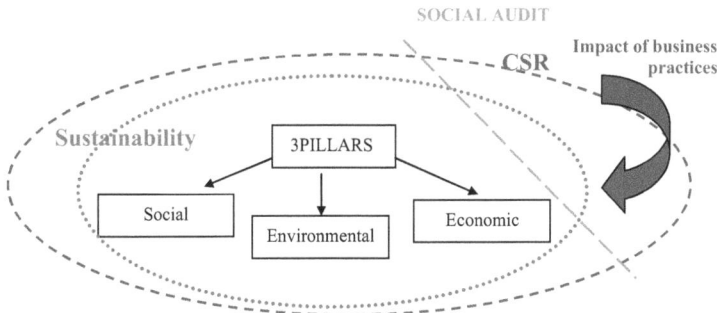

Fig. 4 Social audit at the interplay between sustainability and CSR. *Source*: adapted from Aceituno et al. (2012)

new corporate reporting trends, through which social performance is compulsory for financial performance. Therefore, companies consider reputation risk (Zadek, Evans, & Pruzan, 2013) as one of the top priorities on their agenda.

Social audit was mentioned for the very first time in accounting literature by Kreps (1940), who provides a strong argument for corporations to report on their responsibility towards society: after the U.S. great depression of the 1930s. He perceives a change in reporting trend that should lead to social audit processes. In the author's view, a social audit is intended to serve the external users of the reports (external stakeholders).

Social audit was introduced in the U.K. during the 1960s in the private sector, as a complement to financial audit, and as a result of duties and obligations towards external stakeholders. Similar trends are observed in the rest of Europe, and even the U.S., during the same period. In the U.S., companies started to focus on social performance, by understanding the positive outcome it conveys in terms of profitability. Meanwhile, several national projects involving social reports and social audits were implemented at the European level (in Sweden, Germany, and in the Netherlands). However, in the 1980s, the interest on social audit actions decreased, as companies concentrated on profit maximization. Since 1990 though, we have found some positive signs for social audit developments. This was the period corporations started to disclose information on their social impacts, especially in certain countries such as Denmark, the U.K., and the U.S. (Zadek, Evans, & Pruzan, 2013).

Nevertheless, social audit trends differ from one accounting model to another. For instance, in the Scandinavian countries, social audit is used ore in the public sector, than in the private sector. Moreover, the non-profit organisations from this region, such as government agencies, universities, schools, communities, etc. practice social accounting and reporting, as well as social audit (Percy Smith & Hawtin, 2007; Pearce, Raynard, & Zadek, 1996; Pearce, 1996). There is evidence to support the view that social audit originates from the public sector. The history of social audit practice begins in 1990, when the Policy Research Unit from Leeds Metropolitan University published a manuscript on the practice of social audit as a tool for understanding communities. Seventeen years later, Percy Smith & Hawtin (2007) developed a guide for social audit in identifying community profiling.

When applied to the business environment, social audit represents a process of external assurance, analyzing quantitative and qualitative information (GRI, 2013). The framework proposed in this chapter for conducting a social audit involves only the qualitative dimension of the information. The motivation for our choice is based on the fact that it is difficult to define a common set of quantitative information, performance indicators, and other data that may be introduced in the checklist. In addition, it is difficult to find appropriate measurement scales for qualitative information due to sector/industry specific conditions (GRI, 2013), country legislation, and other factors that do not permit comparability.

In summary, according to international literature, social audit concentrates on a set of elements:

- non-financial information;
- corporate social responsibility;
- corporate sustainability;
- impact on society;
- impact on the environment;
- accountability;
- stakeholder model;
- reputation risk;
- social performance;
- social impacts;
- communities.

3 Verifying Social Audit Compliance Using the SA8000 Standard

3.1 Research Objectives and Methodology

Within this study, we have established a set of objectives, in order to show our intended contribution to the field. The main goal is to propose a standardized framework based on a disclosure checklist simulated on a range of human/social information elements. Starting from how social audit is defined, Percy & Smith, 2007 argue that social audit denotes an understanding of corporate impacts on society and environment, our study focuses only on the second dimension: human/society.

In order to achieve this objective, we have directed our efforts towards specific divisions, with the following subsequent objectives:

- we have selected and consulted the main standards that are related to a non-financial elements' audit (especially, social factors);
- After analysing the SA8000 standard, we have formulated a specific social audit framework, based on the elements found in the standard (which will serve as our content analysis checklist);
- Finally, we have calculated the disclosure index on the checklist for each company from the sample and have interpreted the results by applying various filters.

Our purpose is to check if organisations comply with the information required by SA8000 Table 13.1 which afterwords allows us to define a framework for social audit, starting from the elements presented in the standard. The research question can be formulated as follows:

Do best practice organisations comply with the SA8000 requirements?

Should we obtain a high degree of compliance, we can consider the list of SA8000 elements suitable for a *social audit disclosure framework*.

Currently, there are multiple sets of standards that define the process of assurance for non-financial information (or social audit) such as the ISAE 3000, SA8000 and AA1000 (GRI, 2013). We have chosen to include in our framework the set of elements from the SA8000 checklist, given the fact that it presents a regulation based approach rather than a principle based approach (as included by the other two standards). Also, the SA8000 standard is more suitable for our study because it has a specific social elements' approach and is quite understandable (by contrast, the AA1000 is a standard meant to provide "accountability" regulations, thus having a tangential implication).

In this respect, we have defined an original *Framework for Social Audit*, based on a set of disclosure checklist elements extracted from SA8000 social audit standard, as follows:

Our social audit disclosure framework defines only the social dimension of social audit, as the environmental side that is subject to environmental audit. We also mention the fact that SA8000 is considered an auditable standard for a third party verification system, setting out the voluntary requirements to be met by employers in the workplace. Therefore, the implementation of the standard is at the companies' discretion (SAI, 2008). The outcome of our study, was intending to show whether the companies from our sample (from different countries and industries) were willing to disclose the social components of their activities and to comply with the adequate social regulations (specifically those from SA8000).

The SA8000 Standard was issued by the Social Accountability International (from the Council on Economic Priorities) to support corporate responsibility, ethics, and build trust for the business environment (Leonard & McAdam, 2003). SA8000 is strongly linked to corporate social responsibility. Academics and

Table 13.1 Framework for social audit

Keywords
Guidelines on Social Audit
Child labour
Forced and compulsory labour
Health and safety
Collective bargaining
Discrimination
Disciplinary practices
Working hours
Remuneration
Management systems

Source: authors' design

scholars used this standard to define a management system framework (Castka et al., 2004). Others consider that SA8000 represents a CSR initiative that can improve business reputation (Fombrun, 2005). Göbbels & Jonker (2003) argue that SA8000 is built on principles of "accountability" (Reynolds & Yuthas, 2008) and "social responsibility" while Reynolds & Yuthas (2008) characterize the SA8000 standard as a CSR reporting model. However, the standard refers just to one of the three sustainability pillars: namely the *societal* one. The SA8000 is not related at all to the economic or environmental issues, although it remains an instrument of promoting sustainability reporting and a sustainable society (Lonzano & Huisingh, 2011). In addition, SA8000 can be used as a benchmark for measuring sustainability, social accountability (Gilbert & Rasche, 2007) and even for a successful implementation of social audit (Wallage, 2000).

Using this checklist, we have performed a content analysis on the annual reports of 825 companies that are considered role models in terms of their best corporate citizenship actions. Our investigation used annual reports for the period 2009–2013/2014. We searched for the elements from our framework within the content of these reports and have scored them with '1' if the element was found and '0' if there was no mention of that item.

Afterwards, we calculated the *disclosure index* for each company, by dividing the sum of the available elements (scored with '1') with the total number of elements in the framework. Consequently, the disclosure index can a value of between '0' (non-disclosure) and '1' (full disclosure). We used these results to reveal multiple interpretations, by applying several filters on the aggregate disclosure index values (e.g. industry filters, country filters etc.).

3.2 Sample Selection

For our framework testing, we have used a sample of 825 global corporations. These were selected from an entire population of approximately 1,400 companies, chosen from various sources, namely: the GRI's G4 Participants, GRI Awards, CRRA Winners Best Report 2012, A4S, sustainabilityreports.com, Global 100, 100 Best Corporate Citizens, Corporate Responsibility Magazine, Most admired companies, Forbes Magazine 2011 issue.

The corporations quoted in the above databases are believed to be more oriented towards non-financial information. Therefore, we have performed a content analysis on their annual reports and check if they comply with our framework on social audit. We also seek to compute a scorecard that will generate the degree of compliance for each company.

We conducted an investigation on Annual Reports issued during the period between 2009–2013/2014. We consider that a 5-year analysis period conveys a better perspective on social audit evolution. We seek to investigate the compliance with the framework as an evolution, in order to detect which elements from the framework are constantly disclosed and reported on from 1 year to another.

As we can observe in Fig. 5, the sample is well distributed throughout 20 industry sectors, according to the Industry Classification Benchmark (FTSE, 2012), thus giving relevance to our analysis (by covering a large number of economic activities). The highest number of companies are from the "Industrial Goods and Services", "Personal & Household Goods" and "Technology" sectors. By contrast, the least represented sectors in our sample are "Media", "Real Estate" and "Chemicals". We also mention the fact that a significant number of companies from our sample, given the nature of their activities, have shown great interest in corporate environmental and social responsibility reporting.

3.3 Results and Discussion

The attempt to provide a valid framework for social audit is a difficult task because it implies setting the grounds for an accurate disclosure of several elements that have a rather subjective interpretation in the annual reports. However, by endorsing the elements from SA8000, many companies have revealed the fact that they respect labour regulations, the fair treatment of employees or show a high degree of responsibility towards society as a whole. In this section, we present several results from our study in order to show the extent to which those companies from the sample have endorsed the social aspects mentioned in our proposed framework, as stated in their annual reports.

After defining our version of social audit framework and generating the disclosure checklist, we searched for the existence of the constituent elements within the annual reports and we computed the disclosure index for each company from the sample. We then filtered the results with the aid of several variables in order to have an overview on the compliance with the social audit framework, through different points of view, assuming that there are unquestionable connections between the disclosure level and the social characteristics of the country of origin.

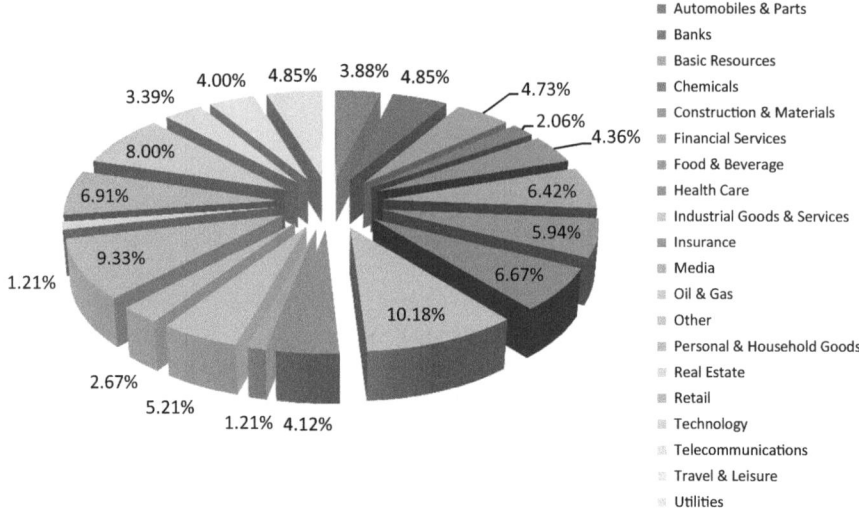

Fig. 5 Sample distribution by industry. *Source*: authors' design

In this respect, we have used four criteria in order to split the sample into categories of companies, namely: the official country language (with two categories: native and non-native speakers), the Human Development Index (UNDP, 2013), the Good Governance Index (World Bank, 2013) and the National Social Responsibility Index (MHCI, 2008). We chose these filters due to their strong social construct and their ability to provide an accurate classification of the sample, by social attributes. Our expectations were to find a superior disclosure level in the case of native English speaking countries (due to the fact that the standard was drafted in English and may have expressions that are better understood by these English speakers). Also, we expected to find a higher average disclosure index for companies from the less developed countries given the fact that these countries have greater social problems and may disclose more information regarding the proposed solutions and outcomes in connection with these problems. In the case of the other two indices, we were expecting to find a direct link between the category's value and the disclosure index. This was because companies from countries with good governance practices and greater social responsibility may disclose more social elements and better integrate them in the corporate overview from the annual reports.

From Table 13.2, we observe that the Disclosure Index (DI) is almost double in Native English speaking countries (0.54 compared to Non-native English speaking countries (0.25). These results can be attributed to the fact that native English speakers have a better understanding of the English-based SA8000 regulation and can better integrate the requirements in the Annual Reports. Another cause could be the fact that many companies from non-English speaking countries issue reports that only partially respect international regulations and/or respect national regulations that do not have social audit requirements.

Table 13.2 Average disclosure index—with specific filters

Criteria	Categories	Average disclosure index
Official language	Native English Speakers	0.54
	Non-native English Speakers	0.25
Human Development Index (UNDP, 2013)	Very High	0.35
	High	0.63
	Medium	0.45
	Low	0.69
Good Governance Index (World Bank, 2013)	Very High	0.34
	High	0.53
	Medium	0.58
	Low	0.48
National Social Responsibility (MHCI, 2008)	Very High	0.34
	High	0.53
	Medium	0.49
	Low	0.11

Source: authors' design

In the case of Human Development Index (HDI) levels, the highest average level of DI was obtained in the sample of companies from countries with a low level of Human Development, which is unsurprising because these companies want to prove they comply with social responsibility standards. A low average DI (0.35) was obtained by companies from countries with a very high level of Human Development, but again, this is understandable because many companies do not consider it is necessary to report their status of social responsibility or human rights; these are considered to be respected implicitly.

Table 13.2 also reveals that the highest average level of DI (0.58) was obtained in the sample of companies from countries with a Medium level in the quality of governance. This is surprising, as we would have expected the highest to be obtained in the "Very high"—instead, in this section we obtained the lowest average DI (0.34). It may be possible that companies are influenced by the level of either too much or too little regulation in these countries. Also, high levels of average DI in the "Medium" level prove that companies from emerging economies have increased their emphasis on human rights and social responsibility, thus having a future effect in increasing their country's quality of governance.

We have also filtered our sample by the National Social Responsibility (NSR) ranking. It is unsurprising to see that companies from countries with a Low NSR ranking have the lowest average DI (0.11). It is, however, surprising to have found that a low average DI (0.34) has been obtained from companies operating in countries with a Very High NSR. Again, this could be explained by these companies not mentioning anything about social responsibility and human rights because

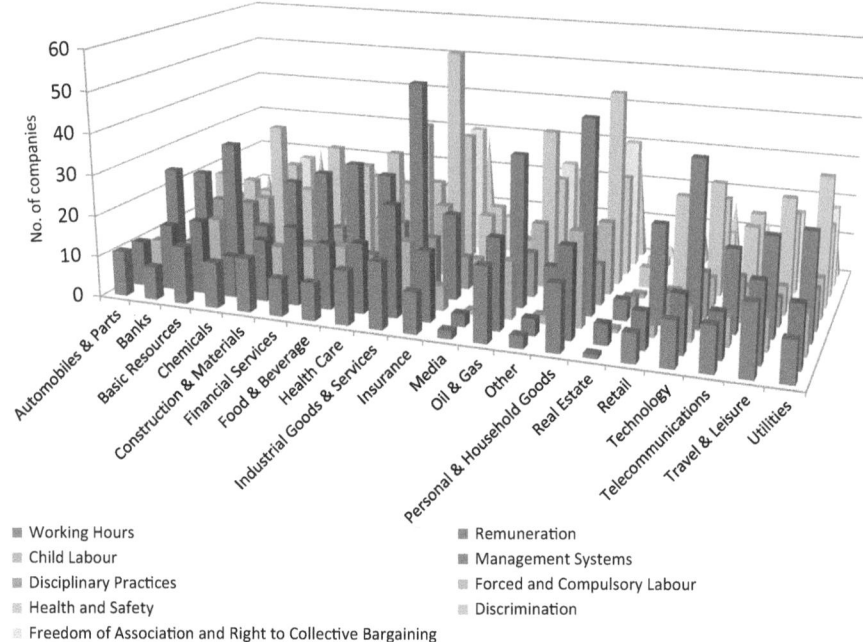

Fig. 6 Social audit—aggregate disclosure levels on industry. *Source*: authors' design

these are implicitly respected—we doubt that companies from Sweden, Denmark, the UK and the USA do not respect Human Rights or CSR frameworks.

Figure 6 presents the aggregate disclosure index per industry. The lowest disclosure levels are registered for the *real estate* companies, while the *industrial goods and services* industry features the highest disclosures on the information requested in a social audit.

From the nine elements of the SA8000 disclosure checklist, *health and safety* has the most prominent frequency. Besides *health and safety*, companies have a preference for *management systems* related disclosures. The maximum disclosure values for *both health and safety* and *management systems* information is registered by corporations in *industrial goods* and *services* industry. *Child labour* and *discrimination* maintain a medium disclosure level for all the industries. *Disciplinary practices, working hours, freedom of association and right to collective bargaining*, as well as *forced and compulsory labour* have a low disclosure tendency, as a few corporations provide information on these elements. *Disciplinary practices* present the highest disclosures in the case of banks and basic resources sectors. Chemical companies disclose less on disciplinary practices and remuneration, while focusing on *management systems* and *health and safety*. Similarly, other industries (construction and materials, financial services, food and beverages, health care) report less on *disciplinary practices, remuneration*, and also *child labour*, leaving more space for *management systems, health and safety,* and even *freedom of association and right to collective bargaining*.

Figure 7 presents the aggregate disclosure level per sector for several industries (which we considered to be relevant for the sample). Companies from construction and materials subsector register lower disclosure values than the industrial goods and services segment. This can be attributed to the fact that this subsector contains a larger sample of companies than the other ones (84–36). The highest disclosure with respect to SA8000 is met for *health and safety* (55 out of a total of 84 organisations operating in the field of industrial goods and services), also mentioning companies' preference to report on *management systems*. The lowest score is recorded by *disciplinary practices* (7 out of 35 companies). The aggregated disclosure level[1] for industrial companies is 16.11 (construction and materials) and 28.89 (industrial goods and services). With regards to the low general disclosure levels (16.11 and 28.89), we argue for the need of more improvements in SA8000 disclosure for industrial companies, because of their impact on society and environment. In addition, social audit should be mandatory for these organisations, and a first step should be the companies' acknowledgment on social aspects, by allocating more space in their reports for disclosure of social information.

In our view, the oil and gas industry has the highest impact towards the environment (and implicitly society). The aggregate disclosure level of 22.56 is the result of 19 lower-medium reporting companies (with DI less than 0.5), and 19 upper-medium reporting entities (DI higher than 0.5). There are four oil and gas companies with a 0 disclosure on SA8000: EnCana, Freeport-McMoRan, Copper & Gold, and SK Holdings. Only four companies (out of 43) report on all the nine elements from SA8000 requirements: ECOPETROL, ENAGAS S.A, Gas Natural and Fenosa Shenhua Group. Similar to industrial companies, oil and gas organisations report most on information related to *health and safety* and *management systems*. At the other extreme is *disciplinary practice*, with the lowest disclosure score of 11.00. We consider that companies operating in the oil and gas sector the report more on SA8000 and social audit should be mandated to do so. Therefore, there is a need for a higher degree of compliance with SA8000 for companies operating in the oil and gas industry, given their impact on planet and people.

Figure 7 also shows the disclosure levels for the basic materials industry. First of all, we have fewer chemical companies than organisations from basic resources subsector in our sample (17 versus 39) and implicitly the disclosure scores are lower. However, the overall level of disclosures for basic materials industry is 20.44, respectively 9.67. 18 companies from the basic resources subsector have disclosure scores of more than 0.5 (upper-medium), while 21 organisations from same subsector record DI of less than 0.5 (lower-medium). Just one company has no compliance with SA8000: Universal Forest Products, while seven have shown full compliance (DI = 100): Resolute Forest Products, BG Group, Fibria Celulose S.A.,

---

[1] The aggregated disclosure level (ADL) means the *average* for the disclosure levels of SA8000 elements: working hours, remuneration, child labour, management systems, disciplinary practices, forced and compulsory labour, health and safety, discrimination, freedom and association and right to collective bargaining. ADL can also be obtained as total disclosures all the companies in a specific industry/subsector.

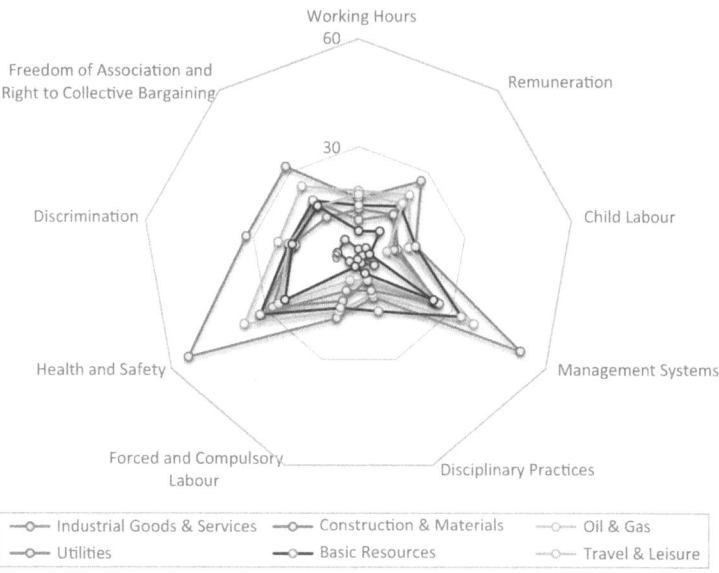

Fig. 7 Disclosure level counts on SA8000. *Source*: authors' design

Gold Fields, POSCO, Stora Enso and Tata Steel. In the chemical sector, Engro Chemical Pakistan Limited fully complies with SA8000 standard, six companies have DI less than 0.5, while eleven corporations have upper-median disclosures (over 0.5). We argue that the disclosure on SA8000 elements can be improved, especially for corporations from the chemical sector (with high impact on society and environment).

For companies from the utilities subsector, *health and safety* and *management systems* are again in their top preferences of companies when reporting on SA8000 requirements. Ten organisations out of 40 report on *child labour, forced and compulsory labour, disciplinary practices,* and *working hours.* The ADL registers 16.87—15 companies are upper medium disclosure organisations, while 25 are lower-medium disclosure ones.

This consumer services industry does not have a significant impact on society and environment, as in oil and gas/chemical sectors. However, companies from this sector report most on *health and safety* and *management systems* issues. In both *travel and leisure* and *retail* subsectors the disclosure trend is the same in the case of certain elements from SA8000: *child labour, forced compulsory labour, health and safety, freedom of association and right to collective bargaining, discrimination, disciplinary services*—and the ADL for management systems records same value 24.00. The media sector has the lowest disclosure values due to the fact that we have fewer companies in our sample in this sector, in contrast to the other two areas.

Conclusion

This chapter was meant to provide an introduction to social audit concepts and evolution. From a literature review perspective, we were able to present the social audit concept as evolving from sustainability and social responsibility practices, and from the need to understand stakeholders' expectations. The changes in corporate reporting has revealed socially responsible investors, who understand and seek the value of the non-financial information. In addition, companies have started to accept their responsibility towards society and environment. All these factors have generated the need for a social audit and specific standards to guide such an audit. From all the current social audit regulations (SA8000, AA1000, ISAE3000), we argue that the SA8000 is the most suitable for being used as the foundation in the implementation process of a social audit framework.

According to the SA8000 Standard, organisations should comply with the following social accountability requirements (SA8000—revised, 2014): child labour, forced and compulsory labour, health and safety, freedom of association and right to collective bargaining, discrimination, disciplinary practices, working hours, remuneration, management systems.

We defined a framework for social audit based on SA8000 requirements, using the nine core elements of the standard. This framework has been tested on 825 global organisations from 20 industry sectors and we have conducted a country-based and industry-based analysis. Our findings have revealed that many organisations prefer to disclose information on their *health and safety* and *management systems* policies. At the industry level, the study revealed that the lowest disclosure levels are registered within the real estate companies, while the industrial goods and services industry has the highest disclosures on the information requested on social audit. From a country's perspective, those whose first language is English have a better understanding of the English-based SA8000 regulation and could better integrate the requirements in the Annual Reports. Going further with our country-based analysis, we found that the highest average level of DI is registered in countries with a low level of Human Development (as they need to demonstrate compliance with social responsibility standards). The lowest DI average is found in countries with a very high level of Human Development (where it might not necessary to report their status of social responsibility or human rights, because these are considered to be respected implicitly). Surprisingly, the highest average level of DI is obtained in countries with a medium level in the quality of governance, and a low average DI was registered in companies operating in countries placed very high on the National Social Responsibility Index.

Finally, we argue that social audit is a relatively new research field that can be further exploited, either by considering the two elements from SA8000

(continued)

that have had the highest disclosure levels: management systems and health and safety, or by testing issues which companies were reluctant to disclose—and even conducting studies on the causes of the disclosure gap.

Acknowledgements This work was co-financed by the European Social Fund through Sectoral Operational Programme Human Resources Development 2007–2013, project number POSDRU/ 159/1.5/S/142115 Performance and excellence in doctoral and postdoctoral research in Romanian economics science domain for which the two authors are grateful.

References

ACCA. (2012). *Integrating material sustainability information into corporate reports should be a key and critical outcome of Rio+20*. Retrieved from http://www.accaglobal.com/en/discover/news/2012/05/sustainability-rio.html
Aceituno, F., Ariza, R., & Sanchez, G. (2012). Is integrated reporting determined by a country's legal system? *Journal of Cleaner Production, 44*, 44–45.
Baker, S. (2006). *Sustainable development*. New York, USA: Routledge.
Balboni A., & Balboni F., (2008). *A duopoly model of corporate social responsibility and location choice*. Dipartimento Scienze Economiche, Universita' di Bologna in its series Working Papers
Benabou, R., & Tirole, J. (2010). Individual and corporate social responsibility. *Economica, 77*, 1–19.
Björn, F. (2012). Development of norms through compliance disclosure. *Journal of Business Ethics, 106*, 73–87.
Castka, P., Bamber, C. J., Bamber, D. J., & Sharp, J. M. (2004). Integrating corporate social responsibility (CSR) into ISO management systems—in search of a feasible CSR management system framework. *The TQM Magazine, 16*(3), 216–224.
Claasen, C., & Roloff, J. (2012). The link between responsibility and legitimacy the case of De Beers in Namibia. *Journal of Business Ethics, 107*, 379–398.
Clements, P.H., & Brown, G., (2012). *Lenses and clocks: Financial stability and systemic risk*. A joint paper presented by the United Nations Environment Programme Finance Initiative (UNEP FI), the International Institute for Sustainable Development (IISD), and The Blended Capital Group (TBCG).
Dahlsrud, A. (2006). How corporate social responsibility is defined: An analysis of 37 definitions. *Corporate Social Responsibility and Environmental Management, 15*, 1–13.
Eccles, R., Cheng, B., & Saltzman, D., (2010a). *The landscape of integrated reporting reflections and next steps*. Harvard Business School, The President and Fellows of Harvard College Cambridge, Massachusetts.
Eccles, R., Krzus, M. P., & Tapscott, D. (2010b). *One report: Integrated reporting for a sustainable strategy* (1st ed.). New York: Wiley.
Eccles, R., Ioannou, I., Serafeim, G., (2012), *The impact of a corporate culture of sustainability on corporate behaviour and performance*. Working Paper, Harvard Business School. Available online at http://www.hbs.edu/research/pdf/12-035.pdf
Fahmi, F. M., & Omar, N. (2005). Corporate reporting on minority shareholders information and its implication on shareholders activism in Malaysia. *Journal of Financial Reporting and Accounting, 3*(1).
Falkenberg, J., & Brunsæl, P. (2011). Corporate social responsibility: A strategic advantage or a strategic necessity? *Journal of Business Ethics, 99*, 9–16.

Fombrun, C. J. (2005). A world of reputation research, analysis and thinking—Building corporate reputation through CSR initiatives: Evolving standards. *Corporate Reputation Review, 8*(1), 7–12.
Fonseca, A., Macdonald, A., Dandy, E., & Valenti, P. (2011). The state of sustainability reporting at Canadian universities. *International Journal of Sustainability in Higher Education, 12*(1).
FTSE International. (2012). *Industry classification benchmark*. ICB Publications, Available online at: http://www.icbenchmark.com/ICBDocs/FTSE_ICB_Corporate_Brochure.pdf.
Garavan, T. N., Heraty, N., Rock, A., & Dalton, E. (2010). Conceptualizing the behavioral barriers to CSR and CS in organizations: A typology of HRD interventions. *Advances in Developing Human Resources, 12*(5), 587–613.
Gray, R. (2001). Thirty years of social accounting, reporting and auditing: What (if anything) have we learnt? *Business Ethics: A European Review, 10*(1), 9–15.
Gray, R. (2006). Social, environmental and sustainability reporting and organisational value creation?: Whose value? Whose creation? *Accounting, Auditing & Accountability Journal, 19*(6), 793–819.
GRI. (2013). The external assurance of. Global Reporting Initiative.
Gyves, S., & O'Higgins, E. (2008). Corporate social responsibility: An avenue for sustainable benefit for society and the firm? *Society and Business Review, 3*(3), 207–223.
Gilbert, D. U., & Rasche, A. (2007). Discourse ethics and social accountability: The ethics of SA 8000. *Business Ethics Quarterly, 17*(2(Apr., 2007)), 187–216.
Göbbels, M., & Jonker, J. (2003). AA1000 and SA8000 compared: A systematic comparison of contemporary accountability standards. *Managerial Auditing Journal, 18*(1), 54–58.
Hopkins, M. (2003). *The Planetary Bargain—CSR Matters*. London: Earthscan. http://www.iso.org/iso/social_responsibility.
Htaybat, K. (2011). Corporate online reporting in 2010: A case study in Jordan. *Journal of Financial Reporting and Accounting, 9*(1).
Kolk, A. (2003). Trends in sustainability reporting by the Fortune Global 250. *Business Strategy and the Environment, 12*(2003), 279–291.
Kolk, A., & Perego, P. (2010). Determinants of the adoption of sustainability assurance statements: An international investigation. *Business Strategy and the Environment, 19*(3), 182–198.
KPMG. (2008). *International survey of corporate responsibility reporting 2008*. KPMG International.
Kreps, T. (1940). *Measurement of the social performance of business*. Washington, DC: United States Government Printing Office.
Kuhlman, T., & Farrington, J. (2010). What is sustainability? *Sustainability, 2*, 3436–3448.
Lanoizelée, Q. F. (2011). Are competition and corporate social responsibility compatible?: The myth of sustainable competitive advantage. *Society and Business Review, 6*(1), 77–98.
Lea, R. (2002). Corporate social responsibility, Institute of Directors (IoD) member opinion survey. London: IoD. http://www.epolitix.com/data/companies/images/Companies/Institute-of-Directors/CSR_Report.pdf [23 June 2003].
Leonard, D., & McAdam, R. (2003). *Corporate social responsibility. Quality progress*. Available online at http://alexandrow.pdforms.texas-quality.org/SiteImages/125/Reference%20Library/Social%20Responsibility%20-%20Leonard.pdf
Lozano, R. (2009). Orchestrating organisational change for corporate sustainability. Strategies to overcome resistance to change and to facilitate institutionalization. (PhD Doctoral thesis, Cardiff University, Cardiff).
Lonzano, R., & Huisingh, D. (2011). *Journal of Cleaner Production, 19*(2–3), 99–107.
Loew, T., Ankele, K., Braun, S, Clausen, J. (2004). *Significance of the CSR debate for sustainability and the requirements for companies*. Institute fo Ecological Economy Research, Available online at http://www.ioew.de/uploads/tx_ukioewdb/future-IOEW_CSR-Study_Summary.pdf

Mandelbaum, M., & Friedman, T. (2009). What are the barriers to mass adoption of sustainability? *SEMIOSIS Communications*. Retrieved from http://www.semiosiscommunications.com/barriers-to-sustainability-adoption/

Marsden, C. (2000). The new corporate citizenship of good business: Part of the solution to sustainability. *Business and Society Review, 105*(1), 9–25.

Martinuzzi, A. (2011). Responsible Competitiveness—Linking CSR and competitive advantage in three European Industrial Sectors. *Zeitschrift für Umweltpolitik & Umweltrecht, 3*, 297–337.

Merce, B., & Simon, A. (2009). Multiple standards: Is this the future for organizations?—Sub-theme 15: Multiplicity and plurality in the World of standards. *Journal of Cleaner Production, 17*(8), 742–750.

MHC International. (2008). *National Social Responsibility Index (NSRI)*. Retrieved from: http://mhcinternational.com/images/stories/national_social_responsibility_index.pdf

Michael, J. (2009). *Best practice CS&R reporting in the Real Estate Industry*. Global MBA Graduate. Ryerson University Toronto.

Montiel, I. (2008). Corporate social responsibility and corporate sustainability, Separate Pasts, Common Futures. *Organization & Environment, 21*(3), 245–269.

Nicol, A. (2000, April). *Adopting a sustainable livelihoods approach to Water Projects: Implications for policy and practice*, Working paper. London, UK: Overseas Development Institute.

Nidumolu, R., Prahalad, C. K., Rangaswami, M. R. (2009). Why sustainability is now the key driver of innovation?. *Harvard Business Review*.

Oba, V. C. (2011). The impact of corporate social responsibility on market value of quoted Conglomerates in Nigeria. *ICAN Journal of Accounting and Finance, 3*, 64–70.

Pearce, J. (1996). *Measuring social wealth: A study of social audit practice for community and cooperative enterprises*. New Economics Foundation.

Pearce, J., Raynard, P., & Zadek, S. (1996). *Workbook on social auditing for small organizations*. New Economics Foundation.

Percy Smith, J., & Hawtin, P. (2007). *Community proofing—A practical guide*. Berkshire: Open University Press McGraw-Hill Education.

Pinney, C., (2001). *Imagine speaks out. How to manage corporate social responsibility and reputation in a global marketplace: The challenge for Canadian business*. http://www.imagine.ca/content/media/team_canada_china_paper.asp?section=media=media [23 June 2003].

Policy Research Unit. (1990). *Finding out about your community: How to do a social audit*. Leeds Metropolitan University.

Radley, Y. (2012). *The value of extra- financial disclosure. What investors and analysts said, Accounting for Sustainability, Global Reporting Initiative*. https://www.globalreporting.org/information/news-and-press-center/Pages/Investors-and-analysts-use-extra-financial-information-in-decision-making-suggests-new-research.aspx

Reynolds, M. A., & Yuthas, K. (2008). Moral discourse and corporate social responsibility reporting. *Journal of Business Ethics, 78*(1–2), 47–64.

Robinson, J., & Tinker, J. (1998). Reconciling ecological, economic and social imperatives. In J. Schnurr & S. Holtz (Eds.), *The Cornerstone of development: Integrating environmental, social and economic policies*. Ottawa: International Development Research Centre.

Selvi, Y., Wagner, E., & Türel, A. (2010). Corporate social responsibility in the time of financial crisis: Evidence from Turkey. *Annales Universitatis Apulensis Series Oeconomica, 12*(1), 281–290.

Signes, P., Segarra, O. M., Jimenez, M., & Vargas, M. (2013). Influence of the environmental, social and corporate governance ratings. *International Journal of Environmental Research, 7*, 105–112.

Social Accountability International. (2008). *Social accountability 8000, International Standard*. Available online: http://www.sa-intl.org/_data/n_0001/resources/live/2008StdEnglishFinal.pdf

Szegedi, K. (2010). The concept and development tendencies of corporate social responsibility. *Club of Economics in Miskolc TMP, 5*, 67–74.

SAI. (2014). *SA8000*. Revised version http://www.sa-intl.org/_data/n_0001/resources/live/SA8000_SideBySide_12_16_%202013.pdf

Tarí, J. J. (2011). Research into quality management and social responsibility. *Journal of Business Ethics, 102*, 623–638.

Tovey, H. (2009). Sustainability: A platform for debate. *Sustainability, 1*(1), 14–18.

UNDP. (2013). *Human development Report 2013*. Retrieved from: http://hdr.undp.org/sites/default/files/reports/14/hdr2013_en_complete.pdf

Unerman, J., Bebbingto, J., & O'Dwyer, B. (2010). *Sustainability accounting and accountability*. London: Routledge.

Veltri, S., & Nardo, M. T. (2013). The Intangible Global Report: An integrated corporate communication framework. *Corporate Communications: An International Journal, 18*(1), 26–51.

Vidal, N.G., & Kozak R.A. (2009, August). *From forest certification to corporate responsibility: Adapting to changing global competitive factors*. Proceedings of the XIII World Forestry Congress 2009: Forests in Development—A Vital Balance, Buenos Aires, Argentina, Proceedings 9 pp.

Wallage, P. (2000). Assurance on sustainability reporting: An auditor's view. *Auditing: A Journal of Practice & Theory, 19*(1), 53–65. Supplement 2000.

World Bank. (2013). *Worldwide Governance Indicators (WGI)*. Retrieved from http://info.worldbank.org/governance/wgi/index.aspx#reports

Zadek, S., Evans, R., & Pruzan, P. (2013). *Building corporate accountability: Emerging practice in social and ethical accounting and auditing*. London: Routledge.

Social Audit Failure: Legal Liability of External Auditors

Larelle Ellie Chapple and Grace Y. Mui

1 Introduction

The core areas of this book relate to social audit and the regulation of this audit practice. Social auditing as used herein refers to the practice of external assurance or evaluation of an organisation's socially responsible reporting assertions. In this regard, we reflect back on the research from the 1970s, such as Bauer and Fenn (1972: p. 2), who refer to "the goal of the social audit movement is the mounting of a comprehensive and objective evaluation of the social performance of firms on a continuing basis." In this chapter, we look at the step beyond reporting, to the external audit or assurance function. The role of any audit engagement is to provide a professional opinion on a set of financial or non-financial assertions reported by an organisation's management, based on an agreed evaluative framework. Any such opinion is not a guarantee that the underlying report is free from fraud or misstatement. Where an audit opinion on financial statements is incorrect, this is referred to as an audit failure. Specifically, the textbook definition of audit failure has two components: that the financial statements contain a serious error and that the auditor has failed to detect the error due to the auditor's failure during the audit process. In this respect a social audit is analogous, it is a professional opinion, and we must concede the possibility that the opinion can be wrong, inaccurate or incorrect, and that the underlying error in the client firm's assertions remained undetected through an error in the audit process. Hence, we pose the question—what is the source of potential legal liability for a social audit failure? By briefly tracing the history of financial reporting audit standards from no regulation, to self-regulation, to

---

L.E. Chapple (✉)
School of Accountancy, QUT Business School, Queensland University of Technology,
2 George Street, 4001 Brisbane, QLD, Australia
e-mail: Larelle.Chapple@qut.edu.au

G.Y. Mui
Thye & Associates, Selangor, Malaysia

© Springer International Publishing Switzerland 2015
M.M. Rahim, S.O. Idowu (eds.), *Social Audit Regulation*, CSR, Sustainability,
Ethics & Governance, DOI 10.1007/978-3-319-15838-9_14

mandatory regulation, this chapter highlights the potential sources of litigation risk faced by auditors in a social audit engagement.

This chapter uses the legal liability framework developed for (financial reporting) audit failure to draw an analogy as to the potential trajectory of the jurisprudence around social audit failure. As such, its conclusions are speculative, based on the experiences documented in the literature as to the developments in the financial reporting and audit sphere. Very little commentary has appeared to date as to the potential for professional legal liability for auditors in social audit engagements; the phenomenon of auditors' legal liability to clients as we now know it developed over a century ago and only in the last couple of decades legal liability expanded to third parties. The ultimate focus of this chapter is to examine the potential sources of legal liability of auditors in social audit engagements.

All audit engagements rely on two sets of agreed evaluative frameworks—first, the reporting framework under which the client firm makes its assertions, and second, the agreed framework under which the auditor agrees to audit those assertions. As the assurance of sustainability reporting itself is relatively new, there is no clear, universal framework for the audit, and it is not regulated in a majority of countries (Junior, Best, & Cotter, 2014). In adopting a comparative methodology in this chapter—comparing the sources of legal liability for financial report audits with the sources of potential legal liability for social audits we acknowledge a major distinction in the underlying regulatory environment. Usually, the social reporting framework organisations use, such as the Global Reporting Index, is a voluntary code for reporting on social performance. For auditors accepting an engagement on social reporting, the appropriate engagement standards may be similarly voluntary. Mock, Rao, and Srivastava (2013: p. 280) report that there is a "worldwide movement toward reporting environmental, social and economic performance beyond what is already reported in the audited financial statements". It is only during this century that financial accounting and reporting has developed a universal set of standards for assurance work on financial reporting engagements, so we acknowledge the incipient nature of the social audit reporting and assurance frameworks.

In addition to the above disclosed focus, i.e. non-financial reports and the voluntary audits thereof, we acknowledge several other limitations in scope. First, the chapter concentrates on the experience of audit and auditors and legal liability from the perspective of the English common law and subsequently the Australian law and practice. Australian case law and statute law and the Australian audit profession and regulators have been at the forefront of global trends and initiatives in setting reporting standards in modern times. Second, as alluded to above, we adopt the perspective that a "social audit" is the non-financial equivalent to the audit of the general purpose financial report and the results of both reports (i.e. the disclosure and the audit) are made generally available by the client firms that commission them. Third, we focus on auditors' potential civil liability arising from assurance engagements, not criminal liability for fraudulent financial reporting (represented famously in the US by *Ernst and Ernst v Hochfelder*[1]: Baker, Bedard, & Hauret (2014). Lastly, the assurance of non-financial reports may be provided by assurers who are either part of the accounting and audit profession, or not (Simnett, Vanstraelen, & Chua, 2009). We focus on the assurance engagements known as the social audit, undertaken by the same profession who also undertake financial report audits.

2 Social Audit Failure: A Definition

As referred to in the Introduction, an audit failure in the sense of a financial report audit is generally defined as follows:

> Audit failure occurs when there is a serious distortion of the financial statements that is not reflected in the audit report, and the auditor has made a serious error in the conduct of the audit. (Arens, Elder, & Beasley, 2002: pp. 109–10).

Accordingly, our first contribution is to propose that social audit failure should be a recognisable phenomenon, and we define a social audit failure as a misstatement in the social reporting assertions made by management in the client firm that are not detected, due to a serious error made by the audits in the conduct of the audit engagement. Financial statement audit failure has four causes (Tackett, Wolf, & Claypool, 2004):

1. auditor unintentional human error;
2. auditor fraud;
3. undue influence caused by financial interests; and
4. undue influence caused by personal auditor-client relationships.

Disregarding auditor fraud, we see that social audit failure can similarly arise from what we might describe as auditor error or auditor independence threats. In the financial report audit arena, we show below that these causes are ameliorated by a combination of professional standards and laws setting best practice in audit procedures, and laws that regulate relationships giving rise to independence threats. Accordingly, our second contribution is to propose that, similarly to financial audit failure, social audit failure can be addressed by the promulgation of internationally accepted standards of best practice. Also later in the chapter, we explain why the incipient process of standard setting in the social audit arena still has some considerable way to go. Finally, our third contribution is to propose the potential consequences of social audit failure as pertain to the audit profession by identifying the sources of legal liability to auditors arising from financial report audits, and

---

[1] 425 U.S. 185 (1976).

comparing this to social audits to identify sources of potential legal liability to auditors arising from these types of engagements.

3 Sources of Legal Liability

In our contemporary setting, professionals such as auditors in a financial report engagement or a social audit engagement, who provide independent, expert advice can be exposed to liability arising from carrying out their professional functions from three legal sources:[2]

1. Contract law (common law);
2. Tort of negligence (common law); and
3. Statutory incursions that impose duties, obligations and remedies, in what otherwise would be private contractual or transactional settings.

The common law is concerned with providing civil remedies where the private ordering arrangements have gone awry. Liability in both contract and tort has requisite hurdles that are described below, with the examples particularly pertaining to the audit setting.

3.1 Contract

Liability in contract depends on establishing the formalities of the contract, primarily in isolating the parties to the contract. Indeed, "for many years, the principle of privity of contract dominated the legal arena" (Samsonova, 2010: p7) and auditors of financial reports were generally seen as fair game in litigation over the causes of audit failure. Privity of contract means only parties to the contract can enforce it, which on the one hand provides a relatively simple liability nexus in the relationship between the client firm and the auditor, via the (usually) written audit engagement letter, but this also means there is no potential for so called third party liability. As described below, case law developments last century as to the scope of the audit and the concept of due care created a real audit litigation risk of client suit.

However, in our analysis, this leads to three complications in applying contract liability in the social audit engagement environment: first, professional standards in the financial report audit arena specifically address the process of agreeing to the terms of the audit engagement (ISA 210). To what extent are auditors in social audits protected by a formal audit engagement standard? The formal engagement

---

[2] We have not included general criminal liability for fraud in this chapter. We are confining the scope to liability of auditors who carry out their routine functions—not liability of auditors who are dishonest, fraudulent or reckless.

letter establishes the terms of the audit engagement, and particularly the agreed upon evaluative framework. Second, contracts can still contain express and implied terms and the common law will always imply a term requiring a professional standard of due care. In the absence of an internationally recognised evaluative framework for social audit, how does a court determine the appropriate level of due care? Third, remedies in contract depend on evidence that one party (the auditor) has breached the contract and the other party (the client firm) can claim that damages that will put them back in the position they would have occupied had the breach not occurred, applying causation tests. In a financial report audit setting, the loss to the client firm is perhaps more easily comprehensible as a quantifiable loss. In a social audit setting, it is more difficult to see what assessable damages a client firm can claim.

Accordingly, we see three difficulties in applying contract law as a source of potential liability in social audit: is there a "standard form" engagement letter that is readily identifiable and widely used to protect social audit engagements, or does each practitioner effectively write their own contract? Second, what evidence will be used to determine the standard of due care in the case of alleged social audit failure? Third, what loss, in the form of measurable damage, flows from the social audit failure as breach of contract? We will examine the other common law source of liability before turning more specifically to these problems.

3.2 Tort of Negligence

Liability in tort also depends on establishing a nexus or relationship between the parties, but tort law has a much more complex set of principles for determining the threshold issue of the "duty of care". Much of the early jurisprudence of tort law in the common law world revolved around the celebrated "neighbour" principle of the duty of care from *Donoghue v Stevenson*,[3] that the law of negligence compensates for harm from activities that we reasonably foresee are likely to injure our neighbours; that is persons closely and directly affected by our acts or omissions. This test appears to serve us well in the case of physical loss or damage from immediate physical, contemporaneous, or geographic nexus between our neighbour and acts or omissions. The tort of negligence also requires evidence that the standard of reasonable care has been breached in the circumstances, and that loss or damage was caused by the breach of the duty of care.

It is now recognised that "pure" economic loss is a form of recoverable loss in negligence, and as this is the basis for auditor liability in damages, it creates a relationship of inter-dependence between the elements of the tort. Identification of the damage depends on the characterisation of the relationship between the parties to the action—the risk of harm becomes the nexus or relationship: "the

---

[3] [1932] AC 562.

characteristic feature of the pure economic loss as a form of harm is that the person's acts [e.g. the audit failure] may bear an indeterminate relationship with the consequences of those acts [e.g. The rendering or manifestation that economic loss occurred] (Corbett, 1994: p. 816). Again, as described below, case law developments last century specifically as to the scope of an auditor's duty of care and liability for economic loss defined some of the indeterminacy of the auditor's litigation risk to client suit.

Accordingly, we see three difficulties in applying the tort of negligence as a source of potential liability in social audit: the duty of care owed by the auditor to the client is somewhat mitigated by the "defence" of contributory negligence, that is management must take responsibility (if they can be proved to be at fault) for the underlying assertions made in the report. It is clear that the contributory negligence applies in financial report audits as a defence for the auditor.[4] Our concern is that in the absence of a strong, international reporting framework, it will be more difficult for an auditor in a social engagement to seek out this defence as management has much more latitude and discretion in a social reporting context as to what they choose to report and how they choose to report it. Providing evidence that management has not acted reasonably according to acceptable standards may be difficult. Second, what evidence will be used to determine the standard of due care in the case of alleged social audit failure? In the absence of a strong internationally recognised social audit engagement framework, social auditors could be exposed to costly litigation in setting the standards expected, based on the experience from financial report audits. Third, what loss in the form of measurable damage flows from the social audit failure as breach of the duty of care?

In each case of the civil law sources of liability we have identified three issues, and in every sense these issues overlap. In our analysis, the development of the auditor's liability in social engagements will be determined by the same three issues:

1. Who defines the scope of the social audit?
2. Who will develop the evaluative frameworks (both for reporting and auditing) that will determine the terms of the engagement as a standard form and determine the standard of due care expected?
3. What loss is reasonably foreseeable from a social audit failure?

By using a comparative methodology, we examine how these issues were resolved in the financial reporting audit context. Before concluding, we will also make some comment on statutory liability as a potential source of auditor liability.

---

[4] In *Daniels t/as Deloitte Haskins & Sells v AWA Ltd* (1995) 16 ACSR 607, the end result was that the Court apportioned damage as 80 % contribution by the auditor and 20 % contribution by client management for the client firm's loss a result of the audit failure.

4 The Defined Scope of Financial Report Audits

The twentieth century witnessed the growth in demand for financial report audits, with a consequent growth in the professional framework to support the assurance function. There are many sources that narrate the history of audit, particularly in the United States and the United Kingdom. In particular, we note that the history of audit enjoyed relatively common origins throughout the nineteenth century, coinciding with the scale of enterprise following the industrial revolution. However, as noted by Brown (1962: p. 699), the American auditing profession diverged from its British origins after the beginning of the twentieth century. The main difference appears to be that the American profession progressed independently in its views of the audit function, moving away from fraud detection to emphasise the objective of ascertaining the financial condition and earnings of the enterprise.[5] This chapter focuses on the development of the audit professional standards and professional liability following the British model; commenting on the experiences of the Commonwealth, common law jurisdictions. The point of following this path is to provide a common narrative of the common law developments, statutory reforms, global financial reporting and audit standards to culminate in the "force of law" audit standards operational in Australia.

The financial report audit today is regarded in Britain as one of the cornerstones of corporate governance (Cadbury Report, 1992). However, it is widely recognised that the British experience of reporting and auditing through the early nineteenth century focused on values of honesty and integrity (Fitzpatrick, 1939; Higson, 2003; Littleton, 1933). The role of the auditor was linked to overseeing management's stewardship, and at various times during the 1800s the financial audit was mandated or not. The nineteenth century marked a period of development of stock markets, but the markets were largely unregulated and highly speculative (Lee & Ali, 2008). The audit function provided some protection for investors where limited liability was introduced by British legislation only in the middle of the century (1855). This culminated with the reintroduction of the mandatory statutory audit in 1900 (The Companies Act 1900 (UK)) and this remains the situation now in all Commonwealth jurisdictions. During that century, the Punishment of Frauds Act 1857 (UK) strengthened the law against fraud by company officers (Higson, 2003), leading to a focus on the financial report audit as a fraud detection device.

However, two cases towards the end of the nineteenth century are attributed with both reinforcing the audit objective of fraud detection (Lee & Ali, 2008); but also signaling the decline in its importance (Higson, 2003): *Re Kingston Cotton Mill (No 2)* [6] and *In re London General Bank Ltd ex parte Theobald (No 2)*[7]. In the latter case, Theobald was the auditor who certified the accounts of a company prior to it

---

[5] Brown (1962) attributes this observation to the author of an early auditing monograph Montgomery (1913).
[6] [1896] 2 Ch 279.
[7] [1895] 2 Ch 673.

paying dividends out of a capital account. Lindley LJ stated that the auditor's "business" is to ascertain and state the true financial position of the company and the auditor must take reasonable care to do so; where suspicion is aroused more care is required, but the auditor is not bound to exercise more than reasonable care even where there is suspicion. Commentators such as Brown (1962) suggest that Lindley LJ's judgment recognises the appropriate use of audit procedures to test assertions rather than implement full scale investigations. Lindley LJ also emphasised that where there is nothing to "excite suspicion" then less inquiry is reasonable and sufficient.

The *Kingston Cotton Mill case* is more directly related to fraud, as it involved fraudulent financial reporting by the managing-director in overstating the value and quantity of inventory. The auditor did not detect the false statements—Lopes LJ found that the auditor did not breach their duty. In the absence of suspicion, the auditor could rely on management assertions.

Around the turn of the nineteenth to twentieth century it was thought that the audit had three objectives: the detection of fraud, the detection of technical errors and the detection of errors in principle (Porter, 2007). Certainly, the role and function of the statutory financial report audit have evolved over the proceeding century such as today, the reverse is almost true.

These early cases are supplemented by cases later in the twentieth century highlighting the audit function somewhere on a continuum between full investigation based on presumed fraud and forbearing to inquire until suspicion arises (Ramsay and Austin, online, [10.550]): *Fomento (Sterling Area) Ltd v Selsdon Fountain Pen Co Ltd* [8]. Sargant LJ in *Re City Fire Insurance Ltd* [9] stated that the duty of an auditor is verification and not detection.

The period after the 1920s witnessed the development of the accounting and auditing professions, and via professional practice and standards, the auditing profession acknowledged progressively less responsibility for detecting fraud. By the 1960s it was denying all but an incidental responsibility in this regard (Porter, 2007). Hence, it has been suggested by more modern commentators that the statutory requirement to provide an opinion on the true and fair view of the financial statements presented by management (e.g. s495 of the UK *Companies Act 2006*; s307 Australian *Corporations Act 2001*) necessitates audit procedures that are not necessarily appropriate to fraud detection (Porter, 2007).

---

[8] [1958] 1 WLR 45.
[9] [1925] Ch 407.

5 The Auditing Framework: The Role of Professional Standards

During the 1950s–1960s, the professional bodies from around the world were well on the way to developing a comprehensive set of standards for the conduct of financial statement audits. For example, by 1961 the Institute of Chartered Accountants in England and Wales had promulgated the General Principles of Auditing. In Australia, the Institute of Chartered Accountants in Australia reissued in 1969 a Statement of General Principles of Professional Auditing Practice that replaced the earlier Trigg statement from the 1950s (Gibson & Arnold, 1981). Earlier that decade, corporate collapses in Australia focused attention on audit failures (Clarke, Dean, & Oliver, 1997) and the audit profession and their professional bodies were being challenged for contributing to the loss (Carnegie & O'Connell, 2012). The profession initially responded with silence (Birkett & Walker, 1971) but in the mid 1960s the professional bodies established the Accountancy Research Foundation, whose role was, inter alia, to develop professional standards of accounting and audit (Boehme & Braddock, 1965; Carnegie & O'Connell, 2012).

Hence, the professional standards as currently derived developed after a period of economic trauma. Since the 1960s, a similar pattern of professional and legal intervention has followed cycles of economic trauma and corporate collapse. In particular, if we now move forward by forty years to the beginning of the current century, Australia similarly experienced corporate collapse of a scale of such magnitude that could not be ignored by policy makers and law makers. Much has been written on the economic failures in Australia of the first part of the decade 2000, and a common theme again was the scrutiny of the role played by gatekeepers such as accountants and auditors in facilitating financial crisis in the firms involved (Clarke & Dean, 2007).

Consequent upon this round of corporate collapse, the law makers waded in more heavily to the debate as to audit standards, in the form of prescriptive amending legislation in 2004 known as the CLERP 9 Act, *Corporate Law Economic Reform Program (Audit Reform and Corporate Disclosure) Act 2004* (Cth) (Chapple & Koh, 2007). Professor Ian Ramsay was commissioned in 2001 by the Commonwealth Minister for Financial Services and Regulation to review the legal regulation of auditor independence. This review lead to the Ramsay Report (2001), which was the precursor to significant legislative intervention in the CLERP 9 Act reforming the law relating to auditor independence (that is, relating to the mandatory audit of financial statements). Prior to CLERP 9 Act, matters of auditor independence had been dealt with at a professional level by voluntary professional ethical standards. It should be noted however that Professor Ramsay cautioned against attributing corporate collapse to audit failure, and much of the contemporaneous literature on the CLERP 9 Act commented that the legislation addressed auditor independence but not corporate governance per se (Fogarty & Lansley, 2002).

The CLERP 9 Act "independence" reforms created direct prohibitions on auditor relationships and conflict of interest situations.[10] In addition, the Act regulated other conduct such as mandatory audit partner rotation after 5 years,[11] a ban on auditor recruitment to client boards,[12] and mandatory disclosure of non-audit fees.[13] Regarding the framework of auditor liability, there are two further aspects of the CLERP 9 Act reforms that did not seem to generate as much attention as the prescriptive reforms, but create a unique auditor liability framework in Australia:

1. Breach of the auditor independence rules as to conflict of interest relationships and situations gives rise to potential statutory criminal liability: section 324CG *Corporations Act 2001*(Cth); and
2. The Australian audit standards are now referred to as "force of law" standards, as s307A *Corporations Act 2001* (Cth) provides that non-compliance with the audit standards is a strict liability offence.

When the force of law provision was first enacted in 2004, there was generally a negative reaction from accounting bodies and it was generally perceived as an overreaction (Hecimovic, Martinov-Bennie, & Roebuck, 2009). To date there have been no actual enforcement actions taken by the regulator Australian Securities and Investments Commission (ASIC) against auditors pursuant to these provisions.

In addition to these strict liability provisions in the *Corporations Act 2001* (Cth), Free (1999) reminds us that auditors are also subject to other legislation such as general consumer protection type liability under misleading and deceptive conduct provisions in Australian state and Commonwealth legislation. For example, the 1999 settled litigation between the insolvent Linter Group and Price Waterhouse; the state of Victoria and KPMG after the Tricontinental collapse, and Southern Cross Holdings against Arthur Andersen, involved actions against auditors under such legislation (Anderson, 1996).

6 The Link Between Evaluative Frameworks and Due Care

The audit and reporting failures in Australia from the 1960s as noted above generated litigation, and the case law from that time has been instrumental in developing the principles of auditor liability. The case *Pacific Acceptance Corporation Ltd v Forsyth & Ors*[14] is a "watershed" case for several reasons pertinent to this chapter. First, as noted, the litigation arose from a controversial period of Australia's economic history, a time when the audit and accounting profession received virulent criticism for their perceived failure in professional standards

---

[10] Part 2 M.4 Div 3 Auditor Independence *Corporations Act 2001* (Cth).
[11] Part 2 M.4 division 5 Auditor rotation for listed companies *Corporations Act 2001* (Cth).
[12] Section 324CI *Corporations Act 2001*(Cth).
[13] Section 300(11B) *Corporations Act 2001*(Cth).
[14] (1970) 92 WN (NSW) 29.

(Carnegie & O'Connell, 2012). Second, the case itself and the judgment of Moffitt J in the New South Wales Supreme Court is instrumental for indicating the role of professional standards in providing evidentiary support for the standards of conduct of due care expected by auditors in professional audits of financial statements. It is the case that the English Courts had previously commented that the professional auditing standards are not determinative of the standard of care expected: *Duple Motor Bodies Ltd v Ostime*.[15] Moffitt J in *Pacific Acceptance* reiterated that the court still determines the standards expected, not the professional statements. The standard of care expected is that of a reasonable expert in audit. The common law of auditor's liability remained reasonably stable thereafter but for two major innovations:

1. The acceptance of the doctrine of contributory negligence as a defence for auditors (against company management): *AWA Ltd v Daniels*;[16]
2. The potential "floodgate" principle in third party auditor liability.

It is generally accepted that the High Court of Australia's decision in *Esanda Finance Corporation Limited v Peat Marwick Hungerfords*[17] is substantially in accordance with the UK law as developed in *Caparo Industries plc v Dickman & Ors*[18] in preventing third party liability (Fogarty & Lansley, 2002). However, this view is perhaps superficial; as identified by Chung, Farrar, Puri, and Thorne (2010), there are potentially four degrees of restrictiveness in identifying third party liability rules: ranging from very restrictive to liberal. These are: (1) privity rule; (2) near-privity standard; (3) restatement rule; and (4) the reasonable foreseeability rule. They position Australia's rule at level 3—that is, the High Court specifically rejected a mere foreseeability rule, but recognised that there may be circumstances where the relationship between the auditor is something more, such that the auditor has created a relationship of reliance between the auditor and the third party. Although in the intervening period there have been no auditor third party liability cases, the High Court has not closed the door completely on third party claims.

7 Analysis of the Development of Financial Report Audits: Where to for Social Audits?

Fundamentally, the role of a financial report audit as described above has several features that have impacted on the sources and doctrines of auditor liability. First, a financial report audit is mandated by statute. Although there are accepted economic arguments that explain why companies might voluntarily provide investors with an

---

[15] [1961] 2 All ER 167.
[16] (1992) 10 ACLC 933.
[17] (1997) 188 CLR 241.
[18] (1990) 8 ACLC 3,011.

audited view of the financial report as presented by management, primarily today we can explain that companies provide audited statements because they have to. Second, in performing the audit, the audit profession is now subject to a global set of professional auditing standards, under the auspices of the International Accounting and Assurance Standards Board. A considerable amount of effort has been expended in the last decade to arrive at a globally harmonised set of standards. Third, the harmonised standards set up a standard template for audit reporting and opinion, such that the audit report itself is a highly structured and technical document. It is not a narrative as such. Fourth, the mandated statutory function is very specific—primarily to form an opinion about whether the financial reports as presented give a true and fair view of the financial position of the company. The concept of "true and fair view" itself is a highly technical one and subject to much debate not reported here, but the concept may be summarised as: "The purpose of financial reporting is to give an understanding, which is not misleading, of the underlying economics of an enterprise": Alexander and Jermakowicz (2006: p. 132). Accordingly, the audit engagement provides reasonable assurance that the financial report presents a true and fair view of the company's financial position and is free from material misstatement or error. It is not a guarantee that the financial report is correct, nor are auditors guardians of good corporate conduct (Fogarty & Lansley, 2002). Fifth, the audit engagement is in the nature of reasonable assurance—there are other forms of audit services such as reviews and limited assurances, as discussed below. Sixth, the sources of legal liability for auditors arising from the statutory financial report audit are the common law (contract or tort duty of care, including potential third party liability); statutory sanctions and penalties for non-compliance with the statutory requirements; statutory sanctions and penalties arising from consumer protection type laws covering misleading and deceptive conduct.

These factors provide a framework for the comparison of the sources and types of liability auditors may be exposed to in the social audit type engagements. This is the topic of the next section, following which some concluding comments are made regarding the possible sources of liability for auditors arising from social audit engagements.

8 The Social Auditing Framework: Developing the Standards for Assurance

There is a wide variety of voluntary sustainability reporting standards[19] that govern the preparation of the sustainability/corporate social responsibility (CSR) reports. The audience and their information needs determine the disclosure (what is

---

[19] Global Reporting Initiative guidelines, Dow Jones Sustainability Index, ISO14000 (environmental) series.

disclosed) and the format of the disclosure (how the disclosure is made) (ICAEW, 2010). Therefore, sustainability/CSR reports vary in format and disclosure according to industry and target audience (ICAEW, 2010). The measures reported in sustainability reports vary according to the standard adopted, the industry the organisation operates in, and the stakeholders. The nature of the disclosure can be qualitative, in the form of narratives, and quantitative (ICAEW, 2010).

The Institute of Chartered Accountants in England and Wales (2010) recommend that sustainability reports are relevant, complete, reliable, neutral, and understandable. Relevant information contributes to conclusions that assist users in making decisions. Complete information ensures that relevant factors that affect conclusions are included. Reliable information allows information to be evaluated in a consistent manner. Neutral information is free from bias. Information that is understandable contributes to clear and comprehensive conclusions that are not subject to significantly different interpretations. External auditors can apply these criteria in social audits.

Given the reporting framework for social reports, the audits thereof are not the same assurance engagement typical in a financial report audit. By way of contrast, a financial report audit is referred to as "reasonable assurance" that according to sufficient appropriate evidence collected; the auditor is able to opine whether the report is materially misstated. A limited assurance engagement provides a lower level of assurance. To describe the difference, we set out below the descriptions from the relevant international standard on assurance engagements:[20]

> a. Reasonable assurance engagement—An assurance engagement in which the practitioner reduces engagement risk to an acceptably low level in the circumstances of the engagement as the basis for the practitioner's conclusion.
>
> The practitioner's conclusion is expressed in a form that conveys the practitioner's opinion on the outcome of the measurement or evaluation of the underlying subject matter against criteria.
>
> b. Limited assurance engagement—An assurance engagement in which the practitioner reduces engagement risk to a level that is acceptable in the circumstances of the engagement but where that risk is greater than for a reasonable assurance engagement as the basis for expressing a conclusion in a form that conveys whether, based on the procedures performed and evidence obtained, a matter(s) has come to the practitioner's attention to cause the practitioner to believe the subject matter information is materially misstated. The nature, timing, and extent of procedures performed in a limited assurance engagement is limited compared with that necessary in a reasonable assurance.

Further, it cannot be assumed that the audit profession per se has a "monopoly" on social audit engagements. Huggins, Green, and Simnett (2011) document the market shares in various social audit engagements and note the diverse nature of the assurance providers. In many of these social audit contexts, the audit profession per se may only have around 50 % of the engagements. The non-audit profession

---

[20] ISAE 3000 (Revised), Assurance Engagements Other than Audits or Reviews of Historical Financial Information International Framework for Assurance Engagements and Related Conforming Amendments, para 12. See also Hassan et al. (2005).

assurance engagements are typically performed by consulting services such as engineering, environmental, risk management services (Simnett, Vanstraelen & Chua, 2009). However, as argued by Simnett, Nugent, and Huggins (2009), the auditing profession is "well-placed" to provide high-quality assurance for a number of reasons, the predominant reason relating to the application of professional standards such as the ISAE quoted above. Simnett, Vanstraelen, and Chua (2009) also document the high reputational capital of the audit profession in enhancing the competency and quality of assurance services in engagements such as social audit.

Social audits differ from compliance audits because they are voluntary in nature. Where an external auditor is engaged to conduct a social audit, the organisation is the primary beneficiary of the social audit. Further, the organisation is also the preparer of the sustainability/CSR report. As in all audit work, the auditor is required to exercise professional judgment in social audits. This is pertinent in the case of social audits because of the absence of mandated auditing standards and guidelines on social audits. A foundational element in exercising professional judgment is the "mindset" of approaching matters objectively and independently, with inquiring and incisive minds (KPMG, 2011). "Mindset" is encompassed by "consultation" with peers, specialists, and other professionals to maintain consistent high judgment quality and enhancing professional skepticism (KPMG, 2011).

Before commencing a social audit, the auditor should determine the objectives of the social audit in the context of the beneficiaries: the organisation (primary beneficiary), regulators, and users of the report (McGladrey, 2012). The primary questions to consider when determining the objective of the social audit are: (1) who is/are the beneficiary/beneficiaries? and (2) What is the beneficiary's interest in the organisation's sustainability/CSR report? Clarifying the objectives of the audit is essential as it sets the framework for the auditor to consider all alternatives on how to conduct the social audit (KPMG, 2011).

We are a long way from promulgating professional standards for social audit. At the moment, the professional pronouncements or prescriptions are very general, in the nature of 'one size fit all' type guidance. For example, ICAS (2012) and KPMG (2011) offer the following advice: after considering all alternatives, gather and evaluate all relevant information. Next, assess the applicable accounting framework, standards; other literature, auditing standards and guidance are assessed. Further, undertake appropriate due process to assess and challenge the client's judgment. Finally, document the judgment, assessment, rationale, and challenge of the preparers' judgment are to be documented (ICAS, 2012; KPMG, 2011).

As in other types of audits, the materiality of qualitative and quantitative information needs to be assessed (ICAEW, 2010: p7). The Global Reporting Initiative's (2013) G4 Sustainability reporting guidelines highlight that "At the core of preparing a sustainability report is a focus on the process of identifying material Aspects—based, among other factors, on the Materiality Principle. Material Aspects are those that reflect the organization's significant economic, environmental and social impacts; or substantively influence the assessments and decisions of stakeholders" (p. 7).

Materiality in social audits falls under ISAE 3000 (Revised) *Assurance Engagements Other than Audits or Reviews of Historical Financial Information* paragraphs 67–70. Materiality of both qualitative and quantitative information should be considered in an audit (paragraph 69). However, materiality is considered only in relation to the subject matter information covered by the audit engagement (paragraph 70). The auditor is to exercise professional judgment when considering materiality (paragraphs 68 and 69) and is to consider the information needs of intended users (paragraph 67 and 68).

Morimoto, Ash, and Hope (2005) used the grounded theory approach in an interview study that aimed to measure CSR for audits. They identified the most significant factors in achieving successful CSR as good stakeholder management; good corporate leadership; greater priority for CSR at board level; the integration of CSR into corporate policy; regulation at national and international level; and the active involvement of and good coordination between government, business, non-governmental organisations, and civil society. The first four factors are especially important for organisations operating in a competitive environment as they can lead to more efficient business, greater share price, and long-term business success.

9 Is There a Framework of Legal Liability for Social Audits?

As discussed throughout this chapter, there are several bases for liability for auditors—either through the common law of negligence or contract (due care), or due to the framework that establishes the basis of engagement, either through standards or legislation; or indirectly through consumer type legislation.

In relation to the common law, the law of professional negligence is grounded as an economic tort, that is, damage for economic loss. Recovery of pure economic loss in tort has been the subject of some controversy and debate in the common law over the last couple of decades (Cane, 1991). The Australian/British common law requires a strong nexus between the subject of the duty of care in negligence and the damage suffered. This has been dealt with doctrinally by the recognition of a relational model of legal responsibility (Corbett, 1994). In more accessible terms, Cardozo CJ in *Ultramares Corp v Touche, Niven & Co*[21] classically stated this in terms of the "floodgate" principle—defendants could not be expected to incur indeterminate liability, as to amount, time and class of claimants. Hence, it is only when damage is suffered and identified are the courts able to assess damages for the loss (Corbett, 1994). The High Court, in the various separate judgments in

---

[21] (1931) 174 NE 441.

the *Esanda* case, emphasised that for third party liability to accrue to the auditor for negligent misstatement on the financial report audit, the third party needs to establish a special relationship directly with the auditor.[22]

If there is no common law liability accruing to the auditor as a result of negligently performed financial statement audit so far as third party investors are concerned, it is difficult to see how under the *Esanda* type formulation, the special relationship will arise under the social audit engagement. Further, the social type reporting matters will not be directly related to the financial investment, so identifying loss to assess damages, is problematic. Whether social reporting "releases value" in monetary terms is the subject of a vast and controversial literature (Gray, 2006).

Now to answer the direct question—what loss will the client company have suffered to have a cause of action in tort against the auditor who performs the social audit? Our speculation leads us to frame further questions, based in doctrine, but for which we currently have no direct answers:

- What damage will have been reasonably foreseeable by the auditor when the social audit engagement is accepted?
- Potentially a negligently performed social audit could expose the client firm to reputation loss or compliance cost—is this loss recoverable in tort?
- Given there are different consulting professionals involved in social audit, would the courts impose differing standards of care based on the profession to which the social auditor belongs?
- What role do the voluntary reporting standards play in setting the standard of care expected?

Conclusion

This chapter has outlined the development of financial report auditing standards and legal liability of auditors who perform these assurance engagements. Based on this experience, we have raised some speculative issues about potential legal liability of auditors who perform social audits. In relation to the framework of the standards for social reporting and assurance, we have established that the reporting and audit is primarily voluntary—hence such codes do not involve legal sanctions.

It is unlikely that the current common law doctrines of negligent misstatement will give rise to potential liability for third parties such as investors. In terms of direct liability to the client, we can only speculate that liability exists, but that there would be significant evidentiary barriers to the client company being able to prove loss. We do not have direct answers to these questions but we hope this chapter inspires further questioning and dialogue.

(continued)

---

[22] (1997) 188 CLR 241, 249–254 (Dawson J), 271–272 (McHugh J), 301 (Gummow J).

However, one final matter involving yet again further speculation—we have noted the unique "power" of s307A *Corporations Act 2001*, which effectively legislates mandatory compliance with the audit standards for financial audits. Given there are no mandatory standards for social audit, this provision may seem irrelevant. Sitting along-side the professional audit standards for financial statement audits is a framework of professional ethical standards. The ethical standards are not specific as to the types of assurance engagement they apply to. The extent to which the professional ethical standards, insofar as they provide guidance on matters, for example, of independence during engagements, are also considered "force of law" standards is certainly not settled in Australia.

Acknowledgements We thank colleagues Scott Hirst and YuYu Zhang for their helpful comments and support, and we are grateful to the anonymous reviewer for their insights that have helped focus the chapter's contribution.

References

Alexander, D., & Jermakowicz, E. (2006). True and fair view of the principles/rules debate. *Abacus, 42*, 132–164.
Anderson, H. (1996). A different solution to the auditors' liability dilemma. *Bond Law Review, 8*, Article 4.
Arens, A., Elder, R., & Beasley, M. (2002). *Auditing and Assurance Services* (9th ed.). New Jersey: Prentice-Hall.
Baker, C. R., Bedard, J., & Hauret, C. P. (2014). The regulation of statutory auditing: An institutional theory approach. *Managerial Auditing Journal, 29*, 371–394.
Bauer, R., & Fenn, D. H., Jr. (1972). *The corporate social audit*. New York: Russell Sage.
Birkett, W. P., & Walker, R. G. (1971). Response of the Australian accounting profession to failures in the 1960s. *Abacus, 7*, 97–136.
Boehme, T. C., & Braddock, L. A. (1965). The Accountancy Research Foundation. *The Australian Accountant*, 573–4.
Brown, R. G. (1962). Changing audit objectives and techniques. *The Accounting Review*, 696–703.
Cane, P. (1991). *Tort law and economic interests*. Oxford: Clarendon.
Carnegie, G., & O'Connell, B. (2012). Understanding responses of professional accounting bodies to crises: The case of the Australian profession in the 1960s. *Accounting, Auditing and Accountability Journal, 25*, 835–875.
Chapple, L. J., & Koh, B. (2007). Regulatory responses to auditor independence dilemmas–who takes the stronger line? *Australian Journal of Corporate Law, 21*, 1–21.
Chung, J., Farrar, J., Puri, P., & Thorne, L. (2010). Auditor liability to third parties after Sarbanes-Oxley: An international comparison of regulatory and legal reforms. *Journal of International Accounting, Auditing and Taxation, 19*, 66–78.
Clarke, F., & Dean, G. (2007). *Indecent disclosure: Gilding the corporate lily*. Cambridge: Cambridge University Press.
Clarke, F., Dean, G., & Oliver, K. (1997). *Corporate collapse*. Cambridge: Cambridge University Press.

Corbett, A. (1994). The rationale for the recovery of economic loss in negligence and the problem of auditors' liability. *Melbourne University Law Review, 19*, 814–867.
Fitzpatrick, L. (1939). The story of bookkeeping, accounting and auditing. *Accounting Digest*, 217
Fogarty, M., & Lansley, A. (2002). Sleepers awake! Future directions for auditing in Australia. *University of New South Wales Law Journal, 25*, 408–433.
Free, C. (1999). Limiting auditors' liability. *Bond Law Review, 11*, article 7.
Gibson, R., & Arnold, R. (1981). The development of auditing standards in Australia. *Accounting Historians Journal, 8*, 51–65.
Global Reporting Initiative. (2013). *G4 Sustainability reporting guidelines: Reporting principles and standard disclosures*. Amsterdam: Global Reporting Initiative.
Gray, R. (2006). Social, environmental and sustainability reporting and organisational value creation? Whose value? Whose creation? *Accounting, Auditing and Accountability Journal, 19*, 793–819.
Hassan, M., Maijoor, S., Mock, T., Roebuck, P., Simnett, R., & Vanstraelen, A. (2005). The different types of assurance services and levels of assurance provided. *International Journal of Auditing, 9*, 91–102.
Hecimovic, A., Martinov-Bennie, N., & Roebuck, P. (2009). The force of law: Australian auditing standards and their impact on the auditing profession. *Australian Accounting Review, 19*, 1–10.
Higson, A. (2003). *Developments in auditing and assurance in corporate financial reporting: Theory and practice*. London: Sage.
Huggins, A., Green, W., & Simnett, R. (2011). The competitive market for assurance engagements on Greenhouse Gas statements: Is there a role for assurers from the accounting profession? *Current Issues in Auditing, 5*, A1–A12.
ICAEW. (2010). *Sustainability assurance: Your choice (re: Assurance initiative)*. London: Institute of Chartered Accountants in England and Wales.
ICAS. (2012). *A professional judgment framework for financial reporting: An international guide for preparers, auditors, regulators and standard setters*. Edinburgh: Institute of Chartered Accountants of Scotland.
Junior, R., Best, P., & Cotter, J. (2014). Sustainability reporting and assurance: A historical analysis on a word-wide phenomenon. *Journal of Business Ethics, 120*, 1–11.
KPMG. (2011). *Elevating professional judgment in auditing and accounting: The KPMG professional judgment framework*. New Jersey: KPMG.
Lee, T. H., & Ali, A. M. (2008). The evolution of auditing: An analysis of the historical development. *Journal of Modern Accounting and Auditing, 4*, 1–8.
Littleton, A. C. (1933). *Accounting evolution to 1900*. New York: American Institute Publishing Co.
McGladrey, L. L. P. (2012). *Using professional judgment in auditing: McGladrey's framework*. Minneapolis: McGladrey.
Mock, T., Rao, S., & Srivastava, R. (2013). The development of worldwide sustainability reporting assurance. *Australian Accounting Review, 23*, 280–294.
Montgomery, R. H. (1913). *Auditing theory and practice*. New York: The Ronald Press Company.
Morimoto, R., Ash, J., & Hope, C. (2005). Corporate social responsibility audit: From theory to practice. *Journal of Business Ethics, 62*, 315–325.
Porter, B. (2007). Auditors' responsibilities with respect to corporate fraud—A controversial issue. In M. Sherer & S. Turley (Eds.), *Current issues in auditing*. London: Sage Publications Ltd.
Ramsay, I. (2001). Independence of Australian company auditors: Review of current Australian requirements and proposals for reform. *Report to the Minister for Financial Services and Regulation*.
Ramsay, I., & Austin, R. (online) *Ford's principles of corporate law*. Sydney: Lexis Nexis Australia.
Report, C. (1992). *The financial aspects of corporate governance*. London: The Committee on the Financial Aspects of Corporate Governance and Gee and Co. Ltd.

Samsonova, A. (2010). *Rethinking auditor liability: The case of the European Union's regulatory reform*. Paper presented at Asia Pacific Interdisciplinary research in Accounting (APIRA) conference, Sydney, Australia.

Simnett, R., Nugent, M., & Huggins, A. (2009). Developing an international assurance standard on Greenhouse Gas statements. *Accounting Horizons, 23*, 347–363.

Simnett, R., Vanstraelen, A., & Chua, W. F. (2009). Assurance on sustainability reports: An international comparison. *The Accounting Review, 84*, 937–967.

Tackett, J., Wolf, F., & Claypool, G. (2004). Sarbanes-Oxley and audit failure: A critical examination. *Managerial Auditing Journal, 19*, 340–350.

Fostering the Adoption of Environmental Management with the Help of Accounting: An Integrated Framework

A.D. Nuwan Gunarathne

1 Introduction

Corporate social accountability has been widely discussed in the recent past though it can be traced back to several centuries (Carroll & Shabana, 2010; Valor, 2005). The evident escalation of corporate power and influence together with recent corporate scandals call for more corporate accountability in regard to the social and environmental impact of organizations (Benn & Bolton, 2011). As a means of being accountable to a wider set of stakeholders, organizations pursue many environmental management strategies in addition to other strategies. The increasing attention of corporations to the environment has been triggered by many factors. Disastrous industrial accidents such as the Fukushima nuclear disaster caused by the recent Tsunami, the Bhopal gas leak, the reactor meltdown at the Chernobyl nuclear plant, the Exxon Valdez oil spill, among many others, have drawn increased media attention and public concern over their harmful impact on the environment as well on society. Climate change, nuclear waste, erosion, routine pollutants, and deforestation have become commonplace concerns while the scope and scale of environmental problems has expanded considerably over the past decades (Colby, 1991; Freeman, 2002; Xiaomei, 2004). This expansion has coincided with unprecedented growth in the scope and scale of human activities. Due to the changing business conditions that demand better environmental performance, corporations have employed many environmental management practices (Banerjee, 2001; Soonawalla, 2006; Sroufe, Montabon, Narasimhan, & Wang, 2003).

In parallel with these developments, there has been growing research on companies' environmental management practices (Delmas & Toffel, 2004). The extant literature has mostly focused on drivers of environmental management practices, development stages of environmental management strategies and the role of

---

A.D.N. Gunarathne (✉)
Department of Accounting, University of Sri Jayewardenepura, Colombo, Sri Lanka
e-mail: adnuwan@gmail.com; nuwan@sjp.ac.lk

(environmental) accounting in managing information, etc. Yet, there is limited guidance on and analysis of how accounting can be integrated with the environmental development stages when pursuing environmental management strategies. This chapter aims to provide an integrated framework that would facilitate the adoption of environmental management with the help of accounting in pursuit of corporate social accountability.

With a view to postulating this integrated framework, the rest of this chapter is organized as follows: The chapter begins with an explanation of the broad drivers of the corporate environmental management agenda in the context of corporate greening. The next section discusses various environmental management strategies and their development. The chapter next moves on to discuss the importance of stakeholder management and engagement in these actions while emphasizing the need for better information management. This is followed by a discussion of the practical approaches available for environmental management. Then the role of accounting, as a means of facilitating and sustaining these practices, is discussed in detail. The integrated framework for the adoption of environmental management with the help of accounting is presented in the next section. The chapter concludes with a practical case study that demonstrates some of aspects of the framework.

2 Greening of Corporations

In the long historical development of corporate greening, the 1980s mark a period that attempted to integrate economic and environmental objectives of corporations (Robbins, 2001). The firms that pursued proactive green markets saw environment as an opportunity while others embraced environmental and social concerns in their business operations. These phenomena collectively gave rise to the greening of corporations. Atmosphere, chemicals and waste, freshwater, land, oceans, biotechnology and bio diversity have been recognized as the key environmental challenges faced by the corporations when greening the organizations. In recent years, firms have increasingly recognized the strategic advantage of superior environmental performance when dealing with environmental challenges (Gouldson & Murphy, 1998; Mross & Rothenberg, 2006; Robbins, 2001). Many organizations now devote substantial time and resources to environmental management practices with the aim of ecologically contributing to sustainable development (Buysse & Verbeke, 2003). In this context, environmental management practices are defined as formal systems that integrate environmental procedures and processes for the training of personnel, for monitoring and controlling environmental impacts and for summarizing, integrating and reporting environmental performance (Sroufe et al., 2003, p. 24). The adoption of these environmental management practices can be identified in relation to specific drivers. The next section describes these drivers.

Various internal and external forces create pressure on corporations to pursue environmental practices (Delmas & Toffel, 2004; Mross & Rothenberg, 2006). Taking this view, researchers have grouped the drivers of environmental management practices broadly into external and internal. External factors include regulations, competitive forces and other stakeholder influences while internal factors include organizational context, design, learning, individual or managerial level, leadership values and managerial attitudes (Berry & Rondinelli, 1998; Mross & Rothenberg, 2006; Qi, Zeng, Tam, Yin, & Zou, 2013; Sroufe et al., 2003). As a crucial external driver of environmental management, environmental-related regulations have received much attention (Henriques & Sadorsky, 1996; Newton & Harte, 1997; Porter & van der Linde, 1995). According to Porter and van der Linde (1995), stringent environmental regulations are needed to introduce better environmental management practices. They suggest that (properly crafted) environmental regulations can direct attention to resource inefficiencies and potential technological improvements, improve corporate awareness through information gathering, reduce the uncertainty in investments, motivate innovation and progress, etc. Highlighting the limitations of volunteerism, Newton and Harte (1997) put forward a similar argument calling for state intervention. Similarly, Henriques and Sadorsky (1996) have empirically identified that government regulation is the single most important source of pressure on firms to consider environmental issues among other forces such as customer pressure, shareholder pressure, community pressure, etc.

The aforementioned forces create a need for corporations to respond to the environmental challenge (or opportunity) in different ways. According to Benn and Bolton (2011), early corporate environmental practices addressed the regulatory requirements, which fall into the command and control category. However, environmental management practices later shifted away from command and control laws to more market-based and voluntary approaches. Driven by these internal and external factors, organizations follow various environmental management practices. In an organization these practices evolve/develop over time. The next section of this chapter describes the models that can be used to analyze the development stages of corporate environmental management practices.

3 Corporate Environmental Management Development Stages

Sroufe et al. (2003), in an attempt to provide a framework for environmental management practices, categorize them into operational, tactical and strategic levels of a firm. In addition to this hierarchy level categorization, there are various models or frameworks which attempt to explain the corporate environmental management practices and their development stages. Among them, the environmental typology of Roome (1992) has received much attention (Buysse and Verbeke, 2003).

Roome (1992) suggests five strategic options available to business in shaping their response to the environmental challenge. These options are non-compliance, compliance, compliance-plus, leading edge, and excellence. The first three of these options are set against the standards of compliance with legal requirements and social pressures. Non-compliance is an option arising from excessive cost, existing liabilities or even management inertia when faced with environmental challenges. However, it characterizes the lack of long term vision for a company and will not be sustainable due to the demands placed by legislation or social pressures. Organizations in the compliance stage will follow environmental management programs in a reactive response to environmental legislation. Hence it is a legislation-push strategy. However, the organizations in the compliance stage will face problems as legislation is more often reactive and deal with one problem at a time. Thus, legislation lags significantly behind the contemporary environmental agenda. These organizations will not therefore achieve any competitive advantage through their environmental management programs. The compliance strategies are based on management techniques and technologies required by legislation. These areas may include waste, pollutants, health and safety and waste water. However, the environmental impact of many organizations is not subject to environmental legislation and, even if it is, legislation will not demand changes in organizational structures and systems. When organizations move to a compliance-plus strategy the organizational response to environmental demand also transforms from reactive to proactive. The compliance-plus organizations will go beyond the current legal requirements and adopt environmental management programs based on a management-pull strategy. Unlike the compliance strategy, the compliance-plus strategy will demand changes in the organizational structures and systems.

The organizations in the next stage of development, i.e., the excellent companies, believe that environmental management is good management. These companies will follow core corporate and managerial values to achieve quality by managing their environmental impact and changing their conventional business concerns. This strategy also highlights a management-pull strategy. Excellent organizations will adopt clean technology techniques while changing their organizational structures and systems. When organizations move beyond excellence and reach leading edge status, they will set the standard for other businesses. Thus, leading edge represents a variant of excellent companies. According to Bhargava and Welford (1996), when organizations pursue/adopt excellent and leading edge strategies they could gain competitive advantage. When one carefully and broadly analyzes Roome's environmental management typologies, these strategies finally boil down to three typologies, i.e, compliance, compliance-plus and excellence.

Providing a somewhat similar analysis, Sakai (2007) suggests three sustainable environmental management views: (a) environmental correspondence, (b) environmental conservation and (c) sustainable environmental management. Organizations in the environmental correspondence stage passively correspond to external environmental pressures while those in the environmental conservation stage attempt to reduce the environmental impact of business activities with noble intent. The organizations in the sustainable environmental management stage actively reduce the environmental impact of a business while creating economic value as a business entity.

Corporate environmental management strategies have also been analyzed according to the natural resource-based view. Accordingly, Rugman and Verbeke (1998) and Buysse and Verbeke (2003) suggest that if the corporate strategy is supported by firm-level competencies (such as physical assets, employee skills, and organizational processes) sustainable competitive advantage can be gained. Taking this view, Hart (1995) suggests three types of resource-based environmental approaches: (1) pollution prevention, (2) product stewardship, and (3) sustainable development. Further, Buysse and Verbeke (2003) suggest a similar classification building upon Hart's (1995) resource-based framework. They suggest three dominant environmental management strategies: (a) reactive, (b) pollution prevention, and (c) environmental leadership. Buysse and Verbeke identify that many firms have already shifted from a reactive to a pollution prevention strategy while only a minority have adopted an environmental leadership strategy, of which many are MNE affiliates.

4 Support and Involvement of Stakeholders and Information Management

Irrespective of the level—early or advanced—of environmental management development, it is necessary to get the commitment of various stakeholders. Buysse and Verbeke (2003) are of the view that effective environmental management requires the identification of important stakeholders. It is therefore necessary to discuss briefly the stakeholder analysis of a firm. It was Freeman's (1984) landmark work that provided a solid and lasting foundation for many models, frameworks, and theories based on stakeholders although older references to the same concept have been found (Clarkson, 1995; Phillips, Berman, Elms, & Johnson-Cramer, 2010; Preston, 1990; Valor, 2005). In this regard, Clarkson (1995) distinguishes between primary and secondary stakeholders. Primary stakeholders are those without whose continuing participation the corporation cannot survive as a going concern. Thus there is a high level of interdependence between an organization and its primary stakeholders. Primary stakeholders include shareholders and investors, employees, customers, and suppliers *and* public stakeholders such as governments and communities. Secondary stakeholders are the stakeholders who influence or affect, or are influenced or affected by, the corporation, but are not engaged in transactions with the corporation. Therefore secondary stakeholders are not essential for the survival of an organization. They include the media and a wide range of special interest groups. However, they have the capacity to mobilize public opinion in favor of or against a corporation. Linking stakeholder analysis to environmental management strategies, Buysse and Verbeke (2003) identify that many companies attach the highest importance to regulators, especially firms with a pollution prevention strategy. Also, they highlight the importance of environmental stakeholder management to the development of green competencies. The green competencies should be focused on making investments in green products and manufacturing technologies, employees, organizational competencies, management systems and procedures.

Some of the extant literature also highlights the importance of information management to improve corporate accountability when dealing with a wide array of stakeholders. Applications such as eco-control have highlighted the importance of information management for better stakeholder management (Schaltegger & Burritt, 2000) when carrying out a corporate environmental strategy. As a specific application of management control systems, eco-control helps organizations to measure, control and disclose their environmental performance by supplying information for decision-making to ensure the attainment of environmental objectives and to provide persuasive evidence in support of the benefits of such actions (Henri & Journeault, 2010). It is a function of the management information system to provide relevant and reliable environmental information for key decision-makers to make better decisions (Ilinitch, Soderstrom, & Thomas, 1998; Wilmshurst & Frost, 2001).

5 Pragmatic Guidelines for Corporate Environmental Management

Along with the growth of literature on the development stages of corporate environmental management and stakeholder management, there is a growing body of guidelines and standards to help managers. These guidelines have taken a pragmatic perspective in improving corporate social and environmental responsibilities and performance (Epstein & Roy, 2003). The guidelines vary widely both in terms of focus and aim. Having analyzed these wide sets of guidelines, Epstein and Roy (2003) suggested nine principles of sustainability which also include protection of the environment. Yet, most these guidelines do not provide a comprehensive and step-by-step approach to the implementation of environmental management strategies. The guidelines offered by Doody (2010) and Certified Management Accountant (CMA) (1995) attempts to fill this need by providing a comprehensive approach to the implementation of environmental management.

Implementing Corporate Environmental Strategies, issued by CMA (1995), provides practical operating principles to implement a corporate environmental strategy. Hence these guiding principles are used to improve environmental performance and to integrate environmental considerations into management decisions as highlighted by Roome (1992). It should also be noted that CMA, being an international management accounting body, attempts to give prominence to the role of the management accountant through these guidelines. Accordingly, CMA emphasizes that the role of a management accountant is integral to planning (development of environmental strategy, policy, objectives and environmental measurements), analysis, and control. Understanding the interdisciplinary nature of environmental management strategies, CMA also stresses the role that a management accountant should be closely associated with other multi-disciplinary groups. CMA also suggests several essential elements to be included when designing an effective environmental strategy. These elements are grouped into three stages, stage one: managing regulatory compliance, stage two: achieving competitive advantage, and stage three: completing environmental integration. According to CMA, companies will move from one stage to the next but in many instances organizations will straddle the boundaries between these stages.

The organizations in stage one develop environmental management measures by acknowledging the financial implications of environmental matters and realizing the risks associated with current practices. Showcasing a similar pattern to what was discussed in the previous models, at this stage the development of environmental management practices is a reactive response to both external and internal pressures. These organizations will adopt measures such as obtaining top management commitment, *and* developing an environmental policy, action program, management system and audit program. In stage two, the focus of environmental management extends beyond simply complying with regulations to improving resource efficiency and profits. These organizations realize that implementing proactive environmental programs can enhance their corporate image, which will in turn increase their market share. In order to derive competitive advantage, organizations should develop an external environmental reporting strategy, design environmentally sensitive products/processes and integrate environmental information into the decision-making process. Thus, these organizations move from cost minimization to cost avoidance through life-cycle management and design for the environment. When the organizations further develop these practices, they will fully integrate the environmental strategy throughout the organization and into all management decisions. The stage three organizations will often introduce performance evaluation systems based on environmental impact considerations, generate revenue generation through environmentally oriented products and services, and follow the principles of sustainable development. These organizations have fully integrated environmental issues into everyone's day-to-day decision-making process since they recognize that long-term economic growth is that of environmental sustainability.

As highlighted in this section and the previous sections, information management is important in implementing environmental management strategies. The next section attempts to provide an overview of information management with the use of accounting in pursuing environmental management strategies.

6 Supporting Tools and Techniques for Environmental Management

As the stakeholders pay a high level of attention to the environmental performance of corporations, environmental information and measurement issues are becoming increasingly important (Ilinitch et al., 1998). Therefore organizations should use effective tools and techniques to support environmental strategies (Sroufe et al., 2003). Further, without greener accounting tools and techniques many environmental initiatives will not succeed (Gray, Bebbington, & Walters, 1993). Hence, in parallel to the increasing interest on the environment, the interest in accounting for the environment has increased (Burnett & Hansen, 2008). Wilmshurst and Frost (2001) suggest that accounting and the accountants can play a significant role in effectively implementing environmental management practices. In order to successfully implement the environmental management strategies it is necessary to bring the traditional functions of accounting to the environmental management process. Hence accounting skills such as measuring, recording, monitoring and verifying data become increasingly important (Gibson & Martin, 2004; Wilmshurst & Frost, 2001). In this context, environmental management accounting (EMA), an accountant's response to environmental challenges, can play a significant role in facilitating the sustenance of an integrated approach to environmental management.

6.1 Environmental Management Accounting

EMA is the identification, collection, analysis and use of physical information on the use, flows and destinies of energy, water and materials (including wastes) and monetary information on environment-related costs, earnings and savings for decision makers (United Nations Division for Sustainable Development (UNDSD), 2001; Burritt, Hahn, & Schaltegger, 2002). It provides physical and monetary information regarding various environmental aspects for internal as well as external decision- makers. This leads to two types of EMA systems: physical EMA (PEMA) and monetary EMA (MEMA) systems. The EMA information provided by these two types of systems may cover three dimensions, i.e., time frame (past, current or future), length of time (short-term vs. long-term) and frequency (ad hoc vs. routine). Accordingly, Burritt et al. (2002) have suggested a comprehensive framework for EMA (refer Burritt et al. (2002) for more details on the framework). According to the framework, EMA encapsulates a wide array of accounting tools and techniques used for internal decision making such as accounting for energy, material flow cost accounting (MFCA), environmental capital budgeting, life cycle analysis, etc. Therefore, EMA is not an environmental management tool among others, but a broad set of principles and approaches that provide information for the successful implementation of environmental strategies (International Federation of Accountants (IFAC), 2005). It acts as an interface between inward focused management accounting and environmental management strategies (Bennett, Bouma, & Walters, 2002). It should also be noted that the same information can be used to report to external stakeholders as well (Schaltegger & Burritt, 2006). Hence, EMA has an external information supply potential to facilitate corporate accountability towards a wider set of stakeholders. As described previously, EMA information is used for internal as well as for external reporting purposes (Deegan, 2003; IFAC, 2005). Due to the importance of the environment in every management activity, EMA is also becoming important for all types of management activities (Gibson & Martin, 2004).

6.2 EMA Techniques

There are many EMA tools and techniques that are continuously developing. Some of them are extensions to or adaptations of conventional management accounting tools and techniques while others are newly developed. In terms of sophistication, these tools and techniques range from simple to advanced. In this chapter, some, yet important, techniques are presented briefly. They are:

- Accounting for energy, materials, water and waste
- Material flow cost accounting
- Environmental capital budgeting
- Life cycle accounting
- Environmental activity-based costing
- Sustainability balanced scorecard

Please refer Appendix for more details on these techniques.

6.3 EMA Benefits and Challenges

When an organization follows EMA, there can be numerous uses and benefits, but they can be identified in relation to three broad areas. They are: ensuring compliance, supporting eco-efficiency and strengthening strategic position (Federal Environmental Agency (UBA), 2003; IFAC, 2005; Doody, 2010). These broad benefits are not mutually exclusive but interdependent. For example, suppose that a hotel installs a waste water treatment plant mainly to comply with environmental law. This is because it cannot simply discharge the waste water to the drainage system or to the nearby environment as per the law. The hotel uses the treated water of the plant for gardening purposes. Re-use of treated water saves water purchased from the municipal council giving economic benefits too. Further, the hotel can use its water purification plant and processes as a marketing tool to generate a favorable public image. It can be portrayed as a green hotel in the eyes of its stakeholders which could strengthen the strategic position. (Refer the practical case study in this chapter for more details).

However, there are many challenges that could reduce the benefits of EMA. The available literature highlights the limitations of conventional management accounting practices as the main impediment to better adoption of EMA (Burritt, 2004; Gray et al., 1993; IFAC, 2005). The assumption of immateriality of environmental costs, lumping of environmental costs with general overheads, too narrow and short term performance appraisal techniques, exclusion of external considerations in investment appraisal, lack of focus on articulation of flow and stock, absence of accounting for externalities and social issues and dominance of financial accounting are some of the problems with conventional management accounting (Burritt, 2004; Fonseka, Manawaduge, & Senarathne, 2005; IFAC, 2005). In addition, underdeveloped communication links between accounting and other functions that collect environmental related information have also been suggested as another barrier (IFAC, 2005).

Having considered the EMA challenges and development stages of environmental management, the next section of this chapter provides the integrated framework that attempts to combine EMA and environmental management.

7 Integrated Framework for the Adoption of Environmental Management

7.1 The Need for an Integrated Framework

The existing literature on development stages of environmental management has so far failed to recognize the importance of environmental information management in propelling an organization to higher levels of development. On the other hand, the challenges to EMA are mainly caused and compounded by the lack of attention paid to the development stages of environmental management. EMA, the so called supporting tool for environmental management, has been developed and discussed in isolation without positioning it within the broad context of different conditions and requirements of environmental management development stages. Therefore the need for an integrated framework for the adoption of environmental management can be emphasized for two reasons: (a) the failure of the existing literature on the development of environmental management stages to recognize the role of accounting in the broader context of information management, and (b) the failure of the existing body of knowledge on EMA to develop EMA along with the environmental management stages. The repercussions of the lack of integration of these two aspects are evident. Many researchers around the world have revealed that the existing corporate environmental management programs and EMA practices have not been systematically and comprehensively implemented internally (Bartolomeo et al., 2000; Gunarathne & Lee, 2015; Lee, 2011). Thus, these existing practices are largely fragmented and have been developed from time to time as a response to various internal and external factors. It is necessary for an organization to follow a structured approach to environmental management to generate competitive advantage. This requires an integrated adoption of EMA to support the environmental strategy of the firm (Godschalk, 2010). However, a systematic and comprehensive adoption of environmental management that fully integrates EMA has to evolve from its own experience after passing some of the early stages of implementation. This chapter aims to postulate a framework that guides companies to continuously develop and systematically adopt environmental strategies with the support of EMA practices over time to generate competitive advantage. The suggested framework is described in below.

7.2 The Integrated Framework

The various analytical models suggested by Roome (1992), Hart (1995), Buysse and Verbeke (2003), Sakai (2007) and practical implementation guidelines such as CMA (1995) finally hold similar views regarding the development stages of environmental management practices. Drawing from the available rich literature this framework also adopts a similar viewpoint suggesting three development stages of environmental management -compliance, conservation, and leading edge. All organizations will be initially compelled to be more environmentally sensitive by an internal or external force/s (compliance stage). Then these organizations will soon realize the conservation (cost saving) potential of these compelling actions and become more active in furthering these initiatives (conservation stage). They are finally propelled into a stage in which these practices become a part of their business through which they enjoy superior environmental and economic performance (leading edge stage).

The integrated framework is intended to be used at the organization level as it provides a micro level analysis of the main environmental management development strategies. The framework captures the development/movement of three key aspects of environmental management according to the development stages to demonstrate corporate social accountability (refer Fig. 1). These three aspects are:

- *Coverage of environmental domains/challenges*
  The environmental domains reflect the subject matter of the environmental management action programs. Electricity, water, solid waste, waste water, emissions, pollution, bio diversity, etc are some of the environmental domains which can also be regarded as areas from which environmental challenges stem.
- *Support and involvement of stakeholders*
  Stakeholder support represents a key success driver of environmental management strategies. Irrespective of whether the stakeholders are internal or external their support or sometimes active involvement will decide the degree of success of environmental management actions.

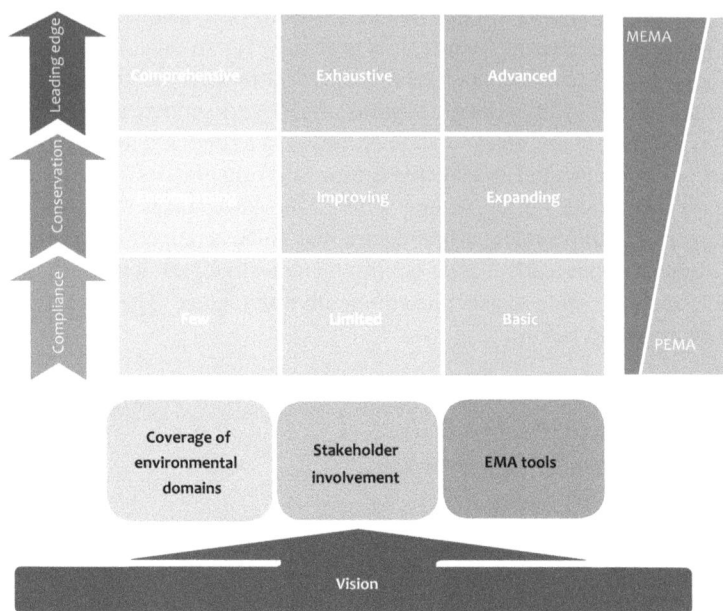

Fig. 1 An integrated framework for the adoption of environmental management

- *EMA techniques being used*

  EMA techniques reflect information provision and management to a wider set of stakeholders. As mentioned previously EMA will be the thread that connects environmental strategies with stakeholders during the different stages of environmental management development.

  The current framework is however not limited to the existing three-stage development models of environmental management. The arrow at the leading edge stage implies the potential for further development as and when the environmental challenges change. Crane and Matten (2007) also emphasize the need for continuous improvement in environmental management, referring specifically to broader sustainability as "the long-term maintenance of systems according to environmental, economic and social considerations (page, 23)". Hence, it should be noted that the leading edge stage is not the most desired ultimate stage of development of environmental management but a desirable status in the current context. However, when the external and internal environment factors change, these practices will have to be adjusted and/or improved to withstand such challenges on an ongoing basis.

  The framework provides an analytical tool for the three key aspects of the adoption of EMA along with the development stages. During the compliance stage, the coverage of environmental domains will be very few and limited to a single, but most important, domain such as energy, waste or even materials. These domains will be determined by the internal or external factor that triggered the compliance stage. Since the environmental domains covered are few and environmental awareness is lacking, an organization will seek the support of key primary stakeholders initially. These key stakeholders can be, mostly, the employees. During this stage, organizations will adopt simple EMA tools such as accounting for energy, water or waste in order to facilitate its limited coverage of environmental strategies. During this stage, EMA system will often provide PEMA information and some monetary values as well.

In the next stage of the development of environmental management, i.e., the conservation stage, an organization will gradually expand its coverage to more environmental domains. For example, if the initial focus has been on energy now the focus will be expanded to incorporate materials, waste or emissions. However, to successfully implement these strategies with an encompassing coverage it is necessary to get the support of a range of stakeholders such as suppliers, distributors, etc. As the coverage of environmental domains is wider, it is necessary to experiment to adopt sophisticated and advanced EMA tools such as ABC, environmental capital budgeting, etc. In this stage, most of the PEMA information will be assigned monetary values. It is through this assignment of monetary values that an organization realizes the resource saving and cost saving potential of environmental strategies pursued. When an organization successfully implements this stage and derives benefits therefrom, it will move on to the next level, i.e., the integration stage. Hence, this stage is only transitional until the next stage is reached, provided the actions taken are fruitful.

In the final stage, the leading edge stage, an organization will have a comprehensive coverage of environmental domains. Hence, its coverage of environmental activities will usually include all aspects that are important for the organization such as pollutants, emissions, bio diversity, etc. which are contained in its environmental policy. In order to implement a wide array of environmental practices, an organization needs the support of all the stakeholders in its value chain such as community, shareholders, upstream suppliers and downstream distributors. Further, the organization will seek to receive the support of secondary stakeholders such as the media and environmental pressure groups. In order to foster the support of various stakeholders the organization will follow internal as well as external environmental reporting systems, which are facilitated through the provision of EMA information. Moreover, in order to foster these various environmental management strategies more advanced EMA tools such as lifecycle analysis and MFCA, will be adopted. When an organization reaches this stage, due to the advanced nature of its EMA system, most of the EMA information will be monetary.

The framework also shows that it is the vision that lays the foundation and drives the environmental management and EMA practices of an organization. Without a clear vision an organization will not progress to the leading edge stage. Lack of a clear vision may dilute the intensity of environmental management and EMA practices when the criticality of the trigger that caused the compliance stage is weakened.

7.3 Contribution of the Framework

The main contribution of the framework is the integration of the current fragmentary discussions on the development of environmental management and EMA. Lack of proper integration of these two strands has posed challenges regarding the practicability and sustainability of environmental management in a dynamic business environment. In addition to this main contribution, the framework mainly provides practical as well as theoretical benefits. From a pragmatic perspective the framework provides many benefits for a practitioner.

- The framework enables practitioners to map where their organization currently stands in terms of the development stages of EMA.

It should be kept in mind that identifying a perfect mapping of the three environmental aspects of an organization according to the development stages is highly unlikely. An organization may display some characteristics at different stages. For example, an organization may have an encompassing coverage of environmental domains but with still limited support from its stakeholders. Hence, the framework emphasizes that what is important is not a perfect mapping of an organization's current status but the identification of lags for future consideration.

- The framework also directs practitioners to understand what the future focus should be in the key aspects of environmental management. Consequently, an organization will be able to move fast to the leading edge stage to derive the full potential of its environmental management strategies.
- In the current dynamic business environment the internal and external factors may change fast posing new environmental challenges and opportunities for organizations. The framework provides insights for the practitioners to develop, or sometimes to retain, an integrated approach to environmental management adoption when environmental challenges change.

From a theoretical perspective, the framework provides a tool for a researcher to analyze the development stages of an organization's environmental management strategies along with the supportive accounting tools (i.e. EMA). Hence, the framework extends the current discussions on the development of environmental management by incorporating vital accounting aspects as a means of pursuing corporate accountability for a wider set of stakeholders.

8 Practical Case Study

Having explained the integrated framework for environmental management adoption, the last part of this chapter uses a case study to demonstrate how the elements described here can be found in a real life situation.

8.1 Scenic Hotel

8.1.1 Background and Drivers of Change

Scenic Hotel belongs to a large hotel chain in the Asian Pacific region. The hotel was performing reasonably well until it faced a critical situation in 2004, which was triggered by the Tsunami that devastated the coastal lines of most of the Asian Pacific countries. Although the hotel was not damaged by the Tsunami, tourist arrivals to the region fell drastically. The hotel experienced a very low level of occupancy amidst increasing operational expenditure. Depressed share prices and the inability to claim dividends frustrated the investors in the hotel. The parent company granted full autonomy for the management of Scenic Hotel to find avenues to reduce its operational costs. In an urgent cost-saving bid, the hotel's management with the help of the accountant carried out a detailed cost analysis. The analysis revealed that the highest cost was energy which accounted for nearly 40 % of the total cost of operations. The next major contributors to costs were materials costs (food, chemicals, etc) and the cost of water. Although the labor cost was significant nothing was possible as the management had promised the hotels' trade unions that no lay off would take place. Also the management did not want to demoralize its employees by reducing the head count.

8.1.2 Internal Compliance Driving Environmental Management

With the primary motive of saving costs to ensure survival, the hotel started devising strategies that were aimed at managing the significant costs identified in the cost analysis. Energy was initially targeted as it offered the greatest and most needed cost savings potential. With a view to saving energy, the hotel introduced solar water heating panels, installed card key switching of room air conditioning, scheduled light switching, colour coded all light switches, and replaced incandescent lights with compact fluorescent lamps (CFL). In order to successfully implement these energy conservation practices, Scenic Hotel sought the support of its key stakeholders -its employees. In this regard, the hotel conducted many awareness programme aimed at employees on the criticality and potential for energy savings. As job security was at risk these energy saving initiatives received overwhelming support from the employees in all departments. The accountant together with the engineer started calculating the energy savings, initially in physical units, which were shared with the employees at regular meetings. The energy savings were identified in terms of energy units by comparing the energy consumption data before and after the initiatives. These accounting aspects reflect the application of simple EMA techniques such as accounting for energy in physical terms. Most of these actions were simple and did not require any significant capital expenditure and represented a bottom-up approach in which most ideas came from the lower level employees. In fact, the hotel was not in a position to make any capital intensive projects due to its financial crisis.

8.1.3 Realization of the Conservation Potential Driving Environmental Management

The cost savings realized by these simple, yet effective, actions were more than anticipated. Encouraged by these savings, the hotel then started to focus on other areas such as water and waste. Many water saving measures were taken such as a sewerage treatment plant, re-use of treated water for garden irrigation, introduction of water-saving cisterns and optional re-use of room linen. In order to minimize waste the hotel started grading garbage, recycling and reusing materials, composting of garden refuse, and reducing the use of environmentally damaging materials. These actions required the support of guests and suppliers in addition to the support already received from employees. To ensure the support of guests the hotel started putting up notices and conducting awareness programme. Most of the guests responded positively and contributed to these initiatives. Similarly, action was taken to educate the suppliers as well. Along with these initiatives, the hotel calculated the water savings and savings of waste. These accounting aspects such as accounting for water and accounting for waste were developed in order to support these initiatives. The accountant then gradually started to assign a monetary value to these savings to convince and encourage the employees and even the management. This expansion of focus of EMA information represents a movement from simple physical EMA tools to somewhat advanced EMA tools that incorporate physical as well as monetary aspects. Some of the savings information was shared with the guests also.

In order to reduce the cost of materials, especially food cost, the hotel started to cultivate vegetable and fruits in its own gardens. Scenic Hotel did not use any chemical fertilizer but used only the compost made from the sewerage treatment plant. In addition to reducing costs, these practices received a lot of attention and were much loved by the guests who started to pick their own vegetables and fruits during their stay.

8.1.4 Integration of Environmental Management to Move on to the Leading Edge Stage

By this time the tourism sector of the country was picking up and tourist arrivals were rising. The hotel was gradually recovering from its financial crisis. However, due to the increased investor confidence in the tourism sector, there were many hotels that were competing in the selected target markets of Scenic Hotel. The hotel, after realizing the potential of the actions, started to revisit these initiatives with a view to developing them. The management soon realized that all these areas relate to the environment and can be used as a competitive tool to generate a greener image for the company. The hotel then devised a comprehensive environmental management policy that encompassed the already established areas such as water, energy and waste as well as new areas such as carbon footprint, pollution, bio diversity, etc. By this period the accounting team of the hotel had developed an advanced toolkit to support these initiatives. Most of these EMA tools had evolved through their own experiments with accounting tools that were developed in the latter two stages. However, in order to view the hotel as a *"mass balance"* and thereby to identify losses (non-product output as discussed in MFCA) it obtained the help of an external consultant. Moreover, it got his support to calculate the carbon footprint of its operations. Especially the advanced EMA tools such as MFCA and carbon footprint calculations enabled the hotel to take a comprehensive look at the site's material flows and to focus attention on areas where carbon savings became possible.

In order to further reduce energy costs and reduce the carbon foot print, the hotel installed a bio mass boiler after carefully carrying out a comprehensive environmental capital budgeting exercise. The firewood for the boiler was purchased from the nearby villagers providing them with a stable livelihood. This initiative ensured their support for the hotel too. The hotel started to conduct awareness and training programs on how to grow and harvest the required firewood. In addition, the hotel maximized the use of indigenous flora in landscaping and eradicated invasive alien species. Most of the flora, fauna and vegetables were supplied by the villagers who became regular suppliers to the hotel. Along with these initiatives the least discussed accounting aspects such as accounting for bio diversity were developed and fully integrated into the corporate information management systems. Moreover, the hotel conducted awareness programs in schools and villages to educate the community about the importance of saving the environment. Some of the waste generated by the hotel was given away to the villagers or to the regular buyers who then use them to produce some ornamental items or for re-use responsibly. During this time the hotel started to use a strategic scorecard that encapsulates the performance of aspects such as financial, environmental, guests (customer), employee, social, etc. This reflects the application of a modified sustainability balanced scorecard to suit the context of the hotel.

Over the years, the green practices followed by the hotel have been recognized by many local and international institutes. Scenic Hotel won many local and international green awards which received a lot of local and foreign media attention. The hotel started sharing its successful green strategies with other industry partners while promoting them in many ways. The hotel started to promote its success story in the media and among environmental groups while reporting the cost savings in its sustainability report. All these actions improved its public image considerably in the local as well as international markets. Today many guests visit the hotel as they love these green practices. In effect, the environmental management practices adopted by the hotel have given it a competitive edge. These environmental management practices of the hotel have been well interwoven with EMA practices such as accounting for material, energy and waste, life cycle design, material flow cost accounting, environmental capital budgeting, etc. Despite the financial recovery, the environmental management practices and supporting accounting tools have become routine and embedded into the daily decision-making process of the hotel. Moreover, all the practices have been implemented with consistent commitment and vision with the support of its stakeholders.

> **Conclusion**
> With a view to furthering the existing discussions on how organizations can meet the environmental challenge, the chapter provided an integrated framework for the adoption of environmental management. In doing so, the framework incorporated an accounting dimension (EMA) along with the development stages of environmental management while highlighting the changes in three key aspects. The framework demonstrates that the development of environmental management is gradual and is initially driven by internal or external compliance. Later, once the business case of environmental management is realized an organization will move forward to leading edge status by incorporating environmental consciousness into its daily decision making. To reach this stage, environmental management strategies, which encompass all the significant environmental domains, should be driven by a clear vision cascading from top to bottom. Furthermore, the continuous engagement of all the primary and secondary stakeholders on a regular basis is warranted. In this process, accounting for environmental management strategies (EMA) will act as the common thread that connects and sustains these practices while ensuring corporate accountability for its stakeholders. Hence, the integrated framework will provide useful insights for organizations to demonstrate corporate social accountability through sustained environmental management strategies with the help of accounting.

Appendix: EMA Techniques

Accounting for Energy, Materials, Water and Waste

According to Bennett and James (1998) accounting for energy and materials is the tracking and analysis of all flows of energy and substances into, through and out of an organization. When organizations realize the importance of energy costs, they can follow a piecemeal approach (in-house initiatives) or a comprehensive approach (top-down approach) or even a combination of them (Gray et al., 1993). Due to the wide range of approaches possible, there is no single hard and fast rule for accounting for energy, but any accounting system should attempt to separately identify different types of energies used, relate these costs to the causes of costs, highlight the energy costs in cost reports, etc. The same is applicable when accounting for materials, water and waste.

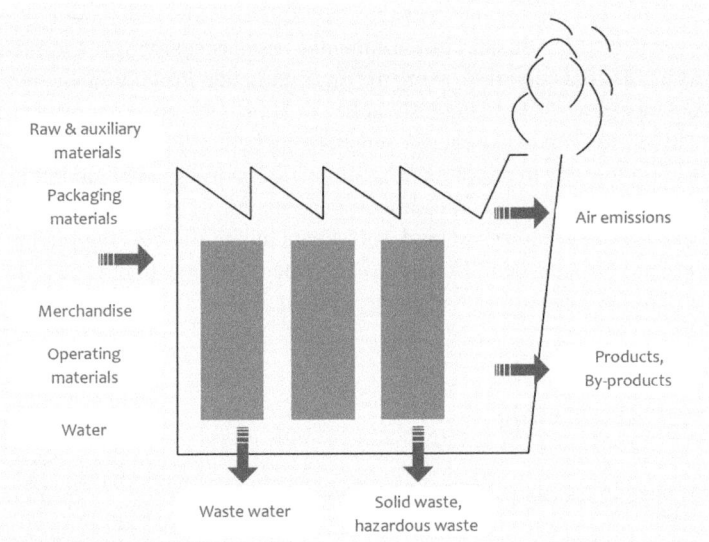

Fig. 2 Mass balance of an organization

Material Flow Cost Accounting

MFCA is a tool for quantifying the flows and stocks of materials in processes or production lines in both physical and monetary units (Kokubu & Kitada, 2012; Strobel & Redmann, 2002). MFCA has been developed based on the principles of mass balance. Accordingly, mass balance implies that the amount of inputs should be consistent with the sum of desirable and no-desirable output (refer Fig. 2). MFCA incorporates both PEMA and MEMA by quantifying material flows and stocks in a process or processes in terms of both physical and monetary units.

In MFCA waste is valued at the same rate as the good output, it thereby brings the cost of waste to the attention of the management immediately for requisite action. The benefits of MFCA are evident from both economic and environmental perspectives. From an economic perspective, MFCA identifies the material in both physical units and in monetary units along with their progress through an organization. From an environmental perspective, the reduction of the consumption of materials and energy reduces the undesirable waste outflows from an organization.

Environmental Capital Budgeting

Capital budgeting is the process of making long-term decisions that involve cash flows beyond the current year (Hilton, Ramesh, & Jaydev, 2008). When environmental considerations are taken into account explicitly in the long-term decision making process, environmental oriented capital budgeting takes place. It is therefore necessary to fully consider environmental costs, cost savings, and revenues in evaluating a potential capital investment (Environmental Protection Agency (EPA), 1995). As Gray et al. (1993) suggest, as the world is becoming more environmentally sensitive, non-environmentally sensitive income streams will be difficult to obtain, leading to early abandonment of projects. Environmental capital budgeting offers financing as well as investment benefits to an organization. Financing benefits such as easy approval and soft financing terms for capital investment projects and investment benefits such as informed decision making are the results of such environmental capital budgeting techniques.

Life Cycle Accounting

Life-cycle costing estimates and accumulates costs over a product's entire life cycle (Drury, 2004). According to EPA (1995), life-cycle accounting assigns and analyzes the product or project-specific costs within a life-cycle framework including usual, hidden, liability, and less tangible costs. However, it will be difficult to conduct a comprehensive life cycle assessment of products or projects (Gray et al., 1993). Yet, accountants and other professionals can contribute to life cycle assessment by bringing in financial implications of existing activities and potential future options.

Environmental Activity Based Costing

Activity-based costing is a two-stage procedure used to assign overhead costs to various cost objects accurately (Hilton et al., 2008; Kaplan & Atkinson, 1998). Despite the fast diffusion of ABC, many companies around the world still use traditional volume-based overhead allocations systems (Fonseka et al., 2005). In a traditional overhead absorption costing system, environmental related costs can be hidden (Burritt, 2004; Gibson & Martin, 2004; IFAC, 2005). It is necessary to bring environmental costs to the attention of corporate stakeholders (EPA, 1995). This necessitates the allocation of environmental costs to the appropriate accounts by allocating them to those who generate them (Soonawalla, 2006).

Sustainability Balanced Scorecard (SBSC)

Balanced Score Card (BSC) has been promoted as a balanced performance measurement system that overcomes the limitations in conventional performance management. BSC has also been suggested as a strategic management tool as well

(Kaplan & Norton, 1992, 1993, Niven, 2002). BSC encompasses four perspectives, namely, financial, customer, internal business and learning and growth. As these perspectives act as only a template, BSC has been suggested as an effective tool to incorporate economic and environmental (and social) dimensions (Dias-Sardinha, Reijnders, & Antunes, 2002, Epstein & Wisner, 2001, Figge, Hahn, Schaltegger, & Wagner, 2002, Moller & Schaltegger, 2005). The environmental (and social) integration into a conventional BSC to make it a sustainable BSC can be achieved in different ways (refer Figge et al., 2002 for more information).

References

Banerjee, S. B. (2001). Managerial perceptions of corporate environmentalism: Interpretations from industry and strategic implications for organizations. *Journal of Management Studies, 38* (4), 489–513.
Bartolomeo, M., Bennett, M., Bouma, J., Heydkamp, P., James, P., & Wolters, T. (2000). Environmental management accounting in Europe: Current practice and future potential. *The European Accounting Review, 9*(1), 31–52.
Benn, S., & Bolton, D. (2011). *Key concepts in corporate social responsibility*. London: Sage.
Bennett, M. J., Bouma, J., & Walters, T. (2002). *Environmental management accounting: Informational and institutional developments*. (Ed), Dordrecht: Kluwer.
Bennett, M., & James, P. (1998). Environment related management accounting: Current practice and future trends. *Greener Management International, 17*, 33–51.
Berry, M. A., & Rondinelli, D. A. (1998). Proactive corporate environmental management: A new industrial revolution. *Academy of Management Executive, 12*(2), 38–50.
Bhargava, S., & Welford, R. (1996). Corporate strategy and the environment: The theory. In R. Welford (Ed.), *Corporate environmental management* (pp. 13–26). London: Earthscan.
Burnett, R. D., & Hansen, D. R. (2008). Eco-efficiency: Defining a role for environmental cost management. *Accounting, Organizations and Society, 33*, 551–581.
Burritt, R. L. (2004). Environmental management accounting: Roadblocks on the way to the green and pleasant land. *Business Strategy and the Environment, 13*, 13–32.
Burritt, R., Hahn, T., & Schaltegger, S. (2002). Towards a comprehensive framework for environmental management accounting: Links between business actors and environmental management accounting tools. *Australian Accounting Review, 12*(2), 39–50.
Buysse, K., & Verbeke, A. (2003). Proactive environmental strategies: A stakeholder management perspective. *Strategic Management Journal, 24*, 453–470.
Carroll, B. A., & Shabana, K. M. (2010). The business case for corporate social responsibility: A review of concepts, research and practice. *International Journal of Management Reviews, 12* (1), 85–105.
Certified Management Accountant (CMA). 1995. *Implementing Corporate Environmental Strategies Management Accounting Guideline (MAG) 37*. Ontario: Society of Management Accountants of Canada.
Clarkson, M. B. E. (1995). A stakeholder framework for analyzing and evaluating corporate social performance. *Academy of Management Review, 20*(1), 92–117.
Colby, M. E. (1991). Environmental management in development: The evolution of paradigms. *Ecological Economics, 3*, 193–213.
Crane, A., & Matten, D. (2007). *Business ethics*. New York: Oxford University Press.
Deegan, C. (2003). *Environmental management accounting: An introduction and case studies for Australia*. Sydney: Institute of Chartered Accountants in Australia.

Delmas, M., & Toffel, M. W. (2004). Stakeholders and environmental management practices: An institutional framework. *Business Strategy and the Environment, 13*, 209–222.
Dias-Sardinha, I., Reijnders, L., & Antunes, P. (2002). From environmental performance evaluation to eco-efficiency and sustainability balanced scorecards. *Environmental Quality Management, 12*(2), 51–64.
Doody, H. (2010). *Environmental sustainability: Tools and Techniques*. The Society of Management Accountants of Canada, the American Institute of Certified Public Accountants and the Chartered Institute of Management Accountants.
Drury, C. (2004). *Management and cost accounting*. New Delhi: Thomson Learning.
Environmental Protection Agency (EPA). (1995). *An introduction to environmental accounting as a business management tool: Key concepts and terms*. EPA: Washington.
Epstein, M. J., & Roy, M. (2003). Improving sustainability performance: Specifying, implementing and measuring key principles. *Journal of General Management, 29*(1), 15–31.
Epstein, M. J., & Wisner, P. S. (2001). Using a balanced scorecard to implement sustainability. *Environmental Quality Management, 11*(Winter), 1–10.
Federal Environmental Agency (UBA). (2003). *Guide to corporate environmental cost management*. Berlin: German Environment Ministry.
Figge, F., Hahn, T., Schaltegger, S., & Wagner, M. (2002). The sustainability balanced scorecard: Linking sustainability management to business strategy. *Business Strategy and the Environment, 11*(5), 269–284.
Fonseka, K. B. M., Manawaduge, A. S. P. G., & Senarathne, D. S. N. P. (2005). *Management accounting practices in quoted public companies in Sri Lanka*. Colombo: Chartered Institute of Management Accountants of Sri Lanka.
Freeman, R. E. (1984). *Strategic management: A stakeholder approach*. Boston: Pitman Publishing.
Freeman, A. M., III. (2002). Environmental Policy Since Earth Day I: What have we gained? *Journal of Economic Perspectives, 16*(1), 125–146.
Gibson, K. C., & Martin, B. A. (2004). Demonstrating value through the use of environmental management accounting. *Environmental Quality Management, 13*, 45–52.
Godschalk, S. K. B. (2010). Does corporate environmental accounting make business sense? In S. Schaltegger, M. Bennett, R. L. Burritt, & C. Jasch (Eds.), *Environmental management accounting for cleaner production* (pp. 249–265). Netherlands: Springer.
Gouldson, A., & Murphy, J. (1998). *Regulatory realities: The implementation and impact of industrial environmental regulation*. London: Earthscan.
Gray, R., Bebbington, J., & Walters, D. (1993). *Accounting for the environment*. London: Paul Chapman Publishing.
Gunarathne, N., & Lee, K. H. (2015). Environmental Management Accounting (EMA) for environmental management and organizational change: An Eco-Control approach. *Journal of Accounting & Organizational Change, 11*(3).
Hart, S. L. (1995). Natural-resource-based view of the firm. *Academy of Management Review, 20* (4), 986–1014.
Henri, J. F., & Journeault, M. (2010). Eco-control: the influence of management control systems on environmental and economic performance. *Accounting, Organisations and Society, 35*, 63–80.
Henriques, I., & Sadorsky, P. (1996). The determinants of an environmentally responsive firm: An empirical approach. *Journal of Environmental Economics and Management, 30*(3), 381–395.
Hilton, R. W., Ramesh, G., & Jaydev, M. (2008). *Managerial accounting: Creating value in a dynamic business environment*. New Delhi: Tata McGraw-Hill.
Ilinitch, A. Y., Soderstrom, N. S., & Thomas, T. E. (1998). Measuring corporate environmental performance. *Journal of Accounting and Public Policy, 17*, 383–408.
International Federation of Accountants (IFAC). (2005). *International guidance document: Environmental management accounting*. New York: IFAC.
Kaplan, R. S., & Atkinson, A. A. (1998). *Advanced management accounting*. New Delhi: Pearson.

Kaplan, R. S., & Norton, D. P. (1992). The balanced scorecard: Measures that drive performance. *Harvard Business Review, 1992*(January-February), 71–79.
Kaplan, R. S., & Norton, D. P. (1993). Putting the Balanced Scorecard to Work. *Harvard Business Review, 1993*(September– October), 134–147.
Kokubu, K., & Kitada, H. (2012). *Material flow cost accounting and conventional management thinking: Introducing, a new environmental management accounting tool into companies.* A paper submitted to the IPA Conference in 2012.
Lee, K. H. (2011). Motivations, barriers, and incentives for adopting environmental management (Cost) accounting and related guidelines: A study of the Republic of Korea. *Corporate Social Responsibility and Environmental Management, 18*, 39–49.
Moller, A., & Schaltegger, S. (2005). The sustainability balanced scorecard as a framework for eco-efficiency analysis. *Journal of Industrial Ecology, 9*(4), 73–83.
Mross, D., & Rothenberg, S. (2006). *Formulation and implementation of environmental strategies: A comparison between U.S. and German printing firms.* New York: Printing Industry Center at RIT.
Newton, T., & Harte, G. (1997). Green business, technicist kitsch. *Journal of Management Studies, 34*(1), 75–98.
Niven, P. (2002). *Balanced Scorecard Step-by-Step: Maximizing performance and maintaining results.* New York: John Wiley & Sons.
Phillips, R. A., Berman, S. L., Elms, H., & Johnson-Cramer, M. E. (2010). Strategy, stakeholders and managerial discretion. *Strategic Organization, 8*(2), 176–183.
Porter, M. E., & van der Linde, C. (1995). Toward a new conception of the environment-competitiveness relationship. *Journal of Economic Perspectives, 9*(4), 97–118.
Preston, L. E. (1990). Stakeholder management and corporate performance. *Journal of Behavioral Economics, 19*(4), 361–375.
Qi, G., Zeng, S., Tam, C., Yin, H., & Zou, H. (2013). Stakeholders' Influences on corporate green innovation strategy: A case study of manufacturing firms in China. *Corporate Social Responsibility and Environmental Management, 20*, 1–14.
Robbins, P. T. (2001). *Greening the corporation: Management strategy and the environmental challenge.* London & Sterling, VA: Earthscan.
Roome, N. (1992). Developing environmental management systems. *Business Strategy and the Environment, 1*, 11–24.
Rugman, A. M., & Verbeke, A. (1998). Corporate strategies and environmental regulations: An organizing framework. *Strategic Management Journal, 19*(4), 363–375.
Sakai, K. (2007). Ricoh's approach to product life cycle management and technology development. In S. Takata & Y. Umeda (Eds.), *Advances in life cycle engineering for sustainable manufacturing businesses* (pp. 5–10). Netherlands: Springer.
Schaltegger, S., & Burritt, R. (2000). *Contemporary environmental accounting.* Sheffield: Greenleaf.
Schaltegger, S., & Burritt, R. (2006). Corporate sustainability accounting. In S. Schaltegger, M. Bennett, & R. Burritt (Eds.), *Sustainability accounting and reporting* (pp. 37–59). Dordrecht: Springer.
Soonawalla, K. (2006). Environmental management accounting. In A. Bihami (Ed.), *Contemporary issues in management accounting* (pp. 380–406). New York: Oxford University Press.
Sroufe, R., Montabon, F., Narasimhan, R., & Wang, X. (2003). Environmental management practices: A framework. *Greener Management International, 40*, 23–44.
Strobel, M., & Redmann, C. (2002). Flow cost accounting, an accounting approach based on the actual flows of materials. In M. Bennett, J. Bouma, & T. Wolters (Eds.), *Environmental management accounting, informational and institutional developments* (pp. 67–82). Dordrecht: Kluwer.
United Nations Division for Sustainable Development (UNDSD). (2001). *Environmental management accounting: Procedures and principles.* New York: UNDSD.

Valor, C. (2005). Corporate social responsibility and corporate citizenship: Towards corporate accountability. *Business and Society Review, 110*(2), 191–212.

Wilmshurst, T. D., & Frost, G. R. (2001). The role of accounting and the accountant in the environmental management system. *Business Strategy and the Environment, 10*, 135–147.

Xiaomei, L. (2004). Theory and practice of environmental management accounting. Experience of implementation in China. *International Journal of Technology Management and Sustainable Development, 3*(1), 47–57.

Index

A
AA1000, 193, 201–214, 249–252, 255, 257, 267, 275
AA1000 Assurance Standard (AA1000AS), 20, 136–138, 142, 143, 151, 176, 178, 250, 251
AA 1000 standard, 7, 8, 193, 222, 250
Accountability, 2, 15, 20, 34, 67, 79–82, 98–99, 107, 120, 121, 131, 169, 190–193, 201–214, 219, 223, 237, 249–251, 256, 264, 267, 301
Accounting, 1, 6, 8, 10, 16, 33–54, 59–77, 111, 135, 142, 151, 172, 176, 187, 197, 205, 220, 221, 224, 261, 282, 292, 301–321
Accounting profession, 10, 36–38, 51, 52, 151, 172, 290
Agency relationship, 135
American Accounting Association (AAA), 60
American Institute of Certified Public Accountants (AICPA), 38–42, 44–48, 52, 54, 60
Analysis, 1, 4, 8, 19, 23, 37, 38, 49, 50, 52, 53, 62, 64, 65, 67, 68, 71, 76, 85, 86, 132, 142, 143, 152, 155, 156, 158, 163, 165, 166, 176, 201–214, 220, 222, 232, 250, 252, 254, 257, 259, 261, 262, 267–269, 275, 284, 286, 291–292, 302, 304, 305, 307, 308, 311, 313, 315, 318
Annual reports, 10, 34, 35, 85, 86, 113, 116, 118, 122, 145, 175–178, 180, 181, 210, 211, 249, 262, 268–270, 275
Answerability, 202
Anti-coal mining activists, 179
ANZ, 170, 177–182

Assurance providers, 7, 19, 50, 52–54, 133, 137, 138, 140–143, 150–152, 293
Audit
 engagement, 281–286, 292, 293, 295, 296
 failure, 6, 281–297
 opinion, 281
Auditing, 2, 8, 15, 22, 33–54, 69, 87, 107, 117–127, 136, 159, 169–183, 187, 189–191, 195–197, 205, 222, 257, 281, 289–290, 292–295
 profession, 8, 33–54, 287, 288, 294
 standards, 6, 22, 170, 182, 227, 291, 292, 294
Auditor liability, 285, 286, 290, 291

B
Bangladesh, 7, 10, 79–102, 107–127, 188, 197
Bangladesh Bank, 109, 110, 113, 114, 117, 122–125
Banking sector, 7, 109–111, 114–117, 121–127
Beneficiary, 79–83, 85, 86, 90–102, 228, 232, 294
Beneficiary-related constraints, 95–96
Benefits, 3, 5–7, 9, 11, 16, 22, 33, 37, 49–52, 60, 62, 66–71, 79–81, 85, 90–99, 101, 102, 108, 115, 116, 118, 122, 159, 189, 197, 208, 225, 226, 232–234, 237–239, 247, 252–254, 259, 261, 262, 306, 309–310, 313, 314, 319, 320
Big 4, 59, 60, 71, 77, 142, 150
Bowen, Howard R., 118, 189, 228
Brazil, 51, 155–157, 161, 162, 166, 209, 211, 260

Business, 1, 9, 15–17, 20, 26, 35, 37, 42, 47, 60, 76, 82, 107, 108, 112, 122, 132, 155–157, 160, 169, 190, 205, 219–257, 288, 301
Business ethics, 231

C
Capacity building, 27, 52
Capital, 18, 35, 36, 38–52, 70, 73, 75, 132, 146, 160, 162, 179, 210, 263, 264, 288, 294, 308, 309, 313, 315–317, 319–320
Capital deployment, 70
Carbon Disclosure Project (CDP), 72, 161, 169, 172–174, 176
Carbon footprint, 73, 177, 317
CAS3101, 136, 140–142, 144, 151
Checklist, 10, 23, 257, 266–269, 272
Child labour, 109, 120, 190, 192, 194, 236, 248, 268, 272–275
China Institute of Certified Public Accountants (CICPA), 136, 140
Chinese listed companies, 145
Chinese standard, 136, 140–142
Climate change, 4, 10, 40, 115, 132, 161, 169–183, 301
Community, 1–3, 6, 8, 16, 17, 20, 22, 25–27, 30, 59, 65, 67, 73, 81–84, 89–93, 96, 99, 111, 112, 118, 120, 124, 127, 134, 144, 155, 157, 160, 161, 163, 169, 172, 174, 175, 178, 179, 181, 189, 190, 193, 195, 207, 209, 211, 212, 219–221, 227–229, 231–233, 235, 236, 240, 242, 245, 246, 248, 254, 263, 266, 303, 305, 313, 317
Community expectations, standards and guidelines, 3, 175
Compact, 9, 17, 20, 158, 162, 175, 211, 212, 233–238, 243, 248, 253, 254, 315
Companies, 1, 15, 17, 18, 33, 42, 61, 70, 72, 80, 107, 131, 144, 145, 155–167, 169–183, 187, 188, 201, 208–212, 219, 223, 224, 228, 233, 235–237, 242, 244, 247, 255, 257, 273, 287, 288, 304
Competitive advantage, 30, 69–71, 112, 157, 191, 261, 262, 304, 305, 307, 311
Compliance, 6, 10, 21–23, 25–28, 60, 67, 71, 81, 108, 111, 149, 170, 171, 177, 182, 193, 209, 237, 241, 248, 250, 253, 258, 262, 266–275, 294, 296, 297, 303, 304, 307, 311–313, 315–316, 318
Conduct, 1, 6, 20, 27, 79, 81, 83, 87–90, 97–99, 119, 120, 124, 140, 144, 177, 178, 194, 205–208, 219, 227, 229, 240, 243–245, 249, 283, 289
Constraints, 7, 81, 86, 94–97, 99, 100, 230
Context-related constraints, 96–97
Contract liability, 284
Corporate communication, 156
Corporate governance, 1, 16–20, 42, 75, 135, 142, 148, 158, 162, 163, 205, 287, 289
Corporate greening, 302
Corporate responses, 170–174
Corporate social accountability, 301, 302, 311, 318
Corporate social and environmental activities, 135
Corporate social responsibility (CSR)
assurance, 6, 19, 21, 132, 134–152
assurance engagements, 137, 138, 141
assurance market in China, 132, 133, 136, 141–152
assurance standards, 144, 150–152
assurance trend, 7, 21, 144–145
reporting, 5–7, 11, 114, 116, 121, 122, 131–133, 138, 140–145, 149, 151, 152, 188, 268
reports, 1, 3, 5–7, 9, 11, 28, 114, 132, 135, 136, 138, 140, 142–149, 152, 177, 178, 180, 188, 196, 263, 292–294
team, 125, 126
Corporations, 2, 7, 10, 21, 59, 70, 71, 107–127, 143, 146, 160, 162, 170, 190, 205, 219, 229, 234, 257–276, 288, 290, 291, 297, 301–303
Corruption, 84, 96, 97, 102, 110, 119, 121, 123, 124, 202, 232, 237, 245, 248
COSO, 6, 59–77
Country, 6, 7, 9, 16, 26, 54, 81, 83–85, 96, 102, 110, 112, 114, 115, 120, 123–125, 132, 134, 146, 155–158, 161–163, 171, 172, 188–190, 196, 197, 204, 209, 211, 212, 227, 232, 235–239, 241–243, 252–254, 258–260, 265–271, 275, 282, 315, 316
CSMAR database, 142, 143
CSR-VRAI, 140, 142, 144

D
Deep pocket, 141
Demand-supply analysis, 132
Democracy, 119, 201–204, 212–214
Development, 4–5, 7–9, 16–19, 33–54, 60, 76, 80, 83, 85, 101, 108, 112, 118–121, 133, 146, 155, 157, 160, 170, 176, 193, 206, 210–212, 220, 221, 229, 232, 235,

Index

242–245, 258, 270, 274, 275, 282, 291–292, 301–305, 308
Dialogue, 79, 84, 97–102, 133, 205, 207, 222, 224, 241, 242, 244, 245, 247, 253, 296
Disaster relief, 113, 115
Disclosure, 4, 20, 33, 40, 51, 59, 79, 109, 131, 143–144, 149, 161, 169–183, 202, 209, 221, 222, 251–252, 257, 259, 270, 274, 282, 289
  content, 149
  index, 257, 267–272
  level, 260, 269, 270, 272–276
Donor, 79–83, 85, 88–92, 98–101, 120, 232
Due care, 284–286, 290–291, 295
Dutch Bangla Bank, 114, 122
Duty of care, 285, 286, 292, 295

E

Eco-control, 306
Economic zone, 146, 147
Electronic Industry Citizenship Coalition (EICC), 194
Emissions trading scheme, 171, 172
Employee, 3, 8, 16–20, 22, 23, 25, 26, 29, 65, 69, 109, 111–113, 116, 118, 120, 144, 149, 163, 165, 166, 187, 192, 209, 211, 227–229, 231, 232, 241, 242, 246, 248, 261, 269, 305, 306, 313, 315–317
Energy accounting, 308, 309, 313, 315, 317, 318
Enterprise risk management (ERM), 60–71
Environment/environmental, 1, 10, 15, 25, 36, 60–63, 73, 96, 111, 113, 116, 143–144, 149, 157, 160–163, 166, 170, 179, 188, 194, 205, 211, 219, 235, 245, 246, 248, 257, 258, 282, 301–318
Environmental accounting, 111, 175, 220–222, 302
Environmental activity based costing, 309, 320
Environmental capital budgeting, 308, 309, 317, 319–320
Environmental challenges, 302–304, 308, 311, 312, 314, 318
Environmental compliance, 71
Environmental conservation, 229, 304
Environmental domains, 311–314, 318
Environmental information management, 310
Environmental legislation, 304
Environmental management
  practices, 301–303, 308, 311, 316, 318
  strategies, 301, 302, 305, 307–309, 314, 318

Environmental management accounting (EMA), 308–314, 316–318
Environmental management accounting (EMA) techniques, 309, 312, 315, 318–321
Environmental, social and governance (ESG), 33, 34, 48, 50, 53, 74, 75, 259, 263, 264
Equator principles, 179–182
Ethical business, 22, 162
Ethical Trading Initiative (ETI), 193–194
External and internal audit, 89, 90, 165, 187
External assurance, 136, 137, 140, 143, 227, 260, 266, 281
External auditors, 6, 9, 93–95, 281–297
Extractive Industries Transparency Initiative (EITI), 189

F

Fair Trade Foundation, 189
Financial audit, 2, 3, 9, 87, 125, 188, 190, 205, 265, 283, 287, 297
Financial auditors, 125
Financial Executives International (FEI), 60
Financial information, 131, 132, 151, 160, 178, 263, 293, 295
Firm, 11, 15–17, 35, 37, 40, 59–77, 108, 109, 111–114, 116–121, 125, 133–135, 138, 141, 142, 145, 150–152, 155, 157, 165, 206, 213, 225–227, 229–231, 233, 236–238, 240, 244, 245, 248, 252, 255, 281–286, 289, 296, 302, 303, 305, 311
Focus group discussions (FGD), 90, 91, 99, 102
Force of law standards, 290, 297
Framework, 8–10, 20, 33–40, 42, 43, 45–48, 50, 51, 53, 60, 63, 70, 81, 113, 132, 140, 170, 173, 176, 219–257, 265, 267, 268, 281, 289–297, 301–321

G

Geographical analysis, 146–148
Global concerns, 107, 170–174
Global Reporting Initiative (GRI), 7, 9, 17, 20, 33–35, 45, 50, 116, 136–141, 143, 144, 149, 156, 161, 165, 166, 175, 176, 178, 206, 222, 233, 244–249, 253–255, 259, 260, 263, 264, 266, 267, 269, 292, 294
Governance, 1, 16–23, 25, 26, 33–36, 40, 42, 43, 47, 49, 60, 74, 75, 108, 125, 135, 142, 148, 158, 162, 163, 165, 205, 207, 232, 234, 236, 240, 246, 253, 258, 259, 261, 263, 264, 270, 271, 275, 287, 289

Government, 2, 4–6, 9, 10, 17, 72, 80, 85, 89–92, 97–99, 102, 110, 114, 118–121, 123–125, 131, 145, 151, 158, 160, 169–171, 176, 182, 190, 209, 210, 219, 223, 226, 229, 234, 236, 237, 239, 241–245, 252, 253, 266, 295, 303
Green accounting, 308
Greenhouse gas emissions, 53
Green image, 316–317

H
Health and Safety, 22, 25, 26, 120, 162, 190, 192, 194, 236, 239, 248, 261, 268, 272–276, 304
History of financial reporting audits, 281

I
Impact, 1, 15, 37, 59, 81, 107, 134, 155, 170, 190, 208, 220, 257, 294, 301
Indonesia, 7, 10, 79–102, 109, 188, 189, 197
Industrial analysis, 146–148
Information, 2, 15, 33, 40, 60, 67–69, 81, 108, 121, 131, 140, 143–144, 160, 170, 177, 188, 202, 219, 245, 257, 263–264, 292, 293, 295, 302, 305–306
Information asymmetry, 135
Institutions, 43, 72, 109, 113–115, 146, 151, 156, 157, 159, 162, 179, 180, 202–204, 209, 213, 223, 227, 228, 232, 237, 241, 246, 247, 253
Institute of Internal Auditors (IIA), 17, 19, 21–27, 38, 40–42, 44–46, 48–50, 52, 54, 60
Institute of Management Accountants (IMA), 60
Integrated framework, 10, 301–321
Integrated reporting, 8, 33–54, 72
Integrated thinking, 36, 51–54
Intergovernmental Panel on Climate Change (IPCC), 169, 170, 173, 174
Internal audit, 10, 15–30, 50, 66, 67, 75, 89, 165
Internal control, 6, 21, 27–29, 41, 47, 60, 66, 77, 187
International CSR reporting framework, 141
International Financial Reporting, 37, 188
International <IR> Framework, 8, 34–36, 38, 39, 47
International Labour Organisation (ILO), 9, 192, 233, 234, 239–242, 253

International Standards, 20, 22, 28, 136–140, 176, 178, 191–195, 197, 206, 223, 239, 293
Interview, 2, 23, 85–87, 90, 96, 97, 125, 261, 262, 295
Irresponsible working practices, 28, 188, 196
ISAE3000, 136–138, 140, 144, 151, 178, 257, 275
ISO 9001, 209
ISO14000, 161, 165, 166, 209, 292

K
Key performance indicators (KPI), 45, 67, 74
Kyoto protocol, 171, 173

L
Latin America, 202, 204, 211
Law enforcement, 235
Leading edge companies, 304
Legal compliance, 111
Legal liability, 6, 45, 141, 142, 144, 148, 281–297
Legal liability framework, 282
Legitimacy theory, 3, 4, 132–135, 261, 262
Liability
  in contract, 284
  in tort, 285
Life cycle accounting, 309, 320
Light SA, 209, 210
Limited-level assurance, 142
Litigation risk, 141, 282, 284, 286
Local administration, 126
Long-term outcomes, 91, 101

M
Mandatory compliance, 297
Mandatory CSR, 188, 197, 262
Mass balance, 317, 319
Material flow cost accounting (MFCA), 308, 309, 313, 317, 319
Mechanism, 2, 24, 28, 79–85, 89, 98–99, 101, 102, 108, 118, 119, 172, 173, 204–208, 212, 214, 219, 222–225, 233, 236–239, 241, 243, 252, 254, 255, 264
Media, 116, 118, 120–124, 152, 157, 158, 170, 172, 174, 175, 177–181, 188, 196, 269, 274, 301, 305, 313, 317
Methodology, 8, 10, 23, 76, 90, 162, 163, 165, 210, 211, 220, 257–276, 282, 286

Index

Microenterprise, 80, 82, 91, 97
Monetary environmental management accounting (MEMA), 308, 319
Monitoring, 2, 5, 9–11, 19, 22, 60, 66–69, 71, 87, 88, 90, 92, 94, 95, 97–99, 101, 107, 158, 161, 170–174, 176, 177, 182, 211, 223, 254, 302, 308
Monitoring and evaluation process, 87, 88, 97, 98

N
Natural resource-based view, 305
Natural resources, 25, 72, 108, 144, 189, 263, 305
Negligence, 141, 284–286, 291, 295
Non-financial information, 7, 257, 259, 260, 263–264, 266, 267, 269, 275
Non-Governmental Organisations (NGOs), 2, 4, 5, 7, 10, 25, 79–102, 125, 156–159, 162, 169, 171, 173, 174, 178, 179, 191, 195, 196, 207, 219, 232, 234, 240, 242, 243, 254
Nvivo, 85

O
Organization, 15–18, 21, 26–28, 34, 36, 39–43, 47, 48, 51, 59–73, 75, 76, 107–109, 111–113, 118–120, 122–124, 127, 134, 143, 156, 157, 159–162, 166, 170, 174, 176, 204–208, 212, 213, 219–229, 231–235, 239–256, 259, 261–263, 294, 301–311, 313, 314, 318–320

P
Participation, 79, 80, 82, 83, 90, 91, 95–99, 102, 121–123, 127, 145, 156, 158, 159, 161, 174, 204, 206–209, 237, 245, 246, 254, 255, 305
Participatory approach, 80, 82, 83, 92, 99–100, 102
Performance, 1–8, 10, 15, 18, 20, 26–28, 30, 34–36, 39, 42, 45, 47, 49, 53, 60, 61, 67–70, 73–76, 79–81, 83, 91–93, 99, 100, 102, 107–127, 134, 136–140, 143, 144, 148, 155, 156, 158, 160, 163–165, 170, 174, 175, 177–180, 195, 206–208, 219, 220, 222, 224–226, 230, 231, 233, 237, 238, 241, 244–248, 250–252, 254, 255, 259, 260, 262–266, 281, 282, 301, 302, 306–308, 310, 311, 317, 320

Performance measure, 28, 91–92, 320
Philanthropy, 27, 111, 113, 117, 156, 166, 227, 228
Physical environmental management accounting (PEMA), 308, 313, 319
Policy, 18, 21, 22, 28, 29, 33, 66, 70, 71, 75, 89, 109, 113, 114, 121, 123, 125, 126, 134, 159, 170, 172–174, 179–181, 190, 192, 196, 225, 226, 230–232, 234, 236–241, 244, 251, 254, 255, 262, 266, 275, 289, 295, 307, 313, 317
Political system, 205
Pollution, 180, 221, 224, 248, 305, 311, 317
Poverty, 76, 79, 80, 82–84, 91, 93, 96, 100, 102, 159, 212, 227, 241
Practice, 1, 3–7, 9–11, 15, 17, 18, 21, 22, 24, 25, 27–29, 34, 40, 59, 60, 67–71, 73–77, 79–102, 107, 109, 111–114, 119, 123–127, 131–152, 155–158, 163, 165–167, 169–183, 188–197, 202, 205, 209–214, 219, 221, 222, 227, 232, 236–238, 241, 245–248, 251, 252, 254, 255, 257–264, 266–268, 270, 272–275, 281–283, 288, 289, 301–303, 307, 308, 310–313, 315–318
Pragmatic approach, 80, 82, 100–102
Private companies, 7, 156, 204, 205, 212, 213
Private interest theory, 37, 50
Proactive environmental management, 307
Professional accounting bodies, 35, 144
Professional audit standards, 297
Professional indemnity insurance, 141
Public consultation, 35, 37, 256
Public interest theory, 37, 50
Public sphere, 203, 205

Q
Quality certifications, 205

R
Reasonable level of assurance, 142
Regulation, 4–7, 9–11, 22, 26, 34, 37, 42, 74, 79–102, 110, 112, 114, 131–152, 158, 165, 166, 170–173, 182, 197, 205, 223, 227, 229, 237, 238, 240, 246, 267, 269–271, 275, 281, 282, 289, 295, 303, 307
  commitment, 5, 101, 112, 135, 171, 307
  of social audits, 6, 7, 9, 281, 289

Regulatory, 4–6, 8–10, 21, 22, 25, 33–35, 37, 45, 49–51, 53, 81, 90, 96, 102, 109, 118, 122–124, 131–133, 144–146, 148, 151, 157, 171, 174, 176, 202, 211, 219–256, 282, 290, 294, 303, 305, 307
   authority, 22, 90, 124
   capture, 37, 50, 51
Report, 6–9, 17, 20, 33–54, 59–77, 116, 121, 123, 136–138, 140, 141, 143, 145, 156, 157, 159, 160, 170–174, 176, 177, 181, 188, 195, 197, 206, 211, 212, 221, 233, 247–249, 251, 254, 256, 258–260, 263, 269, 270, 275, 282, 287–289, 291–292, 294
Reputation management, 67, 70, 75
Review process, 87–89, 93, 96–98
Rio Earth Summit, 170, 172
Risk
   appetite, 61–64, 69
   assessment, 60, 63, 65, 67, 136, 138
   identification, 28, 60, 64–65, 72
   management, 21, 27, 43, 47, 60–71, 158, 179, 294
   response, 60, 65–66, 68, 72

S

SA8000, 56, 166, 191–193, 197, 257, 266–275
Sectorial Inspector, 196
Self-selection adoption, 79, 83
Short-term outcomes, 42
Signaling theory, 132, 133, 135
Social
   and environmental accounting, 111, 175, 220–222
   and environmental reporting, 4, 175
Social audit
   assurance, 2–8, 132, 133, 193, 222, 227, 250–252, 255, 259, 260, 266, 267, 281–283, 287, 292–297
   engagement, 282, 284–286, 292, 293, 296
   failure, 6, 281–297
   framework, 257, 267, 269, 275
   regulation, 7, 10, 79–102, 275
   reporting, 197, 282
   Team, 120, 124–126
Social auditing, 2, 3, 5, 8, 10, 11, 69, 107, 109, 117–127, 170, 175, 182, 187–193, 195–197, 222, 224, 256–258, 281, 292–295
Social Audit Network (SAN), 190
Social auditors, 4, 6–8, 187, 188, 192, 196, 221, 286, 296
Social contract, 108, 134, 135, 229, 262
Socialising accountability, 82, 83, 100
Socially responsible reporting assertions, 281
Social reporting, 4–6, 8, 132, 152, 197, 221, 224, 232, 237, 240, 245, 254, 282, 283, 286, 296
Social reporting framework, 282
Social welfare, 29, 108, 111, 227
Society, 3–5, 16–18, 35, 37, 44, 53, 54, 95, 107–109, 111–114, 118, 119, 121, 123, 125, 134, 138, 157–160, 172, 175, 182, 190, 193, 197, 203–205, 208–214, 219–221, 223, 224, 227–234, 237, 240, 242–245, 247, 248, 253, 257, 261, 262, 264–269, 273–275, 295, 301
Stakeholder
   engagement, 27, 40, 48, 123, 127, 137, 174, 193, 206, 208, 210, 211, 214, 248, 250, 255, 259, 261
   theory, 3, 4, 132–135, 222
Standard, 1–9, 19–23, 26–29, 34, 37, 40, 49, 51, 59, 68, 73, 76, 81, 115, 119–121, 125, 132, 136–144, 150–152, 161, 170, 175–178, 181, 182, 188, 191–195, 197, 201, 206–212, 214, 219, 222, 223, 227, 230, 232–257, 259, 262, 266–275, 281–297, 304, 306
State, 7, 21, 110, 114, 115, 120, 121, 156–160, 166, 171–173, 178, 179, 187, 202–205, 210, 211, 213, 214, 226, 228–230, 240, 259, 261, 262, 288, 290, 303
Strategic plan, 62, 74
Subject matter, 52, 64, 109, 124, 137, 140, 141, 144, 293, 295, 311
Supply chain, 10, 23, 25, 67, 144, 188–189, 193, 235, 238, 249
Sustainability
   lens, 70
   report, 17, 20, 34, 49, 67, 68, 73–75, 77, 134, 136–138, 140, 143, 210–214, 245, 247–249, 251, 252, 254, 255, 259, 260, 263, 264, 268, 282, 292–294, 317
   reporting, 17, 34, 49, 68, 74–75, 77, 137, 138, 140, 143, 214, 245, 247–249, 251, 252, 254, 255, 259, 260, 268, 282, 292, 294
   risk, 6, 59–77

Index

Sustainability balanced scorecard (SBSC), 309, 317, 320–321
Sustainable development, 16, 17, 22, 24, 37, 112, 143, 157, 176, 193, 210, 211, 248, 258, 259, 263, 302, 305, 307, 308
Sustainable poverty alleviation, 80, 82, 91
Suzano Papel e Celulose, 209, 210
Sweatshop, 190
Symbolic legitimacy, 181, 182

T
TBL. *See* Triple bottom line (TBL)
Theory, 3, 4, 6, 34, 37, 50–51, 81, 117, 131–152, 202, 213, 214, 221–223, 231, 259, 261, 262, 295, 305
Tort of negligence, 284–286
Transparency, 3, 18, 25, 48, 54, 68–70, 73, 75, 80, 81, 84, 90, 98, 101, 107, 114, 117–119, 121–123, 125–127, 131, 143, 159–161, 183, 189, 190, 201–205, 208, 212, 214, 237, 245, 246, 248, 262
Triple bottom line (TBL), 18, 34, 60–69, 162

U
UN. *See* United Nations (UN)
UNFCC. *See* United Nations Framework Convention on Climate Change (UNFCC)
UN Global Compact, 9, 211, 233–235, 237, 238, 253, 254
United Nations (UN), 9, 17, 20, 37, 162, 170, 173, 175, 176, 192, 211, 212, 233–239, 245, 246, 253, 254, 308
United Nations Framework Convention on Climate Change (UNFCC), 170, 173, 174
US GAAP, 59, 60

V
Value creation, 34–36, 38, 42, 49, 75, 260
Voluntary
    adoption, 7, 262
    compliance, 209
    CSR, 135, 142–145, 151
    disclosure, 182

W
Waste accounting, 123, 301, 302, 304, 308, 309, 311–313, 316, 317
Water accounting, 37
Website reporting, 27, 85, 98, 100, 113, 177, 178, 180, 181, 246
Whitehaven coal, 170, 178–181
Worker, 131, 134, 158, 189, 191–194, 196, 220, 225, 227, 228, 233, 236, 239, 241, 242, 253
Workplace, 25, 192, 194, 232, 235, 249, 252, 254, 261, 267

Printed by Libri Plureos GmbH
in Hamburg, Germany